18th International Workshop on Magneto-Plasma Aerodynamics (WSMPA 2019)

AA007690

Journal of Physics: Conference Series Volume 1394

Moscow, Russia
9 – 11 April 2019

ISBN: 978-1-7138-1164-0
ISSN: 1742-6588

Printed from e-media with permission by:

Curran Associates, Inc.
57 Morehouse Lane
Red Hook, NY 12571

Some format issues inherent in the e-media version may also appear in this print version.

Institute of Physics
Dirac House, Temple Back
Bristol BS1 6BE UK

Phone: 44 1 17 929 7481
Fax: 44 1 17 920 0979

techtracking@iop.org

Additional copies of this publication are available from:

Curran Associates, Inc.
57 Morehouse Lane
Red Hook, NY 12571 USA
Phone: 845-758-0400
Fax: 845-758-2633
Email: curran@proceedings.com
Web: www.proceedings.com

TABLE OF CONTENTS

Journal of Physics: Conference Series **1394** (2019) 011001 doi:10.1088/1742-6596/1394/1/011001

We are pleased to present the Proceedings of 18th International Workshop on Magneto-Plasma Aerodynamics (The Workshop) which was held in Moscow on April 09 – 11, 2019 at Joint Institute for High Temperatures of the Russian Academy of Sciences (JIHT).

Every spring this Workshop is organized by JIHT under Scientific Council for Direct Energy Conversion, Scientific Council for Fluid Mechanics and Scientific Council for Low Temperature Plasma Physics of the Russian Academy of Sciences.
The objective of the Workshop is to provide forum for presentation and discussion of recent progress in science and technology on magneto-plasma aerodynamics. The results of analytical and experimental studies of physics of gas discharge, physics of shock waves, plasma assisted ignition and combustion, MHD energy conversion, physical gas dynamics of high-speed airflows and their applications are discussed at the meeting.

At the 18th Workshop there were presented 60 works in plenary and profiled sections: shock waves, electrical discharges, plasma assisted ignition and combustion, plasma aerodynamics, MHD&PA flow control and related problems.

In this collection of papers there are published 30 selected articles and we are grateful to IOP Publishing for cooperation.

Valentin Bityurin, PhD, DS, Chairman of WSMPA 2019
Tatiana Kuznetsova, PhD, Scientific Secretary of WSMPA 2019

Journal of Physics: Conference Series **1394** (2019) 011002 doi:10.1088/1742-6596/1394/1/011002

Peer review statement

All papers published in this volume of *Journal of Physics: Conference Series* have been peer reviewed through processes administered by the proceedings Editors. Reviews were conducted by expert referees to the professional and scientific standards expected of a proceedings journal published by IOP Publishing.

Generation of fast electrons in pulsed discharge plasma in helium

Shibkov V. M.

Moscow State University, 119991 Moscow, Russia

shibkov@phys.msu ru

Abstract. The results of experimental studies of the distribution function of fast electrons under conditions of non-equilibrium plasma of a pulsed discharge in helium are presented. It was established by various methods that at the initial stage of the discharge, when the reduced electric field exceeds the critical value, a beam of fast electrons is generated. The beam energy depends on the applied voltage to the discharge gap, and helium pressure.

1. Introduction

Currently, to increase the efficiency of a ramjet engine with a supersonic flow rate in the combustion chamber, new methods are actively developed to increase the burning rate and ensure complete combustion of the fuel inside the engine without the use of various stagnant zones in the form of cavities, ledges, pylons, and poorly streamlined mechanical stabilizers. In our laboratory, for these purposes, along with other plasma sources, in particular, freely localized and surface microwave discharges [1-8], a non-stationary pulsating transverse-longitudinal discharge is used, which is generated under the conditions of subsonic and supersonic flows of air and air-fuel mixtures [9-18]. It is known that in order to realize the mode of formation of a fast electron beam, it is necessary that the reduced electric field exceed a critical value $(E/p)_{cr}$. For a plasma discharge in air $(E/p)_{cr} = 490\ \mathrm{V/(cm{\times}Torr)}$. In the experiments [12, 14-16], the static pressure in the flow was 100–200 Torr, the shortest distance between the electrodes was 0.01 cm, and the maximum voltage value at the discharge gap was 2000 V. Hence, in the breakdown stage, the reduced electric field value can exceed a critical value $(E/p)_{cr}$. It is difficult to measure fast electrons during breakdown of short gaps in a high-speed flow. Therefore, studies of fast electrons generated in the stage of pulsed breakdown of a motionless gas in a tube of a special configuration were conducted under conditions of reduced electric field close to experiments performed in a supersonic air stream.

2. Experimental setup

To obtain information on the kinetics of electrons at the initial stage of a pulsed discharge in motionless gas, various methods were used - the method of retarding potential, the electrons velocity selection method using a transverse magnetic field and the probe method. In experiments a special discharge tube was used. The distance between the anode and the cathode was fixed and equal to 12 cm. The anode was a grid with a transparency coefficient of 0.7. The second grid with a transparency factor of 0.95 was mounted behind the anode in close proximity to it. On the second grid retarding potential was applied. The movable collector was located behind the second grid. The

Journal of Physics: Conference Series **1394** (2019) 012001 doi:10.1088/1742-6596/1394/1/012001

distance between the collector and the second grid could be changed from 0 to 12 cm. A flat movable probe, which could move along the wall of the discharge tube, was also mounted in the tube. In the space between the second grid and the collector, it was possible to create a constant magnetic field transverse with respect to the axis of the discharge tube. The experiments were performed in pure helium at a pressure of 0.02–2.0 Torr. A rectangular impulse voltage of 1 μs duration was applied to the discharge gap. The steepness of the leading edge of the discharge current is 10^8 A/s. Voltage amplitude was varied from 200 to 700 V.

3. Experimental results

It is experimentally observed that in the initial stage of a pulsed discharge with multiple overvoltage's, a generation of a beam of fast electrons is possible. To realize the regime of continuous electron acceleration, it is necessary that the reduced electric field strength exceeds a certain critical value $(E/p)_{cr}$ [19-21]. For weakly ionized helium plasma value of $(E/p)_{cr}$=50 V/(cm×Torr). With a decrease in gas pressure or with an increase in the amplitude of the electric field, the value E/p can exceed the critical value $(E/p)_{cr}$, while the mean free path of the electrons will increase, and the probability that the electron will receive more energy will increase [22]. Under these conditions, therefore, the generation of a beam of fast electrons in a weakly ionized plasma is possible.

Figure 1 shows the dependences of the electron flux to the collector on the distance between the anode and the collector at a helium pressure of 0.15 Torr and various values of the pulse voltage applied to the discharge gap. It can be seen that the flow of electrons passing through the anode and falling into the equipotential space between the second grid and the collector weakens due to elastic and inelastic collisions with atoms.

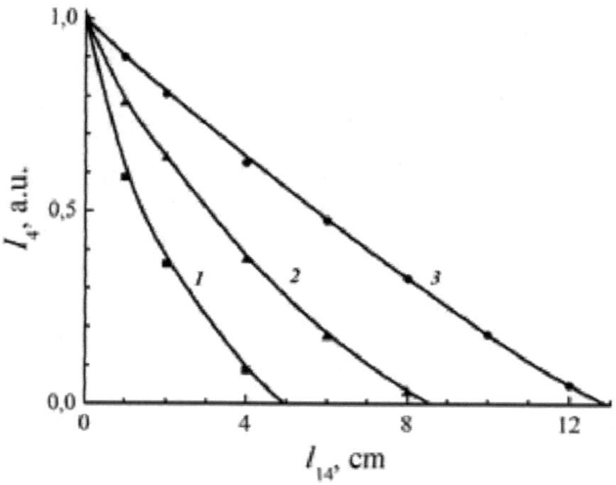

Figure 1. Dependencies of the electron flux to the collector I_4 on the distance l_{14} at the stage of breakdown of a pulsed discharge in helium at p = 0.15 Torr and various voltages between the cathode and anode U_0, V: *1* - 200; *2* - 400; *3* – 700.

From these data, one can find the effective scattering length of electrons and estimate their energy. It follows from such estimates that at the initial stage at threshold values of U_0, electrons with an energy of ~ 20 eV predominate in the discharge, and with an increase in U_0 the electron energy increases and reaches ~ 100 eV at U_0 = 700 V.

Figure 2 shows the dependences of the current to the collector I_4 for a moment of time t~10^{-7} s on the magnitude of the induction B of the transverse magnetic field at p =0.2 Torr and fixed distances from the second grid to the collector l_{24}. It is seen that with an increase in the magnetic field, the value of the current to the collector decreases monotonically. According to the data obtained, knowing the geometry of the discharge tube, we can estimate the energy of a beam of fast electrons. The course of the obtained dependences $I_4 = f(B)$ does not change when the direction of magnetic induction is reversed.

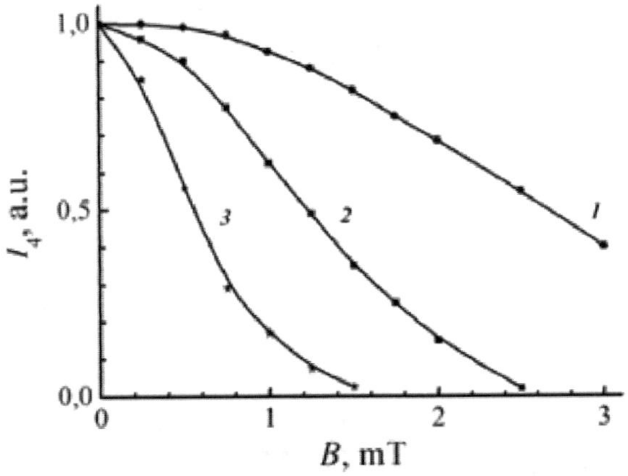

Figure 2. The dependencies of the electron current to the collector I_4 on the magnitude of the transverse magnetic field induction for various distances l_{24}, cm: *1* - 2; *2* - 4; *3* - 8.

In the experiment, the dependences of the current to the wall probe I_6 on the transverse magnetic field for various distances of the probe from the second grid l_{26} were also recorded. Figure 4 shows the obtained dependence $I_6 = f(B)$.

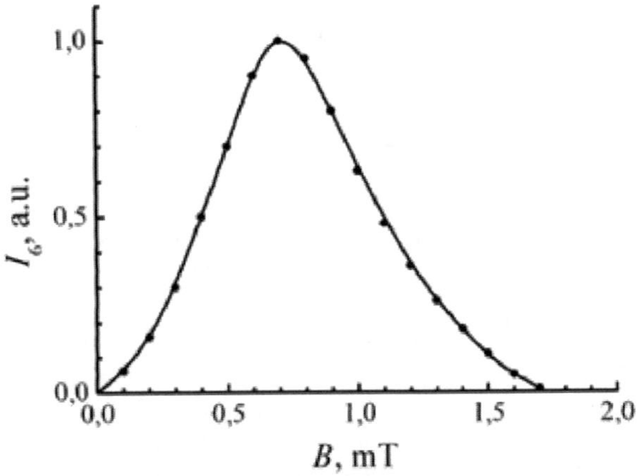

Figure 3. The dependence of the current to the wall probe I_6 on the transverse magnetic field at $l_{26} = 2.3$ cm and $p = 0.4$ Torr.

It can be seen that in the absence of an external magnetic field, the current to the probe is equal to zero. When the magnetic field increases, the current increases and reaches its maximum value at B =0.7 mT. A further increase in B leads to a decrease I_6 and at B =1.7 mT the current to the wall probe is not recorded. Changing the direction of the magnetic field to the opposite, we see that the current to the wall probe is zero at any value of B. In Figure 3 the maximum current corresponds to the electron energy $\varepsilon \sim 30$ eV, and at $B = 1.7$ mT the electron energy is of the order of 100 eV.

Figure 4 shows the dependences of the energy of fast electrons, corresponding to the maximum of the distribution function, on the helium pressure at various values of U_0. It can be seen that the beams of fast electrons are observed over the entire range of investigated helium pressures. Energy ε_m

increases with growth U_0 at all pressures. All curves $\varepsilon_m = f(p)$ pass through a maximum. The obtained result can be explained by the fact that at low pressures the electrons effectively recombine on the walls of the discharge tube, and with increasing p the frequency of inelastic collisions increases, which leads to an effective loss of electrons energy.

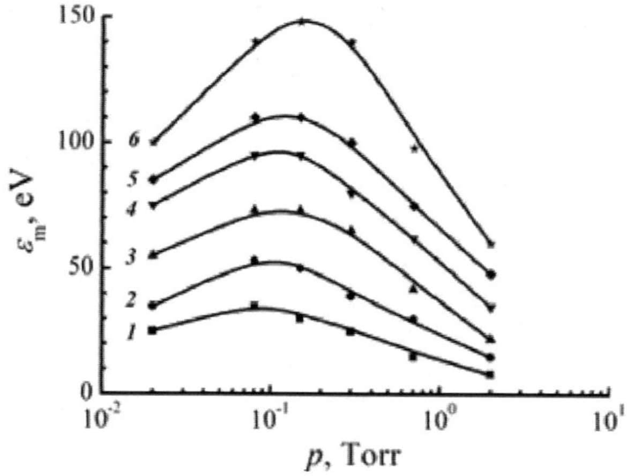

Figure 4. Dependences of energy ε_m on helium pressure for various voltages between the cathode and anode U_0, V: *1* - 200; *2* - 300; *3* - 400; 4 - 500; 5 - 600; 6 - 700.

4. Conclusions

It was experimentally established that in the breakdown stage, when the reduced electric field exceeds the critical value, the regime of continuous electron acceleration is observed. The energy of fast electrons grows with the applied voltage to the discharge gap and no monotonously depends on the gas pressure.

Acknowledgments

This work was supported by the Russian Foundation for Basic Research, project # 18-02-00336-a.

References

[1] Shibkov M V, Aleksandrov A F, Chernikov V A, Ershov A P and Shibkova L V 2009 J. Propul. Power 25 123 DOI: 10.2514/1.24803
[2] Shibkov V M, Aleksandrov A F, Ershov A P, Timofeev I B, Chernikov V A and Shibkova L V 2005 Plasma Phys. Rep. 31 9 795 DOI: 10.1134/1.2048839
[3] Shibkov V M, Dvinin S A, Ershov A P, Konstantinovskii R S, Surkont O S, Chernikov V A and Shibkova L V 2007 Plasma Phys. Rep. 33 1 72 DOI: 10.1134/S1063780X07010096
[4] Shibkov V M, Shibkova L V, Gromov V G, Karachev A A and Konstantinovskii R S 2011 High Temp. 49 2 155 DOI: 10.1134/S0018151X11020143
[5] Shibkov V M, Ershov A P, Chernikov V A and Shibkova L V 2005 Tech. Phys. 50 4 455 DOI: 10.1134/1.1901784
[6] Shibkov V M, Dvinin S A, Ershov A P and Shibkova L V 2005 Tech. Phys. 50 4 462 DOI: 10.1134/1.1901785
[7] Shibkov V M and Shibkova L V 2009 Tech. Phys. 54 10 1467
[8] Shibkov V M and Shibkova L V 2010 Tech. Phys. 55 1 58
[9] Ershov A P, Surkont O S, Timofeev I B, Shibkov V M and Chernikov V A 2004 *High Temp.* **42** 4 516

[10] Ershov A P, Surkont O S, Timofeev I B, Shibkov V M and Chernikov V A 2004 *High Temp.* **42** 5 669
[11] Ershov A P, Kalinin A V, Surkont O S, Timofeev I B, Shibkov V M and Chernikov V A 2004 High Temp. **42** 6 856
[12] Shibkov V M, Shibkova L V, Kopyl P V, Logunov A A 2019 *High Temp.* **57** 2 164 DOI: 10.1134/S0018151X19020196
[13] Shibkov V M 2019 *High Temp.* 57 3 322 DOI: 10.1134/S0018151X19010231
[14] Kopyl P V, Surkont O S, Shibkov V M and Shibkova L V 2012 *Plasma Phys. Rep.* **38** 6 503 DOI: 10.1134/S1063780X12050054
[15] Shibkov V M, Shibkova L V and Logunov A A 2018 *Plasma Phys. Rep.* 44 8 754 DOI: 10.1134/S1063780X18080056
[16] Shibkov V M, Shibkova L V and Logunov A A 2017 *Plasma Phys. Rep.* **43** 3 373 DOI 10.1134/S1063780X17030114
[17] Shibkov V M, Shibkova L V and Logunov A A 2018 *Moscow Univ. Phys. Bulletin* **73** 5 501 DOI: 10.3103/S0027134918050168
[18] Shibkov V M, Shibkova L V and Logunov A A 2017 *Moscow Univ. Phys. Bulletin* **72** 3 294 DOI: 10.3103/S0027134917030109
[19] Mesyats G A, Bychkov Yu I, Kremnev V V 1972 *Uspekhi Fizicheskikh Nauk* **107** 2 201
[20] Bokhan P A, Kolbychev G V 1980 *Letters to the ZhTF* **6** 7 418
[21] Bokhan P A 2018 *Uspekhi Fizicheskikh Nauk* **61** 12 1361
[22] Fon W C, Berrington K A, Hibben A 1981 *J.Phys.B.* **14** 307.

Journal of Physics: Conference Series **1394** (2019) 012002 doi:10.1088/1742-6596/1394/1/012002

Parameters of electron component in a pulsating discharge in a supersonic airflow

Shibkova L.V., Shibkov V.M., Logunov A.A., Andrienko A.A., Kornev K.N., Dolbnya D.S.

Moscow State University, 119991 Moscow, Russia

shibkov@phys.msu ru

Abstract. The plasma parameters of an unsteady pulsating transverse-longitudinal discharge created in subsonic and supersonic air flows are determined. It was shown that in the anode part of the plasma loop, the longitudinal electric field, as well as the electron density and temperature exceed the corresponding values in the cathode part of the loop. The effect of gas flow rate and discharge current on the temperature of electrons in plasma of a pulsating discharge in air and propane-air mixture is also shown.

1. Introduction

For efficient operation of a high-speed ramjet engine, the air-fuel mixture flow should have a supersonic velocity in all areas of its path. The high gas velocity makes it difficult to control the flow inside the combustion chamber. The air-fuel mixture passing with supersonic speed through the direct-flow chamber should have enough time for complete combustion in order to generate maximum thrust. In order to increase the efficiency of a ramjet engine with a supersonic flow rate in the combustion chamber, new methods are actively being developed to increase the burning rate and ensure complete combustion of fuel inside the engine without the use of various mechanical stabilizers. One of such methods is the use of various types of electric discharges [1-19]. In the works presented above, as well as in many other studies, it was shown that the use of gas-discharge plasma in most cases leads to ignition of the fuel.

The induction period varies from a microsecond time scale under condition of high-voltage nanosecond high-current discharges, freely localized and surface microwave discharges to a millisecond range for direct current discharges. Application of nanosecond and microsecond discharge with pulse repetition frequency up to several kilohertz cannot ensure constant mode of fuel combustion in supersonic flow. To stabilize fuel combustion, it is proposed in [10, 19-23] to use quasi-stationary pulsating transverse, longitudinal and transverse-longitudinal discharges produced in high-speed multicomponent chemically active gas flows. For the effective use of a transverse-longitudinal discharge as a fast ignition and combustion stabilization of air-hydrocarbon mixtures, it is necessary to know the main parameters of a low-temperature plasma.

The parameters of freely localized microwave discharges used to ignite hydrocarbon fuels in supersonic air flows are presented in [2, 3, 21, 24, 25]. The main properties and plasma parameters of a surface microwave discharge are given in [2, 3, 26-28]. This article is devoted to determining the parameters of the electronic component in plasma of a non-stationary transverse-longitudinal

Journal of Physics: Conference Series **1394** (2019) 012002 doi:10.1088/1742-6596/1394/1/012002

discharge created in subsonic and supersonic air flows, as well as in chemically active propane-air mixtures.

2. Experimental setup

The experiments were carried out on an installation consisting of a vacuum pressure chamber, a system for creating a supersonic air flow, a high-pressure air receiver, a high-pressure propane receiver, aerodynamic tunnels of rectangular cross section with attached air ducts, a high-voltage power supply for creating a pulsating discharge, a synchronization system and diagnostic equipment. The experimental setup and diagnostic methods are described in detail in [3, 8, 10, 19].

The diagnostic complex consists of spectrographs with digital recording of the spectrum; a probe diagnostics block with digital recording of the volt–ampere characteristics; pressure sensors; photo cameras; a high-speed video camera; oscilloscopes and computers. The discharge was produced using a power source with an output voltage of up to 4.5 kV and a discharge current of up to 16 A, the pulse duration being of up to 2 s. Propane-air mixtures were used as fuel. The second mass air flow rate in the experiment could vary from 25 g/s to 150 g/s, the second mass flow rate of propane varied from 3 g/s to 6 g/s. Aerodynamic channels were located inside a metal pressure chamber with a volume of 3 m^3, the air pressure in which could vary from 10 to 760 Torr.

A pulsed discharge was excited between two well streamlined electrodes [8, 10, 19]. In the experiments an electrode system with the shortest distance between the anode and cathode $d = 0.2$ mm was used. The electrodes were mounted inside the diverging wind tunnel. The air flow velocities varied from 150 to 550 m/s. For synchronizing the operation of electromechanical valves and discharge, a special scheme is used, which provides a time delay between the opening of the valves in the air ducts, which supply air and propane gas, and the start of the discharge.

The emission spectrum of the gas-discharge plasma was recorded using an AvaSpec-2048-2-DT dual-channel spectrograph with an inverse linear dispersion of 0.05 nm/mm for the first channel and 0.32 nm/mm for the second channel.

A distinctive feature of the spectrum is the presence of an intense continuum, which may electron bremsstrahlung from ions or atoms and/or recombination emission. Estimates show that, under these experimental conditions, the intensity of bremsstrahlung caused by electron–ion collisions exceeds the intensity of bremsstrahlung caused by electron-neutral collisions. The electron temperature in the plasma channel was determined by comparing the measured spectrum with the theoretical bremsstrahlung spectrum calculated for different electron temperatures.

To this, it was necessary to correct the measured spectrum with allowance for the spectral sensitivity of the spectrograph. The absolute sensitivity of the spectrograph was calibrated using a tungsten incandescent lamp, whose emission intensity is determined by the temperature of the tungsten ribbon, which depends on the heating current.

The electron density was determined from the Stark broadening of the H_α and H_β spectral lines of the hydrogen Balmer series.

One of the main characteristics of the discharge is the electric field strength in the plasma. In previous studies [15, 19], it was shown that the average electric field along the length of the plasma loop increases with increasing air velocity. Moreover, at any air flow rate, an increase in the discharge current leads to a decrease in the electric field in the plasma.

In the present investigation, the distributions of floating potentials along the anode and cathode parts of the plasma loop were measured by the probe method. The measurements were carried out using a single unit of two probes, the distance between which is 0.7 cm. In our case, the electron energy distribution function in a pulsed discharge plasma is close to the Maxwell function with a temperature of the order of 1 eV. In this case, the difference between the floating potential and the potential of a space of the order of several $kTe = 1$ eV. Therefore, the electric field strength was determined as the difference of floating potentials divided by the distance between the probes.

3. Experimental results

Investigations of the plasma parameters of a pulsating transverse-longitudinal discharge were performed at changing the gas flow velocity from 200 to 500 m/s and the discharge current from 5.5 to 15.5 A.

The dependences of the longitudinal electric field strength in the anode and cathode parts of the plasma loop are shown in Figure 1.

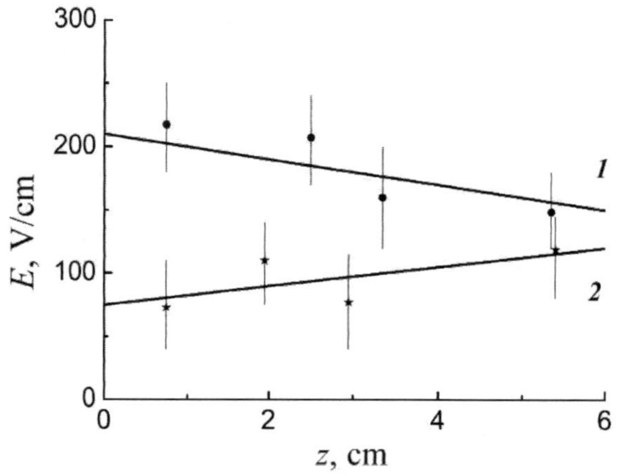

Figure 1. The longitudinal distribution of the electric field strength in the anode (*1*) and cathode (*2*) parts of the plasma loop.

It can see that in the anode part of the loop, the electric field has a maximum value near the anode $E_a = 210$ V/cm and decreases slightly with increasing of distance z from the anode, while the electric field has a minimum value near the cathode $E_c = 77$ V/cm and increases slightly with increasing distance z from the cathode. The excess of the electric field E_a over E_c reaches 60 % near the electrodes and 20 % on distance $z = 6$ cm. Different values of the electric field strengths E_a and E_c lead to different values of the electron concentration and temperature in the anode and cathode parts of the plasma channel. It was too shown that in the anode part of the plasma loop the electron density systematically exceeds n_e at the same distances z in the cathode part of the plasma loop by about 15 %. Similar results were obtained when measuring the longitudinal distribution of electron temperature. It was shown that the electron temperature decreases along the plasma channel downstream as well as the electron concentration, when the distance z from the electrodes increases. In this case, the excess of the measured electron temperature in the plasma channel near the anode over T_e near the cathode reaches 30%.

The dependences of the electron temperature in the plasma on the air flow velocity were shown in Figure 2 at various values of the discharge current. It can be seen that with increasing air flow velocity, the average electron temperature decreases both at a discharge current of 5.5 A and at $i = 15.5$ A. Moreover, at any value of the flow velocity the electron temperature increases with increasing discharge current.

Studies have also been fulfilled on the effect of propane injection into a high-speed air flow on the electron temperature in pulsating transverse-longitudinal discharge plasma. It was experimentally obtained that when propane is added to the air flow, the electron temperature decreases. This is due to the fact that the addition of propane to the air flow leads to an increase in the fraction of electron energy losses due to dissociation, excitation and ionization of impurity molecules. In the case of a propane-air mixture, the dependence of the average electron temperature on the velocity does not qualitatively change as compared to the air flow, i.e. an increase in velocity leads to a decrease in the electron temperature (Figure 2, line *3*).

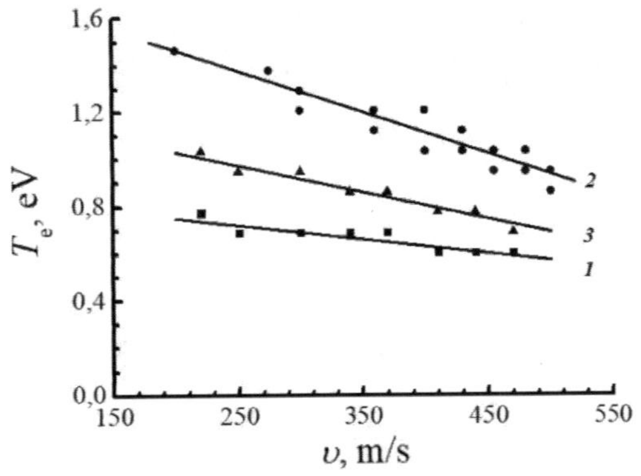

Figure 2. Dependence of the electron temperature on the air (*1* and *2*) and the air-propane (*3*) flow velocity at various values of the discharge current 5.5 A (*1*); 12.5 A (3);15.5 A (*2*).

Mathematical modeling of the electron energy distribution function in a non-equilibrium pulsed plasma was performed to justification of the experimentally obtained results on measuring the temperature of electrons. The electron energy distribution function was calculated based on the solution of the Boltzmann equation for the spherically symmetric component of the EEDF.

$$\frac{1}{n_e n_0}\sqrt{\frac{m\varepsilon}{2}}\frac{\partial(n_e f(\varepsilon,t))}{\partial t} = \frac{\partial}{\partial\varepsilon}\left(\left(\frac{1}{3}\left(\frac{eE}{n_0}\right)^2\frac{\varepsilon}{\sigma_c} + \delta k T_0 \varepsilon^2 \sigma_c\right)\frac{\partial f}{\partial\varepsilon}\right) +$$

$$+\frac{\partial}{\partial\varepsilon}\left((\delta\varepsilon^2\sigma_c + 4\varepsilon B_0 \sigma_{re})f\right) - \begin{cases}\sum\limits_{i\in\square_e}\gamma_i^+ \varepsilon\sigma_i(\varepsilon)f(\varepsilon), \varepsilon \ge \varepsilon_i^{nop}\\ 0, \qquad \varepsilon < \varepsilon_i^{nop}\end{cases}$$

$$+\sum_{i\in\square_n}\gamma_i^+(\varepsilon+\varepsilon_i^{nop})\sigma_i(\varepsilon+\varepsilon_i^{nop})f(\varepsilon+\varepsilon_i^{nop})$$

$$++\sum_{i\in\square_i}\gamma_i^+(\varepsilon+\varepsilon_i^{nop})\sigma_i(\varepsilon+\varepsilon_i^{nop})f(\varepsilon+\varepsilon_i^{nop})$$

$$+\begin{cases}\sum\limits_{i\in\square_{обр}}\gamma_i^- \varepsilon\sigma_i(\varepsilon)f(\varepsilon-\varepsilon_i^{nop}), \quad \varepsilon \ge \varepsilon_i^{nop}\\ 0, \qquad \varepsilon < \varepsilon_i^{nop}\end{cases} +$$

$$+2\pi e^4\gamma_e \ln\Lambda\frac{\partial}{\partial\varepsilon}\left(f(\varepsilon)\int_0^\varepsilon \varepsilon'^{1/2}f(\varepsilon')d\varepsilon' + \frac{2}{3}\frac{\partial f}{\partial\varepsilon}\left(\int_0^\varepsilon \varepsilon'^{3/2}f(\varepsilon')d\varepsilon' + \varepsilon^{3/2}\int_0^{¥} f(\varepsilon')d\varepsilon'\right)\right),$$

$$\ln\Lambda = \ln\left(\frac{T_e^3}{e^6 N\gamma_e}\right)$$

Journal of Physics: Conference Series **1394** (2019) 012002 doi:10.1088/1742-6596/1394/1/012002

Here $f(\varepsilon, t)\varepsilon^{1/2}$ is the spherically symmetric component of the electron energy distribution function, $\sigma_{c,i}(\varepsilon)$ is transport cross-section of the *i*-th component of the mixture, $\sigma_c(\varepsilon, t)$ is total transport section of the mixture, γ_i is mole fraction of the corresponding component, N – total concentration (mol/cm^3), E is the electric field (V/cm), $C_{p,i}$ – the molar heat capacity of the *i*-th component at constant pressure, α_{iq}^- and α_{iq}^+ – stoichiometric coefficients, $\ln \Lambda$ – Coulomb logarithm.

It was found that a close to Maxwell electron energy distribution function with an effective electron temperature of the order of one electron-volt is formed at high degrees of plasma ionization, which is realized under conditions of an unsteady transverse-longitudinal discharge in a gas stream (Figure 3, dash line *2*).

For the same value of the reduced electric field E/n= 30 Td, but in a weakly ionized plasma with an ionization degree of 10^{-6}, a non-equilibrium electron energy distribution function is realized, the form of which strongly depends on the degree of vibrational excitation of air molecules (Figure 3, curve *1*).

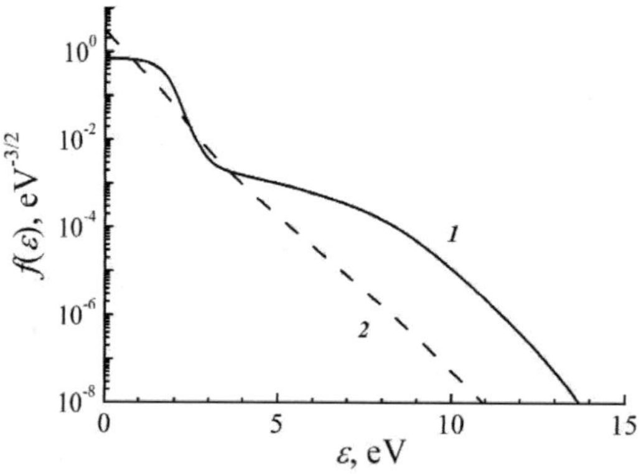

Figure 3. Electron energy distribution function at E/n = 30 Td and the plasma ionization degree of 10^{-6} (solid curve *1*) and 10^{-3} (dashed curve *2*)

4. Conclusions

The spatial distributions of the plasma parameters of a pulsating discharge as a function of air velocity and discharge current are obtained. Significant excess of the longitudinal electric field in the anode part of the plasma loop of the pulsating transverse-longitudinal discharge over the field in the cathode part of the loop was revealed.

Acknowledgments

The work was supported by the Russian Foundation for Basic Research (grant No. 18-02-00336-a).

References

[1] Starikovskaya S M 2006 *J. Phys.* D **39** R265.
[2] Konstantinovskii R S, Shibkov V M and Shibkova L V 2005 *Kinet. Catal.* 46 6 775
[3] Shibkov M V, Aleksandrov A F, Chernikov V A, Ershov A P and Shibkova L V 2009 *J. Propuls. Power* **25** 123
[4] Jacobsen L, Carter C, Baurle R, Jackson T, Williams S, Barnett J, Tam C-J and Bivolaru D 2008 *J. Propuls. Power* **24** 4 641
[5] Starikovskii A Y, Anikin N B, Kosarev I N, Mintoussov E I, Nudnova M M, Rakitin A E, Roupassov D V, Starikovskaia S M and Zhukov V P 2008 *J. Propuls. Power* **24** 1182.
[6] Adamovich I V, Lempert W R Rich J W and Utkin Y G 2008 *J. Propuls. Power* **24** 1198
[7] Jacobsen L, Carter C, Baurle R, Jackson T, Williams S, Barnett J, Tam C-J and Bivolaru D 2008 *J. Propuls. Power* **24** 4 641
[8] Shibkov V M, Shibkova L V, Kopyl P V, Logunov A A 2019 *High Temp.* **57** 2 164

[9] Shibkov V M 2019 *High Temp.* **57** 3 322
[10] Kopyl P V, Surkont O S, Shibkov V M and Shibkova L V 2012 *Plasma Phys. Rep.* **38** 6 503
[11] Shibkov V M and Shibkova L V 2009 *Tech. Phys.* **54** 10 1467
[12] Shibkov V M, Shibkova L V 2010 *Tech. Phys.* **55** 1 58
[13] Aleksandrov N L, Kindysheva S V, Kochetov I V 2014 *Plasma Sources Science and Technology* **23** 1
[14] Aleksandrov N L, Kindysheva S V, Kukaev E N, Starikovskaya S M, Starikovskii A Yu 2009 Plasma Phys. Rep. **35** 10 867
[15] Kosarev I N, Aleksandrov N L, Kindysheva SV, Starikovskaia S M, Starikovskii A Yu 2009 *Combust. Flame*, **156** 221
[16] Alekseev A I, Vaulin D N, Stepanov A I, Chernikov VA 2018 *High Temp.* **56** 1 20
[17] Alekseev A I, Vaulin D N, Deshko K I, Chernikov V A 2018 *Plasma Phys. Rep.* **44** 8 766
[18] *Leonov S, Yarantsev D, Carter C* 2009. J. of Propuls. Power. **25** 289
[19] Shibkov V M, Shibkova L V and Logunov A A 2018 *Plasma Phys. Rep.* **44** 8 754
[20] Shibkov V M, Shibkova L V and Logunov A A 2017 *Plasma Phys. Rep.* **43** 3 373
[21] Shibkov V M, Aleksandrov A F, Ershov A P, Timofeev I B, Chernikov V A and Shibkova L V 2005 *Plasma Phys. Rep.* **31** 9 795
[22] Shibkov V M, Shibkova L V and Logunov A A 2018 *Moscow Univ. Phys. Bulletin* **73** 5 501
[23] Shibkov V M, Shibkova L V and Logunov A A 2017 *Moscow Univ. Phys. Bulletin* **72** 3 294
[24] Shibkov V M 1997 High Temperature 35 5 681
[25] Shibkov V M 1997 High Temperature 35 6 858
[26] Shibkov V M, Dvinin S A, Ershov A P, Konstantinovskii R S, Surkont O S, Chernikov V A and Shibkova L V 2007 *Plasma Phys. Rep.* **33** 1 72
[27] Shibkov V M, Ershov A P, Chernikov V A and Shibkova L V 2005 *Tech. Phys.* **50** 4 455
[28] Shibkov V M, Dvinin S A, Ershov A P and Shibkova L V 2005 *Tech. Phys.* **50** 4 462

Journal of Physics: Conference Series **1394** (2019) 012003 doi:10.1088/1742-6596/1394/1/012003

Ignition and combustion stabilization of multicomponent supersonic chemically active flows in low-temperature plasma conditions

Shibkov V. M.

Moscow State University, 119991 Moscow, Russia

shibkov@phys.msu ru

Abstract. An electrodeless freely localized microwave discharge, a surface microwave discharge, a non-equilibrium non-stationary transverse-longitudinal pulsating discharge were used for ignition and combustion stabilization. The induction periods for various discharges and the possibilities of their application for supersonic combustion stabilization inside the direct-flow aerodynamic channels are experimentally determined. A comparison of experimental results with the data of mathematical modeling is presented.

1. Introduction

Low-temperature plasma in a gas flow is widely used in scientific research and various technological applications [1-10]. The possible means of effective control of combustion processes via different physical actions has been discussed in the scientific literature for several decades. In recent years, the interest in the intensification of the combustion of gas-phase systems with the use of different forms of gas discharge considerably increased. The methods of control of the combustion of air–hydrocarbon supersonic flows based on the generation of electric discharges seem to be most promising at present.

The aim of this work is an experimental study of supersonic ignition and combustion of air-hydrocarbon mixtures initiated by low-temperature gas-discharge plasma. In the article transverse-longitudinal discharges [11-15], freely localized microwave discharges [2, 6, 16] and surface microwave discharges excited on a dielectric antenna [17-19] are used for the initiation of ignition and combustion stabilization of hydrocarbon fuels in supersonic air flows.

2. Experimental setup

Experiments were carried out on the installation consisting of a vacuum chamber, a receiver of a high pressure of air, a receiver of a high pressure of propane, a system for mixing propane with air, a system for producing a supersonic gas flow, two magnetron generators, two systems for delivering microwave power to the chamber, cylindrical and rectangular aerodynamic channels, two sources of high-voltage pulses, a synchronization unit, and a diagnostic system. A detailed description of the experimental setup is presented in [6, 9, 11, 18].

The basic component of the experimental setup is an evacuated metal cylindrical chamber, which serves simultaneously for supersonic flow creation, and as a tank for the expiration of gases or combustion products. The inner diameter of the vacuum chamber is 1 m, and its length is 3 m. A supersonic flow was produced by filling the vacuum chamber with air through a specially profiled Laval nozzle mounted on the outlet tube of the electromechanical valve. In our experiments, we used cylindrical and rectangular nozzles.

The microwave source is a pulsed magnetron generator operating in the centimeter wavelength range. The parameters of the magnetron generator are as follows: the wavelength is $\lambda = 2.4$ cm, the pulsed microwave power is $W_p < 200$ kW, the pulse duration is $\tau = 1$–100 μs, and the period-to-pulse duration ratio is $Q = 1000$. Microwave power was delivered to the discharge chamber through a 9.5×19-mm rectangular waveguide. All the components of the microwave transmission line were sealed. To avoid electric breakdowns inside the waveguide, it was filled with an insulating gas (SF_6) at a pressure of 4 atm. The vacuum system of the chamber allows us to vary the pressure over a wide range from 10^{-3} to 10^3 Torr.

A transverse-longitudinal pulsating discharge was excited between two well streamlined electrodes. The design of the electrode unit was described in [9, 11, 12]. The minimum distance between the electrodes was varied from 0.1 to 3 mm. The electrodes were mounted inside the diverging wind tunnel. The discharge was produced using a power source with an output voltage of up to 4.5 kV and a discharge current of up to 20 A, the pulse duration being of up to 2 s. The air flow speed varied from 150 to 550 m/s. The second mass air flow rate in the experiment could vary from 25 g/s to 150 g/s, the second mass flow rate of propane varied from 3 g/s to 6 g/s.

Combustion was realized inside smooth (without stagnant zones) wind tunnels of rectangular cross section. Variable cross-section channels were used to prevent their thermal blocking. The electrode system of a special configuration was used to create a pulsating transverse-longitudinal discharge without additional initiation in a wide range of flow velocities of $\upsilon = 100$-600 m/s and in a wide range of air pressures of $p = 40$–760 Torr. The electrical circuit consists of a series-connected stationary power source with an output voltage of $U_0 = 2.0$-4.5 kV and an internal resistance of $r = 100\ \Omega$, the ballast resistance, the value of which could be gradually switched to five fixed positions $R_4 = 145$-$675\ \Omega$, non-inductive resistance $R_5 = 0.32\ \Omega$ for measuring the discharge current. A non-inductive voltage divider with a division factor of $k = 8760$ was used to measure the voltage drop on the discharge.

The diagnostic complex consists of monochromators and spectrographs with digital recording of the spectrum; a probe diagnostics block with digital recording of the volt–ampere characteristics; pressure sensors; thermocouples; strain sensors; a shadow system; refracting laser sensors; an electric probe incandesced in the flame flow; a system of flame conductivity measurement; electron sensors measuring the concentration of propane and carbon dioxide, temperature, and absolute and relative humidity; digital photo cameras; a high-speed digital video camera; digital oscilloscopes; and computers.

The completeness of hydrocarbon fuel combustion under plasma-assisted combustion was determined via measurements of the gas temperature in the closed chamber immediately after switch off of discharge with help of a calibrated thermocouple and electron thermometer placed near walls of the chamber far from the hot flow.

Simultaneously with the temperature measurement, the time behavior of the pressure change in the closed chamber in the process of air injection without a discharge, with a discharge without combustion, and as a result of hydrocarbon fuel combustion were also recorded. This made it possible to determine the fraction of the burned propane.

The concentration of water vapors in the closed chamber after propane burning was also recorded. Knowing the amount of propane introduced into the flow, one can calculate the water vapor concentration that must be formed in the process of complete fuel combustion. Comparison of this value with the measured water vapor concentration that forms in the experiment makes it possible to calculate the completeness of propane combustion.

The combustion completeness was also estimated by the time dependence of the thrust force appearing under the plasma-assisted combustion of a supersonic propane–air flow in a 60-cm-long aerodynamic channel equipped with an outlet nozzle.

Journal of Physics: Conference Series **1394** (2019) 012003 doi:10.1088/1742-6596/1394/1/012003

3. Experimental results

It is known, that self-sustained microwave discharges, as well as high-voltage nanosecond electrode discharges, exist at high values of the reduced electric field. However as against the nanosecond high-voltage discharge, high value of the reduced electric field under conditions of a microwave discharge can be supported during long (hundreds microseconds) time. Therefore, features of self-sustained microwave discharges are very perspective from the point of view of the fast reliable ignition of high speed air-hydrocarbon flows. However, as well as for other types of the gas discharges, combustion of high speed combustible flow under condition of microwave discharges stops, as soon as the supply of microwave energy is switched off.

For stationary work of the engine under condition of non-stationary low-temperature plasma for ignition of hydrocarbon-air mixture it is necessary to optimize a mode of initiation of the pulsed discharge, i.e. the value of energy put into plasma, pulse duration and pulse repetition rate. This can be done using the discharge in a programmable mode. In a programmable pulse mode gas breakdown and creation of plasma is carried out with the help of a short powerful pulse, and maintenance of the formed plasma and energy input occur during the long low-power pulse.

A gas discharge is usually considered only as a source of thermal energy introduced into the system. But a different degree of gas ionization is achieved for different types of gas discharges at the same input power. In this case, the electrical energy is redistributed in various ways along the internal degrees of freedom of the molecular gas. This redistribution in a very strong degree depends on the reduced electric field, which, in turn, is determined by the electrodynamics of the discharge. Different results can be achieved using the same power source. For example, to ignite a high-speed propane-air flow with a Mach number $M = 2$, we used a direct current electrode discharge, a freely localized electrodeless microwave discharge, a surface microwave discharge, a transverse-longitudinal discharge.

Figure 1 presents a photograph of the general views of the ignition and combustion of fuels in high-speed air flows under conditions of freely localized microwave discharge (*1*), surface microwave discharge (*2*), (*3*) and transverse-longitudinal pulsating discharge (*4*), (*5*).

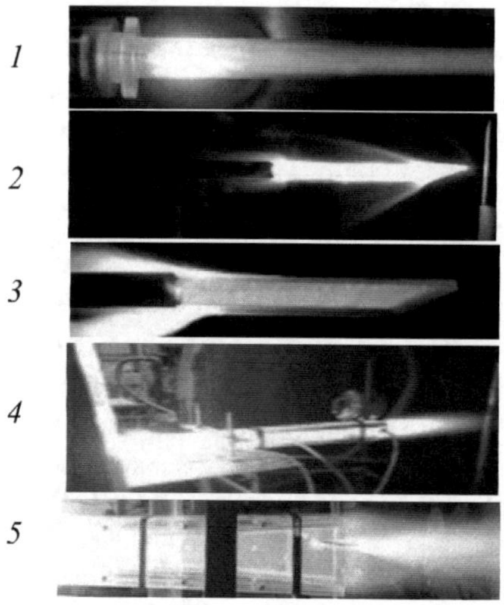

Figure 1. General view of the ignition and combustion of fuels in high-speed air flows under conditions of freely localized microwave discharge (*1*), surface microwave discharge (*2*), (*3*) and transverse-longitudinal pulsating discharge (*4*), (*5*).

The experiments show that under conditions of pulsed microwave discharges with pulse duration of 100 μs (photo *1*, *2*, *3* in Figure 1) the combustion completeness of propane is low, since when the pulsed discharge is turned off, combustion stops. Pulsed discharges can be used to quickly ignite; however, in order to maintaining stationary combustion it is necessary to use mechanical stabilizers. The maximum completeness of fuel combustion reaches ~90% in the conditions of a quasi-stationary pulsating transverse longitudinal discharge (photo *4*, *5* in Figure 1).

Figure 2 shows the experimentally obtained dependence of the induction period on the reduced electric field for a supersonic propane-air flow $M = 2$.

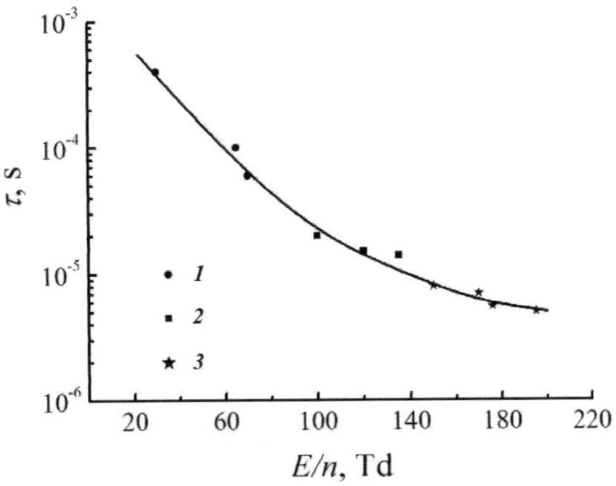

Figure 2. Dependence of the induction period on the reduced electric field. Transverse-longitudinal discharge (*1*), freely localized microwave discharge (*2*), surface microwave discharge (*3*).

4. Mathematical modeling

Numerical model has been developed to study C_3H_8-air plasma-stimulated combustion. In the article, the ignition of gas fuel mixtures was simulated under the conditions of non-equilibrium gas discharge plasma of microwave discharges. Simulation was performed for a stationary propane-air mixture.

The kinetic model includes 166 components and 1168 direct and reverse reactions. In the simulation takes into account the following components: neutral particles: H, H_2, N_2, N, NH, NH_2, NH_3, N_2H, N_2H_2, N_2H_3, N_2H_4, N_2O, NO, NO_2, NO_3, NHO, HNO, HNO_2, HNO_3, O, O_2, OH, H_2O, HO_2, H_2O_2, O_3, C, C_2, CH, CH_2, CH_3, CH_4, C_2H, C_2H_2, C_2H_3, C_2H_4, C_2H_5, C_2H_6, CO, CO_2, HCO, CH_2O, CH_3O, CH_2OH, CH_3OH, CH_3O_2, CH_3OOH, C_2HO, C_2H_2O, CH_2CO, CH_3CO, CH_3CHO, C_2H_5O, $C_2H_5O_2$, $C_2H_5O_2H$, CN, C_2N, C_2N_2, HCN, NCO, C_3H_4, C_3H_5, C_3H_6, н-C_3H_7, изо-C_3H_7, C_3H_8, C_3H_5O, $C_3H_5O_2$, $C_3H_5O_2H$, C_3H_6O, н-C_3H_7O, изо-C_3H_7O, н-$C_3H_7O_2$, изо-$C_3H_7O_2$, н-$C_3H_7O_2H$, изо-$C_3H_7O_2H$, C_4H, C_4H_2, C_4H_3, C_4H_6, C_4H_7, C_4H_8, н-C_4H_9, втор-C_4H_9, C_4H_{10}, electronically excited states: H(2s+2p), $H_2(c^3\Pi_u)$, $H_2(a^3\Sigma_g^+)$, $H_2(e^3\Sigma_u^+)$, $H_2(d^3\Pi_u)$, $N_2(A^3\Sigma_u^+)$, $N_2(B^3\Pi_g)$, $N_2(C^3\Pi_u)$, $N_2(D^3\Sigma_u)$, $N_2(a^1\pi_g)$, $N_2(b^1\pi_u)$, $N_2(a^1\Sigma_g^+)$, $N_2(11.87)$, $N_2(a'^1\sigma_u^-)$, $N(^2D)$, $N(^2P)$, $O(^1D)$, $O(^1S)$, $O_2(a^1\Delta_g)$, $O_2(b^1\Sigma_g^+)$, positive ions: H_2^+, H_3^+, H_5^+, N^+, N_2^+, HN_2^+, N_3^+, N_4^+, O^+, O_2^+, OH^+, H_2O^+, HO_2^+, C^+, CH^+, CH_2^+, CH_3^+, C_2^+, C_2H^+, $C_2H_2^+$, $C_2H_3^+$, $C_2H_4^+$, $C_2H_5^+$, $C_2H_6^+$, C_3^+, $C_3H_2^+$, $C_3H_3^+$, $C_3H_4^+$, $C_3H_5^+$, $C_3H_6^+$, $C_3H_7^+$, $C_3H_8^+$, H^+, CO^+, CH_4^+, O_3^+, CO_2^+, H_3O^+, $O_3H_2^+$, O_4^+, $O_2^+\cdot N_2$, $NO^+\cdot N_2$, $NO^+\cdot O_2$, $NO^+\cdot NO$, negative ions: O^-, OH^-, CH_4^-, H^-, O_2^-, O_3^-, O_4^-, CH_2^-, NO_3^-, NO_3^{-*} and electrons e.

The model of ignition of fuel mixtures was based on a system of equations describing the oxidation processes and including the equation of energy balance, the equation of the balance of density (concentration) of particles and the equation of state:

$$\frac{d\gamma_i}{dt} = G_i - \gamma_i \sum_{j\in\kappa} G_j\,,\quad G_i = \sum_{q\in\Re} \frac{\alpha_{iq}^- - \alpha_{iq}^+}{N}\left[R_q^+ - R_q^-\right]\,,$$

$$R_q^\pm = k_q^\pm \prod_{j\in\kappa}\left(N\gamma_{n_q^\pm}\right)^{\alpha_q^\pm}\,,\quad \frac{dN}{dt} N \sum_{k\in\kappa} G_k\,, p = NRT_0\,.$$

The calculation of the kinetic coefficients of the electrons was carried out based on the solution of the Boltzmann equation for the spherically symmetric component of the EEDF.

The dependences of the propane-air induction periods on the reduced electric field are presented in Figure 3 under conditions of microwave discharges.

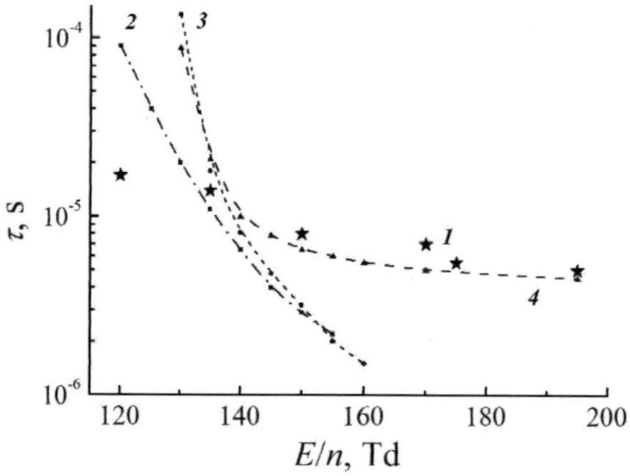

Figure 3. Dependences of the induction period on the reduced electric field.

In addition to the experimental data (*1*), the graph shows the results of calculating the dependence of the induction time of the propane-air mixture on the reduced field for several cases. First, the discharge was calculated at a constant value of the amplitude of the electric field (*2*). Secondly, a similar calculation was performed in the case of a preliminary breakdown of the gas by a microwave pulse with a reduced field of 150 Td, and the pulse duration 1.3 μs, which led to an increase in the electron concentration up to 10^{12} cm^{-3} (*3*). Thirdly, the discharge was calculated with a constant amplitude of the electric field with the electron concentration limited to 4.2×10^{12} cm^{-3} (*4*). It can be seen from the Figure 3 that such a limitation affects the ignition times for large values of the reduced electric field and gives dependence better consistent with the experiment. In plasma of microwave discharges, generated in a focused beam of electromagnetic radiation, the electric field is screened. This electrodynamic effect leads to a decrease in the electron concentration behind the leading front of the propagating discharge.

5. Conclusions

It is shown, that the all types of the discharges result in a reliable ignition of hydrocarbon fuel. The stationary plasma-assisted combustion of a propane–air mixture in aerodynamic channels without the use of different dead zones is realized under condition of a nonstationary transvers-longitudinal pulsating discharge. Mathematical modeling confirmed the strong influence of the reduced electric field on the induction period

Acknowledgments

This work was supported by the Russian Foundation for Basic Research, project # 18-02-00336-a.

References

[1] Starikovskaya S M 2006 *J. Phys.* D **39** R265.
[2] Konstantinovskii R S, Shibkov V M and Shibkova L V 2005 *Kinet. Catal.* 46 6 775
[3] Starikovskii A Y, Anikin N B, Kosarev I N, Mintoussov E I, Nudnova M M, Rakitin A E, Roupassov D V, Starikovskaia S M and Zhukov V P 2008 *J. Propuls. Power* **24** 1182.
[4] Adamovich I V, Lempert W R Rich J W and Utkin Y G 2008 *J. Propuls. Power* **24** 1198
[5] Jacobsen L, Carter C, Baurle R, Jackson T, Williams S, Barnett J, Tam C-J and Bivolaru D 2008 *J. Propuls. Power* **24** 4 641
[6] Shibkov M V, Aleksandrov A F, Chernikov V A, Ershov A P and Shibkova L V 2009 *J. Propuls. Power* **25** 123

[7] Fridman A, Nester S, Kennedy L, Saveliev A and Mutaf-Yardimci O 1999 *Prog. Energy Combust. Sci.* **25** 211
[8] Pashchenko N T and Raizer Yu P 1982 *Sov. J. Plasma Phys.* **8** 617
[9] Shibkov V M, Shibkova L V, Kopyl P V, Logunov A A 2019 *High Temp.* **57** 2 164
[10] Shibkov V M 2019 *High Temp.* 57 3 322
[11] Kopyl P V, Surkont O S, Shibkov V M and Shibkova L V 2012 *Plasma Phys. Rep.* **38** 6 503
[12] Shibkov V M, Shibkova L V and Logunov A A 2018 *Plasma Phys. Rep.* **44** 8 754
[13] Shibkov V M, Shibkova L V and Logunov A A 2017 *Plasma Phys. Rep.* **43** 3 373
[14] Shibkov V M, Shibkova L V and Logunov A A 2018 *Moscow Univ. Phys. Bulletin* **73** 5 501
[15] Shibkov V M, Shibkova L V and Logunov A A 2017 *Moscow Univ. Phys. Bulletin* **72** 3 294
[16] Shibkov V M, Aleksandrov A F, Ershov A P, Timofeev I B, Chernikov V A and Shibkova L V 2005 *Plasma Phys. Rep.* **31** 9 795
[17] Shibkov V M, Shibkova L V, Gromov V G, Karachev A A and Konstantinovskii R S 2011 *High Temp.* **49** 2 155
[18] Shibkov V M and Shibkova L V 2009 *Tech. Phys.* **54** 10 1467
[19] Shibkov V M and Shibkova L V 2010 *Tech. Phys.* **55** 1 58

Influence of the sign and magnitude of a surface charge on the breakdown voltage of a barrier corona discharge in Ar

Yu S Akishev[1,2], A A Balakirev[1], M A Medvedev[3], A V Petryakov[1], N I Trushkin[1]

[1]SRC RF TRINITI, Moscow, Troitsk, Pushkovykh str. 12, 108840, Russia
[2]NRN University MEPhI, Moscow, Kashirskoe shosse 32, 115409, Russia.
[3]P. N. Lebedev Physical Institute, Leninsky Avenue 53, 119991, Moscow, Russia

akishev@triniti.ru

Abstract. The topic of this article is related to the breakdown phenomena in the surface dielectric barrier discharge (SDBD). The surface charge, locally deposited on the dielectric barrier, changes markedly the configuration and intensity of the electric field both on the barrier and in the gas gap. In turn, a change in the local electric field affects the rate of ionization multiplication of electrons in this area and, accordingly, influences the breakdown conditions of both volume and surface barrier discharges. In the case of sinusoidal SDBD, the first breakdown in each next half-cycle will be controlled by the charge deposited on the barrier during the previous half-cycle. Contrariwise, if there are several successive breakdowns in the half-cycle, each next breakdown after the first one will be controlled predominantly by the charge and plasma created by the previous breakdown in this half-cycle but not only by the charge deposited in the previous half-cycle. These features of the SDBD breakdown were investigated by the example of the barrier corona in argon at atmospheric pressure in the pin-to-plane configuration. This discharge allows us the controlling of the location of the deposited charge. Surface charge was deposited in advance and by the SDBD itself.

1. Introduction

The monopulse or periodic gas discharges created in electrode systems with one or two electrodes covered with a thin dielectric (for example, corona discharge in the gap between a metallic pin and a flat dielectric barrier or the discharge between two dielectric barriers), are widely used in scientific research and numerous practical applications as well [1-3]. Barrier discharges are the self-extinguished ones. A reason is that the electric current arising after every breakdown of the gap deposits on the barrier(s) the charge(s) creating the additional electric field $E_{opposite}$ directed against the initial E-field in the gap. The $E_{opposite}$ quickly grows with time and, eventually, reduces the resultant field in the gap down to such low magnitude which is not able further to support the discharge. The new current pulse can arise if the voltage across the gap continues to increase and the resultant electric field between electrodes will increase again to the breakdown magnitude. For this reason, in many cases (especially in gases at atmospheric pressure and with the use of AC voltage of high amplitude) barrier discharges exhibit themselves in the form of irregular current pulses (microdischarges) arising one after another in each half-cycle on a phase of the increasing voltage [4-7].

One of the important questions in physics of the microdischarges formation in the periodical barrier discharge is the question about the influence of the amount and sign of the surface charge deposited in the previous half-cycle on the first breakdown in each subsequent half-cycle of the sinusoidal voltage. At the qualitative level, it is a well-known fact that the surface charge, locally deposited on the dielectric barrier, changes markedly around the configuration and intensity of the electric field both on the barrier and in the gas gap. In turn, a change in the local electric field affects the rate of ionization multiplication of electrons in this area and, accordingly, influences the breakdown conditions of both volume and surface barrier discharges. In the case of sinusoidal DBD, the first breakdown in each next half-cycle will be controlled by the charge deposited on the barrier during the previous half-cycle. However, if there are several successive breakdowns in each half-cycle, all breakdowns being happened in the half-cycle after the first breakdown will be controlled by the charge and plasma that were created by the previous breakdowns just in this half-cycle but not by the charge deposited in the previous half-cycle. These features of the DBD breakdown were investigated experimentally by the example of the barrier corona in argon at atmospheric pressure in the pin-to-plane geometry. This discharge allows us controlling the location of both the microdischarges and charges deposited by them. Besides, the manipulation of the value of the electric charge pre-posited on the barrier surface was done in the experiments; this charge was formed by an auxiliary discharge.

2. Experimental setup

The pin-to-plane electrode system consists of the following elements. The plane electrode, a copper disc with a diameter of 110 mm, was covered with a thin dielectric film (lavsan, 80 μm thick, $\varepsilon = 3.5$) (Figure 1). High-voltage (HV) electrode (the pin) is the stainless steel wire with a diameter of 1 mm, the end of which was sharpened on a cone with a radius of 0.1 mm. The distance between the wire tip and the film is 1 mm (at smaller distances, there was an electric breakdown of a thin film by a high voltage being applied to the pin electrode). Before applying the sinusoidal voltage powering the barrier corona, an electric charge Q was deposited in advance on the film. It was done by applying a DC high voltage to the pin. This voltage initiates a short-term discharge both in the tip-film gas gap and on the film surface as well. Variation of both the amount and sign of the pre-posited charge Q was performed by changing the amplitude and polarity of the auxiliar DC voltage. The Q amount was being determined from the measured voltage on the capacitor C connected between the copper disk and the ground (Figure 1). After the charge deposition, the auxiliar DC voltage was disconnected, and the sinusoidal generator was connected to the HV pin.

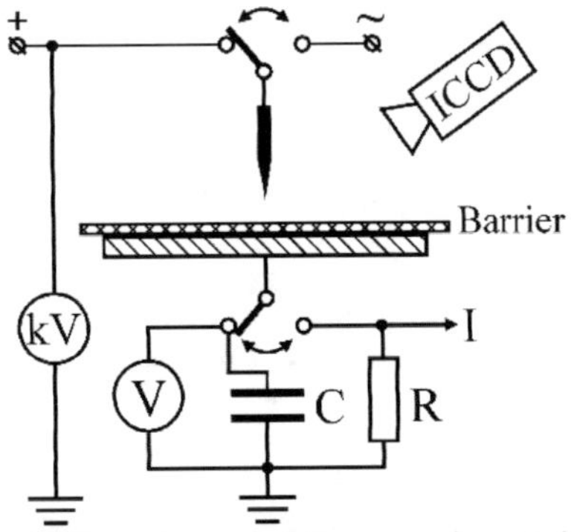

Figure 1. The scheme of the experimental setup.

The generator had two strengths useful for our experiments. First, the generator enables to vary the frequency and the amplitude U_0 of the output sinusoidal voltage within the range of 50 Hz - 100 kHz and 0 - 4.5 kV respectively and can be turned on with any preset phase of the sinusoidal voltage. The phase was always chosen in the way that the polarity of the first half-cycle of voltage was the opposite to the sign of the pre-posited charge Q. In such a case, the E-field of the pre-posited charge promotes the breakdown in the corona gap. Second, the generator sharply dumped the output voltage after each corona breakdown but after that, the output voltage slowly was increased. On the one hand, the voltage sharp dumping stops quickly the corona current after the breakdown. On the other hand, the subsequent slow increase of the output voltage gives the opportunity for the next breakdown to happen.

The voltage applied to the HV electrode was measured by a high voltage probe (1:2000). The voltage and current of the barrier corona were recorded with anoscilloscope (C8-17). The breakdown moments were identified by the appearance of the sharp pulses on the voltage and current waveforms. The images of the barrier corona during its breakdown development are taken by a fast multi-frame intensified camera (LV-03). The images were synchronized with the voltage and current waveforms. Before each deposition of a pre-charge Q, the surface of the dielectric barrier was being "cleaned" from the residual charges of the previous experiment. The charge erasing was done with the use of the grounded soft brush made of many thin copper wires. All experiments were conducted in the gas discharge chamber under argon pressure slightly exceeding the ambient atmospheric one and with weak pumping of gas.

Close examination of the parameters characterizing the first and subsequent breakdowns in each half-cycle allowed us to reveal the role of the surface charge and plasma covering the charge in the development of primary and all other secondary breakdowns in the barrier corona.

3. Results and discussion

Figure 2 presents the results on the first corona breakdown when the sinusoidal voltage was being switched on starting from zero magnitude. The polarity of the chosen half-cycle (positive or negative) was opposite in sign to the sign of pre-posited charge (negative or positive). As expected, in such a case the pre-posited surface charge Q decreases the breakdown voltage U_b. One may see in Figure 2, this effect exhibits itself the stronger, the greater the charge Q irregardless of its sign. But the strong dependence of this effect on the frequency of the sinusoidal voltage (the higher frequency, the stronger effect) was really unexpected. One of the possible reasons for this effect can be attributed to the formation of corona around the HV pin at the pre-breakdown stage. The current of this pre-breakdown corona can deposit the charge of opposite sign than that of the pre-posited charge. The discharging the pre-posited charge will increase with duration of the pre-breakdown corona (i.e. with diminishing the frequency of the applied voltage) that leads to an increase in the breakdown voltage.

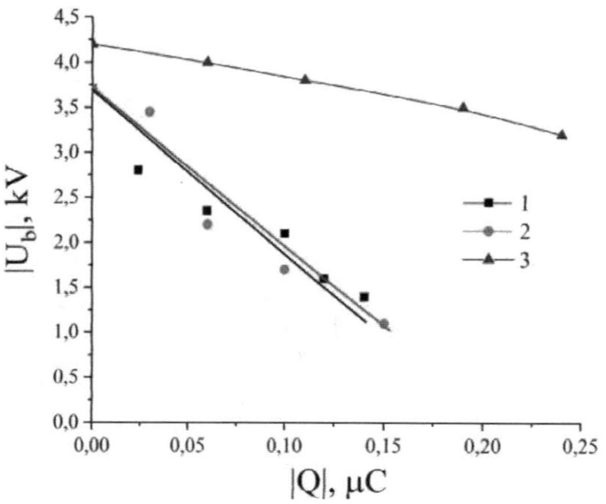

Figure 2. The relation between the voltage amplitude U_b of the first breakdown in sinusoidal barrier corona and the magnitude of the pre-deposited charge +Q or –Q for two different frequencies of sinusoidal voltage. Ar, P=770 Torr. 1) 3 kHz, –Q, $+U_b$; 2) 3 kHz, +Q, $-U_b$; 3) 400 Hz, –Q, $+U_b$.

At the beginning, we present in Figure 3 the results on the barrier corona breakdown in absence of the pre-posited charge Q. The experiment has revealed the following feature of such breakdown: under

Journal of Physics: Conference Series **1394** (2019) 012004 doi:10.1088/1742-6596/1394/1/012004

sinusoidal voltage with the same amplitude, the first breakdown can occur in any half-cycle and not always immediately after switching on the sinusoidal voltage. As a rule, the breakdown occurs after several half-cycles of the sinusoidal voltage. Besides, in absence of the pre-posited charge Q, the first breakdown always occurs near the maximum of the sinusoidal voltage, therefore the current pulse and the transversal size of the plasma sheet on the barrier exceed those formed during the breakdown with the presence of pre-posited charge Q (see the proper images in Figures 3 and 4).

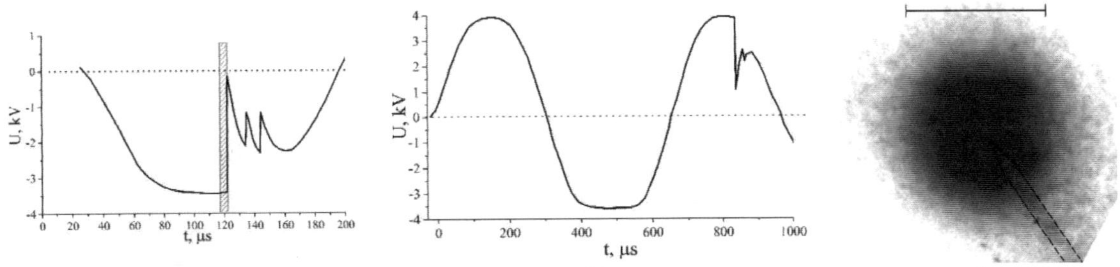

a) b) c)

Figure 3. The voltage waveforms for the sinusoidal barrier corona breakdown at f=3 kHz with no pre-posited charge: a) the breakdown happened in the first negative half-cycle; b) the breakdown happened in the second positive half-cycle; c) the image of the plasma sheet formed on the barrier by breakdown current pulse; this photo corresponds to the waveform presented in Figure 4a. Exposure time is 5 μs. The HV electrode is shown with a dashed line. The above-given scale is equal to 60 mm.

Before the examination of data presented in Figure 4 note the following. The executed by us comparison of the images of plasma sheets $S1$ and $S2$ formed respectively by the auxiliar discharge providing the pre-posited charge and by the first breakdown current pulse revealed that the visual diameter of $S1$ is always less than that of $S2$. It means the pre-posited charge Q can be fully or partially eliminated by a current pulse of the first breakdown which deposits the opposite sign's own charge on the barrier. In other words, the pre-posited charge Q influences mainly the first breakdown but all breakdowns happening after the first one are being influenced also by the charge and plasma created by a current pulse of the previous breakdown.

One may see in Figures 3 and 4 that the typical visual diameter of the plasma sheet on the barrier ranges within 20-70 mm that is much larger compared to the interelectrode gas gap of 1 mm. It means the corona breakdown is being determined predominantly by the processes controlling the surface breakdown rather than processes controlling the volume breakdown in the short pin-to-plane gap. The difference between these breakdowns is that the volume breakdown is accompanied by the formation of plasma in the pin-to-plane gap but the surface breakdown includes the deposition of the electric charge on the barrier and formation of plasma covering the surface charge. The plasma propagation along the barrier and the surface charge deposition interplay with each other and form in total the self-consistent breakdown process.

The results presented in Figure 4 show that both the maximum current pulse (correlated with the maximum sharp diminishing the applied voltage) and the maximum transversal size of the plasma sheet on the barrier happen at the first breakdown in each half-cycle. Figure 4 shows also the increase in the value of the pre-posited charge Q leads to an increase in the number of repeated breakdowns in the half-cycle. Besides, both the breakdown voltage U_b and the transversal size of the plasma sheet formed by every secondary breakdown are smaller than those corresponding to the first breakdown.

Note the plasma sheet formed in half-cycle by the first breakdown is always located symmetrically around the axis of pin-to-plane barrier corona, while the plasma sheets formed by secondary breakdowns are not axisymmetrical. One may see in Figure 4 that every secondary plasma sheet is being formed only within a certain sector around the axis. As a rule, every subsequent plasma sheet is not being overlapped appreciably with the plasma sheet formed by the previous breakdown.

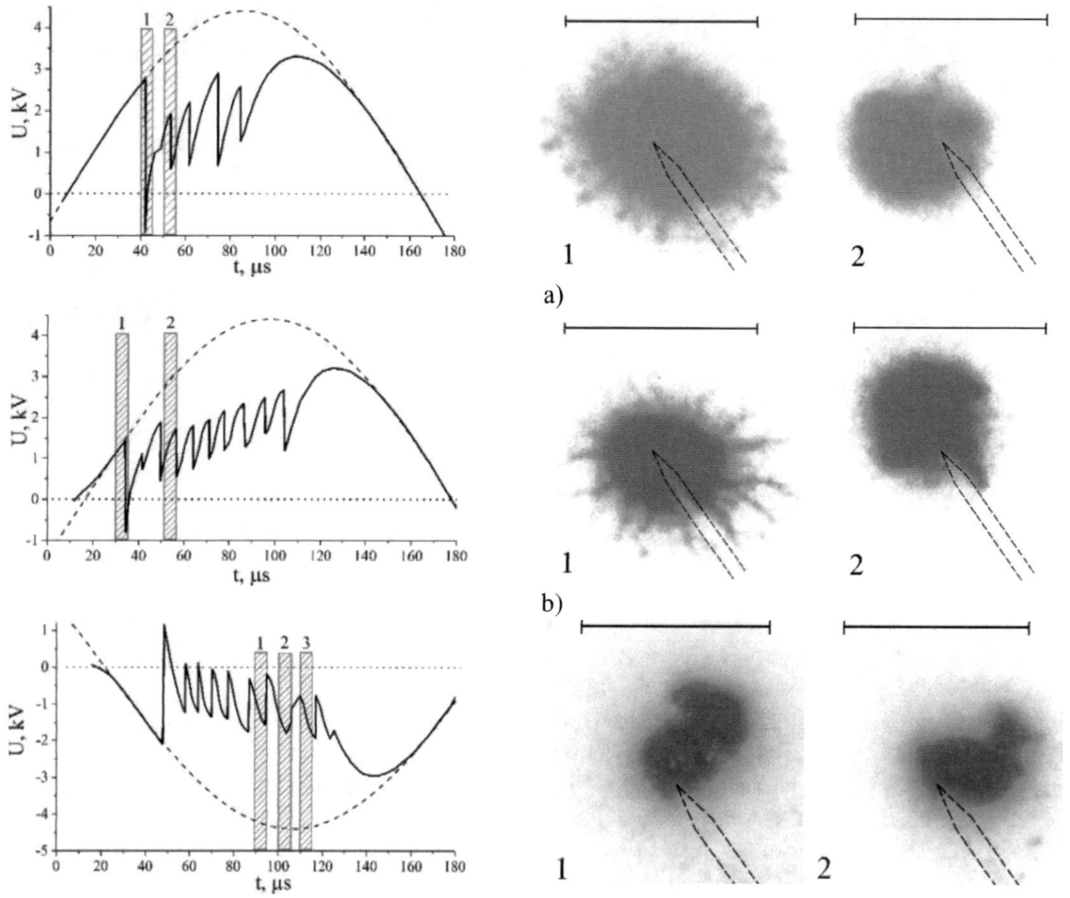

The shaded vertical strips on all waveforms indicate the moments where the images were taken. The numerals above the strips correspond to those pointed beneath the discharge images shown in the right. The dashed line in each figure shows the sinusoidal voltage formed by the generator in the absence of the barrier corona.

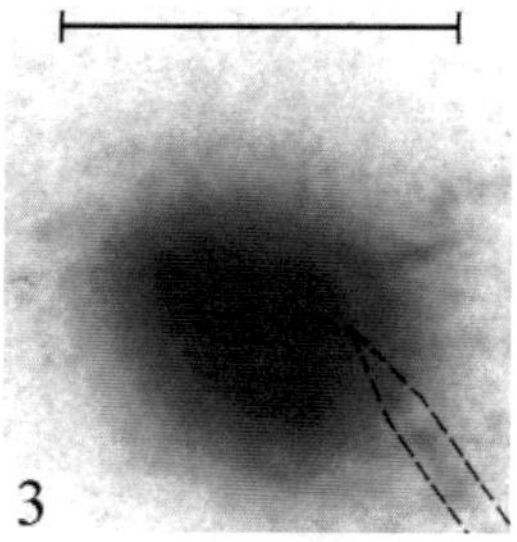

c)

Figure 4. To the left – the set of the voltage waveforms of the sinusoidal barrier corona in Ar with multiple breakdowns during the half-cycle at frequency f=3 kHz happening under different magnitudes and sign of the pre-posited charges: Q= $-2.4\cdot10^{-8}$C (a); Q = $-1.4\cdot10^{-7}$C (b); Q = $+6\cdot10^{-8}$ C (c).
To the right – the set of the plasma sheet images formed on the barrier due to the barrier corona breakdown (shots are taken at angle of 45^0 to the barrier surface). These images are synchronized with the marked moments on the proper voltage waveforms: the numerals at the photo correspond to the numerals on the waveform. The scale line is given at the top of each picture and equal to 60 mm. Exposure time is 5 μs. The HV pin electrode is shown with a dashed line.

Close examination of the results presented in Figure 4 has revealed the following features in the sinusoidal barrier corona breakdown. With a positive pre-posited charge Q, the first breakdown of occurs in the first negative half-cycle without delay after switching on the sinusoidal voltage; in the case of negative Q, the first breakdown happens in the first positive half-cycle after switching on the sinusoidal voltage.

4. Conclusion

The breakdown formation of the sinusoidal barrier discharge in every half-cycle is accompanied by the formation of microdischarges developing in the gap and on the surface of a barrier. This process leads to the deposition of the electric charge on the dielectric. It is clear the surface charge deposited in the previous half-cycle influences the first breakdown in each subsequent half-cycle of the sinusoidal voltage. The situation is more complicated when many breakdowns happen one after another during alone half-cycle. This issue is one of the important questions in physics of the microdischarges and was investigated in this work experimentally by the example of a barrier corona in pin-to-plane configuration. In order to clarify the surface charge influence in more detail, the experiments with the pre-posited charge on the barrier were performed.

It was found out the area on the dielectric barrier occupied by a pre-posited charge was always smaller than the area of the plasma sheet formed by the first breakdown in a half-cycle. It means that the first breakdown covers the whole area occupied by a pre-posited charge and deposits the own charge of the opposite sign. In other words, the first breakdown can diminish the resultant pre-posited charge (or completely neutralize it and even change the sign of the resultant surface charge). In any case, the experiment showed that the resultant charge does not eliminate (or impede) the occurrence of the next repeated breakdowns, which still occur even at a lower magnitude of the applied voltage compared to the first breakdown. We associate the possibility for repeated breakdowns in a half-cycle either with the incomplete neutralization of the pre-posited charge by the first and next breakdowns or (if complete neutralization of pre-posited charge and change the sign of the resultant surface charge is happened) with the plasma sheet formed on the barrier by each previous breakdown. This plasma can survive up to the next breakdown and shield the negative influence of the surface charge and therefore promote the next breakdown.

So, one may conclude that the surface breakdown in SDBD is controlled not only by the surface charge but the plasma sheet formed by the previous breakdown as well.

Acknowledgement
The work is supported by the RFBR (grant No 19-52-53003-a).

References

[1] Eliasson B. and Kogelshatz U. 1991 Modeling and applications of silent discharge plasmas *IEEE Trans. Plasma Sci.* **19** 309–23
[2] Kogelschatz U. 2007 Applications of microplasmas and microreactor technology *Contrib. Plasma Phys.* **47** 80–8
[3] Bruggeman P. and Brandenburg R. 2013 Atmospheric pressure discharge filaments and microplasmas: physics, chemistry and diagnostics *J. Phys. D: Appl. Phys.* **46** 464001
[4] Akishev Yu. S., Demyanov A. V., Karalnik V. B., Monich A. E. and Trushkin N. I. 2003 Comparison of the AC barrier corona with DC positive and negative coronas and barrier discharge *Plasma Phys. Rep.* **29** 82–91
[5] Akishev Y., Aponin G., Balakirev A., Grushin M., Karalnik V., Petryakov A. and Trushkin N. 2011 'Memory' and sustention of microdischarges in a steady-state DBD: volume plasmaor surface charge? *Plasma Sources Sci. Technol.* **20** 024005
[6] Wu K., Pan C., Meng Y. and Cheng Y. 2013 Dynamic behavior of surface charge distribution during partial discharge sequences. *IEEE Trans. Dielectr. Electr.Ins.* **20** 612–9
[7] Emelyanov O.A. and Shemet M.V. Single dielectric barrier discharge characteristics in needle-to-plane configuration *J. Phys. D: Appl. Phys.* **47** (2014) 315201

Increase of electric field on streamer/surfaces contact area with using external electrode

I V Schweigert[1,2], S Vagapov[1], L Lin[2], M Keidar[2]

[1] Khristianovich Institute of Theoretical and Applied Mechanics, 630090, Institutskaya str 4, Bld. 1, Novosibirsk, Russia
[2] George Washington University, D.C. 20052, 2121 I St NW, Washington, USA

ischweig@yahoo.com

Abstract. Nowadays, a low-temperature atmospheric plasma jets are widely used in the anticancer therapy. Enhancement of electric field near the surface is one of possible ways to increase an efficiency of plasma jet treatment. Another mechanism of plasma jet's influence on living cells is generating reactive oxygen nitrogen species (RONS). In this work effect of presence of external biased electrode was studied. Electric fields near the dielectric surface and zone of interaction were much higher than that near the conductive one due to accumulation of charge on the surface. The evolution of ionization rate near the surface was studied. It was observed experimentally and in the simulation that there are two main stages of evolution – the stage of fast ionization which is few hundreds of ns long and the stage of slow ionization with characteristic time of ~10 μs. A 0D modelling was applied in the boundary layer near the surface in order to obtain mixture composition of RONS after several pulses.

1. Introduction

Cold atmospheric-pressure plasma jets attract attention of researchers from various areas of science for about few decades. The reason of such interest is a wide spectrum of applications of plasma jets, especially in medicine [1] as a promising tool for cancer treatment. Due to presence of strong electric field, high energy electrons, ions and metastable atoms and molecules in the head of the streamer a lot of various reactive oxygen nitrogen radicals (RONS) are being generated in the vicinity of the surface. Also it is believed that plasma jets can influence on cells due to delivery of electric field to the biological tissue and this, along with RONS, may have a synergetic effect.

In this work we provided increasing of electric field with ring external electrode installed between tube's nozzle and the surface. The main streamer's characteristics such as rate of ionization and electric field were investigated with and without external electrode. Streamer's interaction with conductive and dielectric surfaces was also studied. Using obtained characteristics of the streamer above the surface we implemented 0D modelling to estimate amounts of RONS generated during plasma jet treatment.

2. Experimental setup and model description

Plasma jet and computation domain are illustrated in figure 1. Plasma is generated between two electrodes powered with sinusoidal voltage at 12.5 kHz by feeding 13.5 sl/min helium through dielectric tube. The tube has nozzle diameter of 0.46 cm. The copper or dielectric targets are placed at 3 cm from tube's nozzle. External electrode is placed at 1 cm from tube's nozzle. Its external and internal radii are 0.65 cm and 0.185 cm, respectively. The propagation of plasma bullet is recorded by an Andor iStar ICCD camera with exposure time – 10 ns. The experimental setup is described in details in [2].

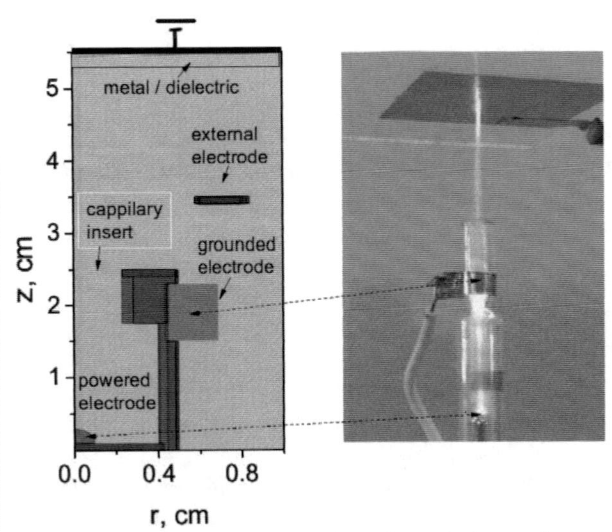

Figure 1. Computational domain and plasma jet.

2D modelling includes Poisson's equation, continuity equations for electrons, ions and mean electron energy. Molar fractions of He, O_2, N_2, H_2O during the mixing of helium flow with ambient air were pre-calculated using ANSYS Fluent. Transport coefficients in the continuity equations for electrons and their energy were calculated from the solution of Boltzmann's equation. For ions transport coefficients were given by the next analytical expressions:

$$\mu_i = \frac{3e}{16 n_e \sigma_i}\left(\frac{\pi}{m_i kT}\right)^{1/2}\left(1+\frac{9\pi^2 eE}{512 n_g kT\sigma_i}\right)^{-1/2}$$

$$D_i = \mu_i kT$$

where μ_i is the mobility of ions, D_i is the coefficient of diffusion of ions, e is an elementary charge, n_e is the concentration of electrons, k is the Boltzmann constant, T is the gas temperature, E is the electric field, m_i is the ion's mass and σ_i is its transport cross-section.

The coefficient of recombination rate β for helium at atmospheric pressure and 300K is $\beta = 2\times10^{-7}$ cm^3/s [3]. The secondary emission coefficient γ from copper target is 0.02 [4]. For the dielectric surface γ was taken 0 and 0.1.

The calculation domain has a cylindrical symmetry with $R_{max} = 5$ cm and $H_{max} = 5.5$ cm. Three values of voltage amplitude U_0 were investigated – 3.9 kV, 5 kV and 7 kV.

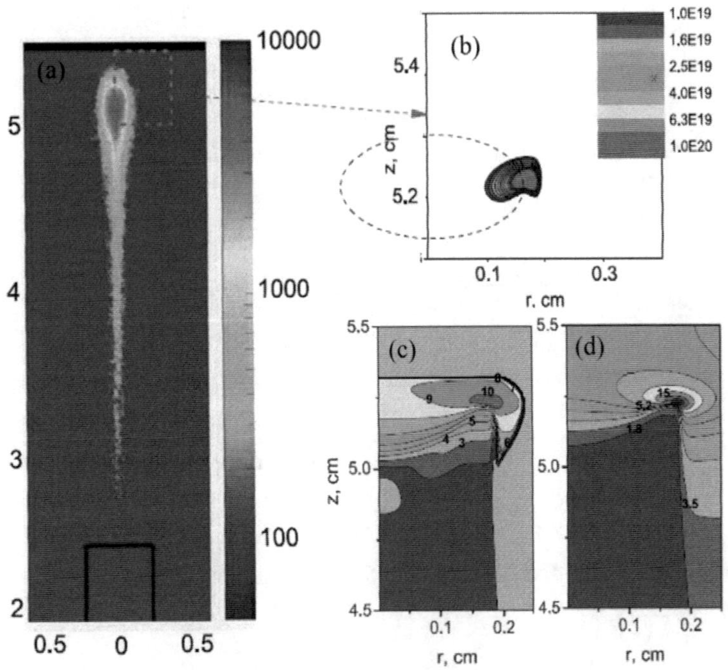

Figure 2. Image of the streamer (a), calculated ionization rate in cm^{-3}/s for U_0 = 5 kV (b), spatial distributions of electron energy in eV (c) and electric field in kV/cm (d) in linear scale.

Journal of Physics: Conference Series **1394** (2019) 012005 doi:10.1088/1742-6596/1394/1/012005

3. Results and discussion

3.1. Main streamer characteristics

When voltage on the powered electrode reaches threshold value wave of ionization starts to propagate from the anode over the dielectric tube. The optical emission (see figure 2 (a)) indicates an approximate radius of the streamer channel, which is essentially smaller than the radius of a wider part of the tube. It is seen from figure 2 (b) that streamer has toroidal shape with external and internal radii of 0.185 cm and 0.14 cm, respectively. Figures 2 (c) and (d) illustrate spatial distributions of electron energy and electric field. The maximum values are 11 eV and 17 kV/cm, respectively. The electron density in the streamer head is 2×10^{12} cm^{-3}.

3.2. Effect of external electrode

As was mentioned above the external ring electrode was maintained at 1 cm from tube's nozzle. We applied positive and negative (+1.5 kV and -1.5 kV) voltages U_r to the ring and studied a variation of streamer characteristics.

Figure 3 illustrates spatial potential distributions for cases with +1.5 and -1.5 kV at the moment when the streamer arrives to the surface. In the case of negatively biased ring there is larger electric field near the surface. For both cases a spot of ionization has the shape of the ring of 0.7 mm wide with external radius of 0.19 cm. The rate of ionization exponentially decays to the center of the ring as it can be seen in figure 3 (c).

Evolution of the ionization rate near the surface for various cases is illustrated in figure 4. It can be seen that for U_0 = 5 kV and with the external electrode ionization rate was larger by factor of two. Electric field in this case was also larger by 30%.

For all cases the ionization rate has the same character of evolution and two stages can be distinguished. In the first one which is 100-150 ns long ionization rate reaches its maximum when the streamer approaches the surface. The second one lasts approximately 10 μs and it is characterized by much less intensive ionization.

This evolution of ionization agrees with experimental observations. In figure 5 experimental images of streamer over the Cu-plate at various moments of time are presented. The first image shows much more intensive optical emission what corresponds to the first stage. The next images show lower optical emission which remains constant for the next ~10 μs which can indicate the occurrence of the second part.

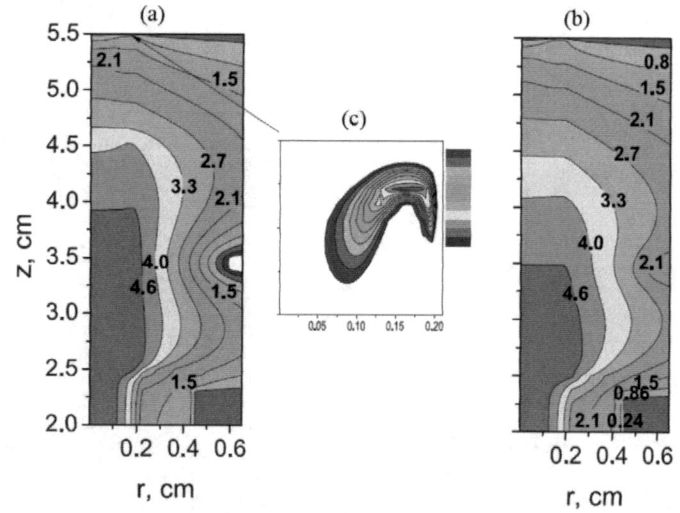

Figure 3. Spatial potential distribution in kV for U_r = -1.5 kV (a) and 1.5 kV (b) for U_0 = 5 kV. Ionization rate (c) with colour palette (0.01-2)$\times10^{22}$ cm^{-3} s^{-1} for U_r = -1.5 kV.

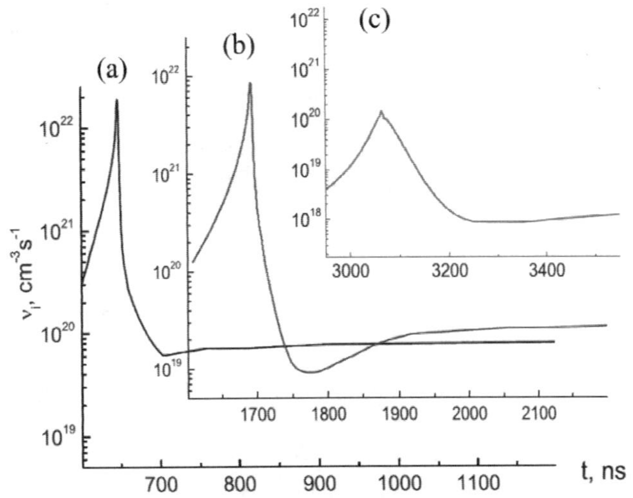

Figure 4. Evolution of ionization rate near the surface for $U_0 = 5$ kV, $U_r = -1.5$ kV (a), for $U_0 = 5$ kV, no ring (b) and for $U_0 = 3.9$ kV, no ring (c).

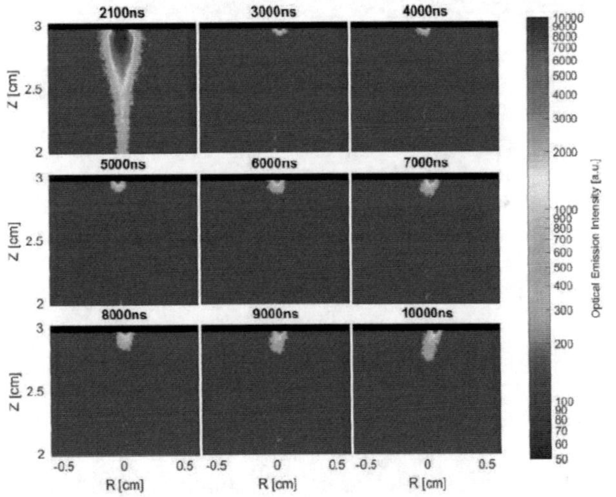

Figure 5. Images of the streamer near the metal surface

3.3. Streamer interaction with dielectric surface

Streamer interaction with metal and dielectric surfaces with various value of ε/d (d is dielectric's thickness) and various configurations already was studied in [5, 6].

In this work we studied streamer interaction with dielectric with $\varepsilon = 10$ in the presence of negatively biased ring electrode. A dielectric which is 0.3 cm thick was placed on the grounded metal plate. Spatially non-uniform surface charge accumulates when streamer approaches to the surface. In the case with $\gamma = 0.1$ ionization wave starts to propagate along the surface. In the case with non-emissive surface there are not enough electrons crossing the leading front of the ionization wave to maintain its propagation.

Axial and radial electric field components E_z and E_r over the emissive surface are illustrated in figure 6. E_z can achieve 80 kV/cm near the surface. The radial component E_r appears due to non-uniform surface charge distribution. Its maximum is 22 kV/cm. Due to presence of E_r the value of electric field is much higher than that over the conductive surface. Inside of dielectric, the E_z ranges

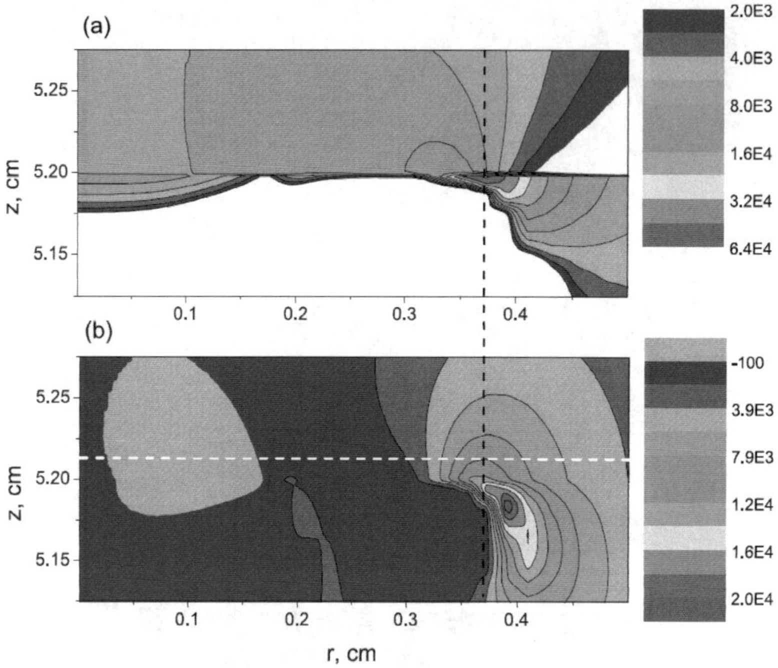

Figure 6. Spatial distribution of electric field (a) and (b) in V/cm over an emissive dielectric surface for U_0 = 5 kV and U_r = -1.5 kV.

from 10 kV/cm to 6 kV/cm and the E_r varies from 12 kV/cm to 4 kV/cm. It is seen that close to the center the E_r changes its sign.

4. 0D model of plasma chemistry

We use 0D model for calculation of plasma enhanced chemical reactions in the thin boundary layer near the surface where speed of the gas flow is almost zero. So there is no need to consider the convection of species.

Kinetic scheme consists of 872 chemical reactions and 98 corresponding components. The scheme includes reactions of ionization (including Penning ionization), dissociation, recombination, charge exchange, excitation and quenching of electronic, vibrational and rotational levels. 190 of them are electron-impact processes. Rate coefficients were taken from literature for ion-ion, neutral-ion and neutral-neutral reactions and were calculated from the solution of Boltzmann equation using cross-sections for electron-impact processes and a mixture composition. Boltzmann equation has been solved periodically with BOLSIG+ solver in order to obtain tables of rate coefficients which were interpolated for current value of the electron temperature.

2D simulation and experiment revealed that after streamer reached the surface during t_s = 1-10 μs (corresponding to the positive cycle of voltage) rates of ionization and excitation are almost constant. As seen from figure 5, the light emission spot has the same intensity for several μs after the streamer approaching the surface. In our simulation of plasma chemistry, we chose t_s = 5 μs and assume that the electron concentration and temperature are constant. For t_s > 5 μs the ionization and excitation rates are zero until the next streamer comes to the surface during a positive cycle of the alternative applied voltage. The electron temperature was chosen as a function of time as follows. Simulation begins with the maximal electron temperature (5 eV) which remains constant for the next 4 μs. Then in 0.1 μs electron temperature exponentially decreases to the gas temperature (300 K) and remains constant for 76 μs until the next positive cycle. After that electron temperature exponentially increases to the maximum value of 5 eV in 0.1 μs.

Due to presence of the sheath near the surface the ion fluxes were added in the model. Ions' velocity is considered to be equal to Bohm velocity, electrons' flux equals to the sum of ion's fluxes to satisfy the quasi-neutrality condition. Also it is considered that all ions recombine after their contact with surface and return in the form of corresponding neutrals. The evolution of concentration of species is described with the following equations

$$\frac{\partial n_\alpha}{\partial t} = \sum_{\lambda=1}^{N_\alpha} (\nu_\lambda^R - \nu_\lambda^L) k_\lambda \prod_{l=1}^{M} n_l^{\nu_l^L} - \frac{\Gamma_\alpha}{d}$$

where n_α is number density of αth component, ν_λ^R and ν_λ^L are right- and left-handed stochastic coefficients of λth reaction, N_α is total number of reactions where αth component was participated, k_λ is rate coefficient of λth reaction, Γ_α is the flux of αth component (positive for ions and electrons, negative for neutral components obtained after ions' recombination, zero for others), d is estimated thickness of boundary layer (~0.1 mm).

The initial values of n_{He}, n_{O_2}, n_{N_2} and n_{H_2O} were taken from the solution of Navier-Stokes equations for the point corresponding to the point of streamer contact with surface. In our case they equal to $He/N_2/O_2/H_2O$ = 0.9767/0.0176/0.0050/0.0007.

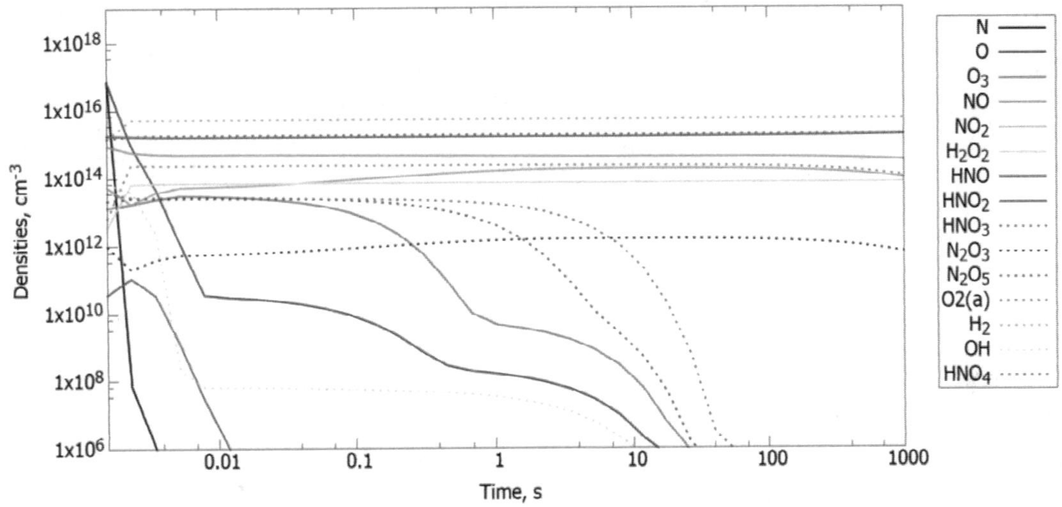

Figure 7. Evolution of some radicals after 20 pulses of electron temperature.

Figure 7 shows the evolution of several species for long-time scale (about 16 minutes) after 20 pulses. Concentrations of H_2O_2, NO, HNO_x, NO_2, $O_2(a)$, O_3 remain high. Concentrations of OH sharply decreases at first ~0.01 s as OH is converting into H_2O_2. O_3, N_2O_5 and HNO_4 decays for 1-10 s. HNO_2, H_2O_2, NO and some other radicals stay constant at this time scale.

5. Conclusion

In this work streamer's interaction with conductive and dielectric surfaces and effect of external electrode were studied. The rate of ionization and electric field distribution near the surface were calculated using 2D modelling. Two main stages the ionization rate's evolution were distinguished – the stage of fast intensive ionization (~100 ns) and the stage of slow equilibrium ionization (~10 μs). Obtained results are in the agreement with the experimental observations. It was found that in case with negatively biased external electrode the rate of ionization is higher by factor of 2 and electric field is larger by 30% than that in case without external electrode. Also electric field was much higher over the dielectric surface compared to the conductive one due to presence of surface charge. 0D plasma chemistry calculation was also conducted where the temperature of electrons was taken as a periodical function of the time. The evolution of concentrations of RONS were calculated in the framework of 0D chemical model.

Acknowledgements
Two of the authors, I. Schweigert and S. Vagapov were supported by Russian Science Foundation N 9-19-00255 and by Russian Foundation for Basic Research N18-08-00510 (plasma chemistry modeling).

References

[1] Keidar M, Walk R, Shashurin A, Srinivasan P, Sandler A, Dasgupta S, Ravi R, Guerrero-Preston R and Trink R 2011 *Br. J. Cancer* **105** 1295–301
[2] Lin L and Keidar M 2016 *Phys. Plasmas* **23** 083529
[3] Raizer Y P 1991 *Gas Discharge Physics* (Berlin: Springer)
[4] Kaminsky M 1965 *Atomic and Ionic Impact Phenomena on Metal Surfaces* (New York: Academic)
[5] Akishev Yu, Karalnik V, Medvedev M, Petryakov A, Trushkin N and Shafikov A 2017 J. of Physics: Conf. Series 927 012040
[6] Akishev Yu, Karalnik V, Medvedev M, Petryakov A, Trushkin N and Shafikov A 2017 J. of Physics: Conf. Series 927 012039
[7] Phelps database (www.lxcat.net) (accessed 16 November 2018)
[8] Norberg S A, Johnsen E and Kushner M J 2015 *J. Appl. Phys.* **118** 013301
[9] Lazarou C, Anastassiou C, Topala I, Chiper A S, Mihaila I, Pohoata V and Georghiou G E 2018 *Plasma Sources Sci. Technol.* **27** 105007
[10] Shashurin A and Keidar M 2015 *Phys. Plasmas* **22** 122002

Study of a discharge with a liquid cathode by the methods of high-speed visualization and emission spectroscopy

A V Chistolinov[1]*, R V Yakushin[2], M A Sargsyan[1], M A Khromov[1], A S Tyuftyaev[1]

[1]Joint Institute for High Temperatures of the Russian Academy of Sciences
[2]Dmitry Mendeleev University of Chemical Technology of Russia

* E-mail: a-chi@yandex.ru

Abstract. In this work the detailed structure of a direct current discharge with a liquid cathode was studied by high-speed imaging and emission spectroscopy. The movement of filaments into which the discharge channel is divided is studied. The processes of formation and destruction of filaments are investigated. Spectra were recorded at different distances from the liquids surface on the discharge axis. The dependences of the luminescence intensity of the main spectral lines and bands on the distance to the liquid surface at different discharge currents are found.

Introduction

Systems in which plasma interacts with the surface of a liquid are interesting objects in terms of both applied and fundamental research.

Among the areas of application of systems in which plasma interacts with the surface of a liquid, we can distinguish: destruction of organic pollutants in water, modification of the surface of polymeric materials, spectral analysis of the content of metal ions in water and aqueous solutions, obtaining ultrafine powders in aqueous solutions, biomedical applications.

The simplest way to organize a system in which plasma interacts with the surface of a liquid is a discharge cell with a liquid electrode [1]. The most intense plasma-chemical processes in a liquid occur when the liquid electrode is a cathode. It is assumed that the bombardment of high-energy ions in this case leads to non-equilibrium dissociation, first of all, of solvent molecules, which triggers the subsequent chain of physicochemical processes in the liquid [2, 3].

Another feature of a DC discharge with a liquid cathode is that in this type of discharge there is a non-equilibrium transfer of solvents and solutes molecules into the gas phase under the action of ionic bombardment of a liquid cathode, similar to the process of cathode sputtering in a classical glow discharge [4]. The analysis of the processes taking place is hampered by the fact that the composition of the gas phase is substantially determined by the processes of transfer of the solution components to the gas phase [2, 3]. These complex processes that occur near the surface of a liquid in a DC discharge with a liquid cathode make it difficult to build a detailed mathematical model of such a discharge. In addition, the detailed structure of such a discharge with a high temporal and spatial resolution remains poorly studied experimentally.

The aim of this work was to study the detailed structure of a DC discharge with a liquid cathode. In particular, the study of the movement of filaments and cathode spots, the study of the formation and death of filaments and cathode spots. As well as the study of plasma composition in different discharge zones at different discharge currents by emission spectroscopy.

Experimental setup and measurement methods

In this work, we investigated a direct current discharge, in which an aqueous solution was used as the cathode. The discharge occurred at atmospheric pressure in the air. A tungsten electrode with a diameter of 2 mm was used as the anode; the tip radius was 1 mm. Sodium hydroxide solution in deionized water with a concentration of 60 mg/l and an electrical conductivity of 330 μS/cm was used as a liquid cathode. The discharge was performed in a flow discharge cell.

The high-speed video recording of the discharge was made with a colored video camera Phantom VEO 410S and a black and white video camera Phantom MIRO M110, with a recording speed of up to 10,000 frames per second. AvaSpec 2048 three-channel fiber optic spectrometer with a spectral resolution of 0.15 nm and a MS-5204i high-resolution spectrograph with a spectral resolution of 0.03 nm with an Andor CCD matrix at its outlet were used to study the discharge plasma by the methods of emission spectroscopy. The image of the discharge was formed by a quartz lens in the plane in which the entrance slit of the spectrometer or the inlet of the optical fiber was placed, which made it possible to study the spatial distribution of the spectra.

Results

According to the data of high-speed video filming, near the surface of the liquid, the discharge channel is divided into separate filaments (figure 1).

a) b) c)

Figure 1. Frames of high-speed video recording of the discharge with a liquid cathode. Exposure 250 μs, discharge current 100 mA, the distance between the electrode and the surface of the liquid: a) 1 mm, b) 2 mm, c) 4 mm

High-speed video shows that the filaments and cathode spots randomly move along the surface of the liquid. Characteristic speeds are on the order of several meters per second. At the same time, the number of filaments and the area occupied by cathode spots on the surface of the solution do not remain constant, but fluctuate in time around some average values.

As shown by high-speed video, the main mechanism for the formation of new filaments and cathode spots is the separation of filaments and cathode spots. The main mechanism of the destruction of filaments and cathode spots is their displacement to the periphery zone and extinction over the course of hundreds of microseconds.

The study of high-speed video frames also showed that each cathode spot bends the surface of the liquid and forms a small deepening (recess) on it. When the cathode spots move along the surface of the liquid, the recesses follow them. As a result, capillary waves are formed on the surface of the liquid, which diverge concentrically from the discharge zone (figure 1).

The average number of filaments and the average area occupied by cathode spots on the surface of the solution increase with increasing discharge current [5]. Also, with an increase in the discharge

Journal of Physics: Conference Series **1394** (2019) 012006 doi:10.1088/1742-6596/1394/1/012006

current, the distance between the solution surface and the branch point (point where discharge channel is broken down into the filaments) increases. At a certain critical current, the position of the branch point coincides with the electrode. When the current is equal to or greater than the critical filamentation of the discharge begins directly from the electrode (figure 1a).

Interpretation of the emission spectra of the discharge plasma showed the presence of emission bands of OH - radicals, molecular nitrogen N_2 (2^+) and N_2 (1^+), molecular ion N_2^+, the bands of which are very weak against the background of molecular nitrogen bands and visible only near the surface of the liquid figure 2, as well as atomic hydrogen lines HI (H_α and H_β) and oxygen OI. The sodium line 589 nm is completely absent in the spectrum. This suggests that sodium ions are not transferred to the gas phase under these conditions.

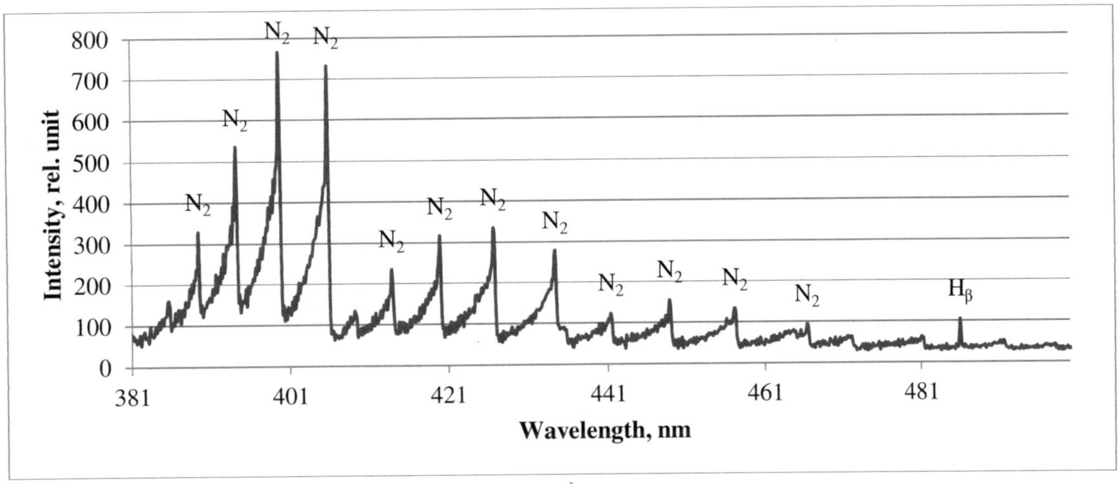

a)

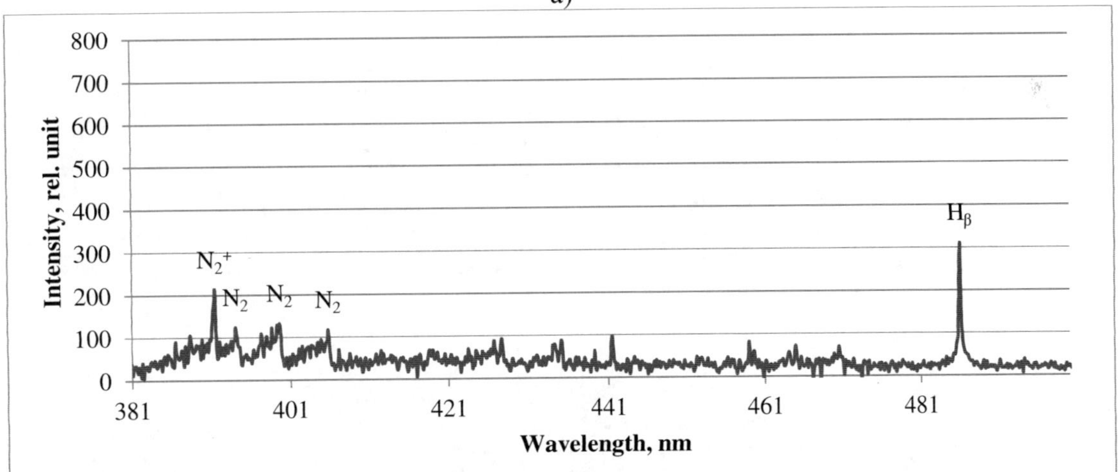

b)

Figure 2. Fragment of the discharge spectrum: a) near the electrode, b) near the surface of the liquid. The distance between the electrode and the surface of the liquid is 2 mm, the current is 100 mA.

Cathode spots are the brightest points on high-speed images figure 1. As shown by spectral measurements, a sharp increase in the intensity of the emission of atomic hydrogen lines H_β and H_α is observed near the surface of the liquid (figures 2 and 3), and only these lines; therefore, we associate the bright luminescence of the cathode spots with the luminescence of atomic hydrogen.

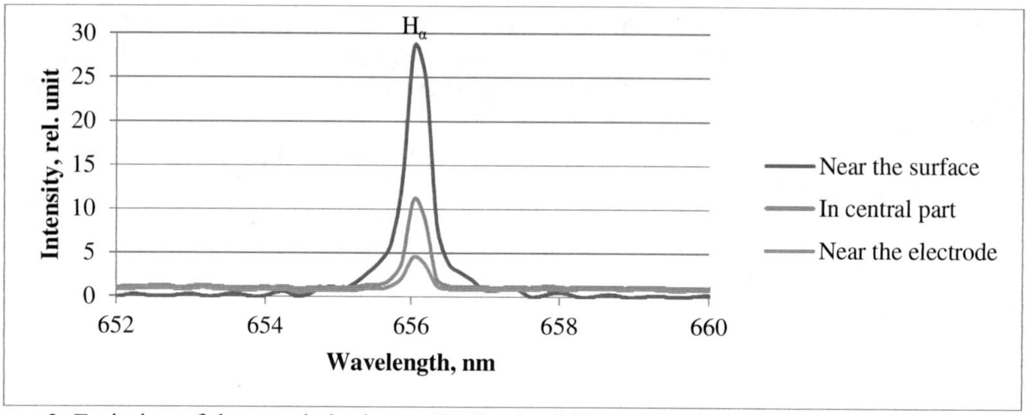

Figure 3. Emission of the atomic hydrogen H_α line in different discharge zones. The distance between the electrode and the surface of the liquid is 2 mm, the current is 100 mA.

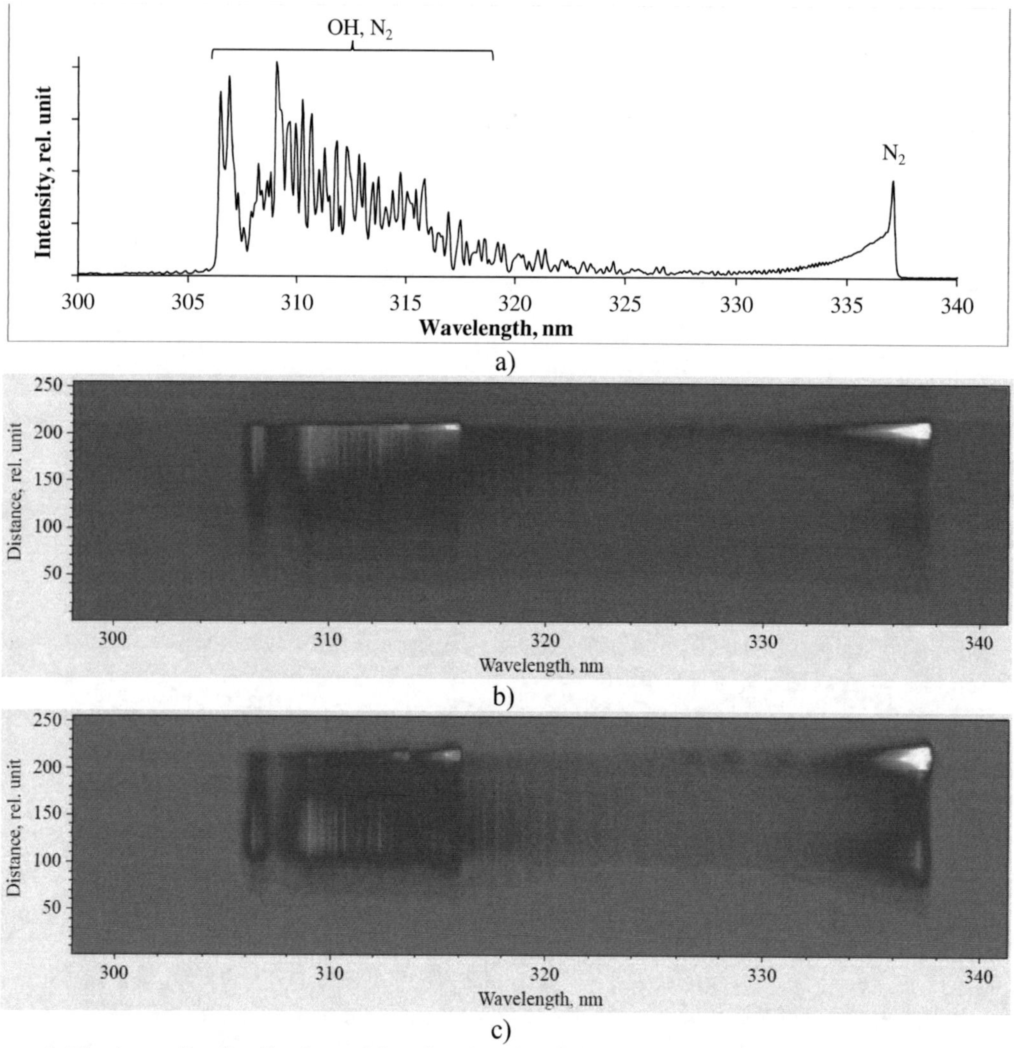

Figure 4. The intensity distribution of the glow bands of OH radicals and N_2 molecules in a discharge with a liquid cathode, a discharge gap of 2 mm, an electrode is on the top: a) spectrum interpretation, b) discharge current 100 mA, c) discharge current 50 mA.

According to spectral measurements, the emission maximum of the second positive system of molecular nitrogen N_2 (2^+) is always located near the electrode, regardless of the discharge current, in the range of studied currents from 10 to 100 mA. The position of the maximum radiation of OH - radicals significantly depends on the discharge current. At a current of 100 mA, the OH - radical emission maximum is located near the electrode. When the discharge current decreases, the radiation maximum shifts to the surface of the liquid (figure 4).

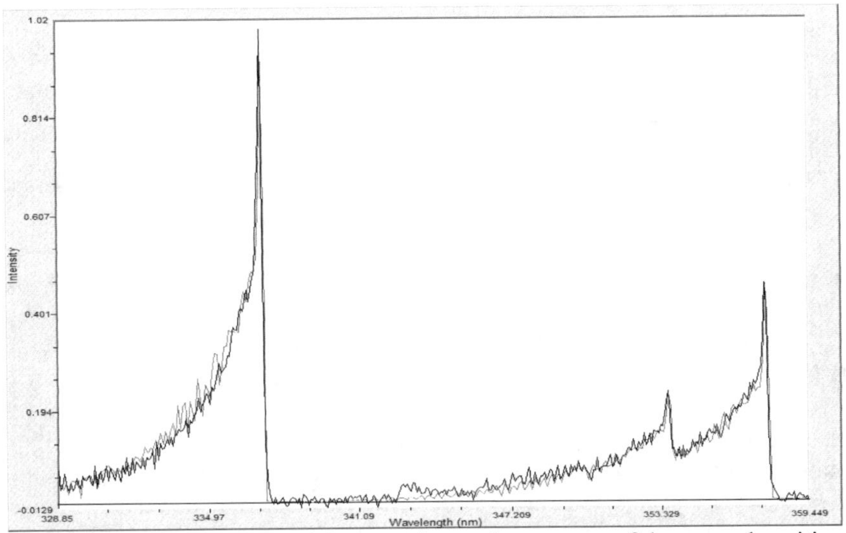

Figure 5. Comparison of a fragment of the experimental spectrum of the second positive molecular nitrogen system N_2 (2^+) with the model one built in the Specair program. Black is an experiment, red is the model.

At a current I = 100 mA and an interelectrode distance of 2 mm, near the surface of the liquid, the vibrational and rotational temperatures of the plasma were determined from the emission spectrum of the second positive molecular nitrogen system N_2 (2^+). The determination of the vibrational and rotational temperatures was carried out by comparing the recorded spectrum with the model spectrum built in the Specair program (figure 5).

For the axial part of the discharge, near the surface of the liquid, at a current I = 100 mA and an interelectrode distance of 2 mm, rotational and vibrational temperatures were obtained, respectively: T_r = 2800K ± 10%, T_v = 3800K ± 20%. Which agrees well with the data [6] obtained under similar conditions.

Discussion

It is known that in the DC discharges with a liquid cathode there is a non-equilibrium transfer of the solvent and solute substances into the gas phase under the action of ion bombardment of a liquid cathode [2, 3]. In this case, one incident ion, according to direct measurements carried out in [4], has from 300 to 500 molecules of sprayed water associated with it. Such a large number of sprayed water molecules per incident ion indicates a high energy of ions bombarding the surface of the liquid. That, in turn, is associated with a large amount of cathode potential voltage drop, which, according to [4], ranges from 500 to 700 V.

Simple estimates show that the rate of outflow of non-equilibrium steam from cathode spots is several meters per second. At the same time, a reactive force acts on the surface of the liquid from the jet of a nonequilibrium vapor, which is balanced by the forces of the surface tension of the liquid. This leads to the formation of characteristic deepenings under the cathode spots.

The steam jet flowing from the cathode spot along the filament axis reduces the temperature on its axis, thus making its position unstable. This is supposed to cause the filament and the associated cathode spot to move along the surface of the liquid.

Conclusion

Near the surface of the solution in a discharge with a liquid cathode, the discharge channel is divided into separate filaments that are in constant chaotic motion. Each filament on the surface of the solution ends with a separate cathode spot. In this case, near the anode, the discharge channel may not have a filamentous structure. The point where discharge channel breaks apart into filaments (branching point) is determined by the discharge current. With increasing discharge current, the distance between the surface of the solution and the branching point on the discharge channel increases.

The number of filaments and cathode spots do not remain constant, but fluctuate over time around some average values. The main mechanism of formation of new filaments and cathode spots is the division of filaments and cathode spots. The main mechanism of death of filaments and cathode spots is their movement to the periphery and extinction.

Interpretation of emission spectra of discharge plasma shows the presence of radiation bands of OH - radical, molecular nitrogen $N_2(2+)$ and $N_2(1+)$, molecular ion N_2^+, as well as lines of atomic hydrogen HI (H_α and H_β) and oxygen OI. The emission of most molecular bands and atomic lines on the discharge axis is highly inhomogeneous in height. In the entire range of the studied currents from 10 to 100 mA, the atomic hydrogen lines have a sharp maximum near the liquid surface, and the intensity of the molecular nitrogen bands $N_2(2+)$ reaches a maximum near the metal anode. The position of the OH – radical emission maximum significantly depends on the discharge current. At a current equal to 100 mA, the maximum emission of OH-radicals is near the metal electrode. When the discharge current decreases to 50 mA and below, the emission maximum is shifted to the liquid surface.

Acknowledgments

This work was partially supported by the Russian Foundation for Basic Research grant No. 19-08-00592.

References

[1] Kutepov A.M., Zakharov A.G., Maximov A.I. Vacuum-plasma and plasma-solution modification of polymeric materials. // Moscow. Nauka, 2004. 497 p.

[2] Bruggeman P., Leys C. Non-thermal plasmas in and in contact with liquids // J. Phys. D: Appl. Phys. 2009. V. 42. № 5. P. 1-28.

[3] Bruggeman P., Kushner M., Locke B. et al. Plasma–liquid interactions: a review and roadmap // Plasma Sources Sci. Technol., 2016, 25, 053002.

[4] Sirotkin N.A., Gurina D.L. The influence of the transfer processes of the components of a liquid cathode on the properties of a dc discharge at atmospheric pressure // International Journal of Applied and Fundamental Research. 2016, №11, pp.94-99.

[5] Khlyustova A. V., Maksimov A. I. Double Electrical Layer at the Plasma-Solution Interface. // Contrib. Plasma Phys. 2013. V. 53. №6. P. 481-489.

[6] Bruggeman P., Liu J.J., Degroote J., Kong M.G., Vierendeels J., Leys C. Dc excited glow discharges in atmospheric pressure air in pin-to-water electrode systems // J. Phys. D: Appl. Phys. 2008. V. 41.215201.

Boundary layer and the structure

L N Pyatnitsky

Joint Institute for High Temperatures RAS Russia. Moscow, Izhorskaja 13.

pyat7@mail.ru

Abstract. L.D. Landau's analysis showed, the plane harmonic wave used in the small perturbation method is inapplicable as the turbulent disturbance motion form. The search for the movement form carried out in accordance with the considerations of H. Lorentz, led to replace the longitudinal wave with a spherical wave layer arising on the wall as a result of friction. This form of disturbing motion formed the basis of the turbulence wave model which reproduced the space-time field of the flow parameter pulsations. In this paper, the model is applied to describe the formation of the structure of the boundary layer and of the viscous sublayer.

Keywords: laminar and turbulent flows, acoustics, spherical wave layer, parameters pulsations, boundary layer, viscous sublayer, flow structure.

1. Introduction

The accepted description of the laminar flow in the channels (pipes) does not cause general objections. As for the turbulent flow, it is appropriate to refer to the figurative assessment of T. Karman [1]: "When I have to appear before the Creator, the first revelation I ask of Him is to reveal the secrets of turbulence." The absence of a complete theory of the turbulence was repeatedly noted by Landau and Lifshitz in [2], §31, §33, §35.

The reason for this strange situation is explained by the lack of a correct idea about the form of the disturbing movement. Two approaches were determined from the very beginning Rayleigh [3] and Reynolds [4] took one dimensional harmonic wave $\boldsymbol{A(t) = A_0 \exp(\gamma t + i\omega t)}$ of small amplitude as a perturbing motion. ***H***. Lorenz [5], on the contrary, believed that the primary perturbation can have any form compatible with the continuity equation. Since Lorenz did not find the proper form, it was the first assumption which got spread. However, it was not possible to build a field of pulsations on this path. It was only shown that the perturbation wave amplitude tends to a finite limit [6].

$$\boldsymbol{|A|^2_{max} = 2\kappa/\alpha} \qquad (1)$$

where $\boldsymbol{\alpha}$ is the Landau constant, $\boldsymbol{\kappa}$ is the attenuation index of the disturbance wave.

It follows that the advantage of the first assumption is not justified. In the meantime, there appeared numerous theories based on different assumptions, which did not bring the desired result, since they had no correct disturbing motion, and accordingly, they did not describe the turbulence occurrence. An exception is the vortex model proposed by Richardson [7], in which large vortices can be qualified as primary perturbations. However, even in this case there is no exact analytical solution of the equations for the turbulent flow region and the boundary layer in which vortices form.

2. Flow core structure

The incompleteness of the turbulence theory when we deal with the flow of a finite aperture, for example in a pipe, lies in the fact that the actual conditions of the flow are not taken into account. Also the process is often considered in the established mode. Meanwhile, these circumstances significantly change the flow properties, moreover, the formation of the flow and its pulsations occur, most likely, simultaneously.

Indeed, every movement is created by a force. In the current environment, the force is the pressure gradient. A standard example is the Riemann wave in a pipe, running in front of the piston which compensates for frictional energy losses. The flow parameters in Riemann wave are usually considered as monotonic functions. But Riemann wave of a finite aperture does not exist as such [8], since friction and diffraction divergence of flow waves, being inevitable in the flow of an aperture bounded by walls, drastically change the wave propagation and the flow properties. Both of these factors form acoustic disturbances – spherical waves. Moving in the pipe from the walls, they create turbulent pulsations. The process of their formation and the structure of the flow core are described in [9].

In short, the process is as follows. The Reynolds criterion is the ratio of the longitudinal pulse flow to the flow of the transverse pulse. The latter means the transverse pressure gradient appearance in the medium stream, i.e.

$$-\frac{\partial P}{\partial z} > 0 \tag{2}$$

The greatest increase in pressure occurs at the wall. At the same time, the reflection of diverging waves (elementary simple waves that support the flow) forms local pressure maxima at the wall, generating spherical wave layers that carry oscillations of the medium parameters. The propagation of the layer creates parameter pulsations in the flow, which are shown in scheme of Fig. 1. The set of such waves forms the turbulent flow structure.

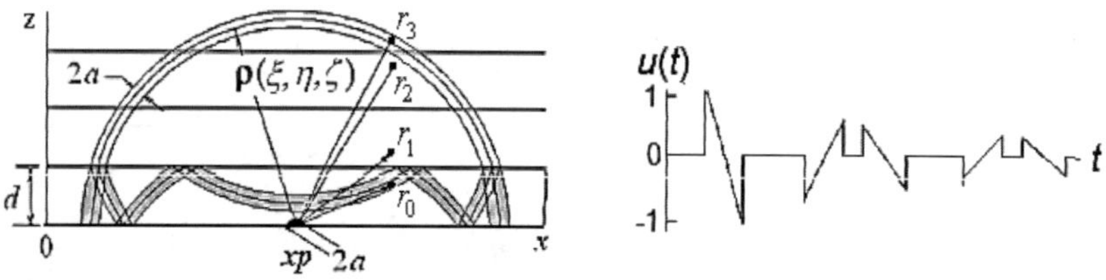

Figure 1 Distribution of a perturbation of radius a in the form of a wave layer ρ(ξ, ή, ζ) in a flat channel with a height d from point xp and velocity fluctuations in an arbitrary point $\boldsymbol{r}_0$, $\boldsymbol{r}_{i}$ – its imaginary positions.

3. Boundary layer structure

The boundary layer is strongly affected by the wall, which requires special consideration. The task of the boundary layer structure and, thus, completing the construction of the wave model of the turbultnce was solved by the example of air flow in a channel formed by two flat plates with a gap $\boldsymbol{d} = 0.1$ m. The velocity profile has been choosen in the form

$$M = M_0\,(4\,(1-z)\,z)^{1/m}. \tag{3}$$

Here $M_0 = U_0/c$ is the dimensionless flow velocity in the middle of the channel, $z = 0.5$, and m is the degree of turbulence of the flow. Formula (3) satisfactorily reproduces the velocity profile and is convenient in that it is applicable for laminar ($m = 1$) and turbulent ($m = 5$) flow regimes. Profiles of the velocities and their vorticity are shown in Fig. 2 for $m = 5$.

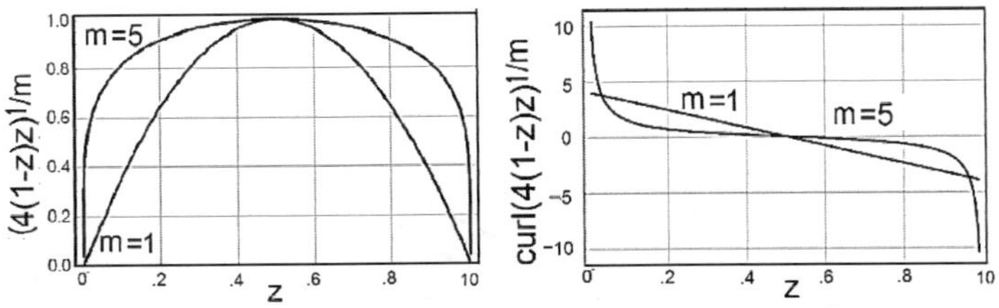

Figure 2 Flow velocity profiles (3) and profiles of the vorticity laminar m=1 and turbulent m=5 flows

4. Laminar flow features

The specificity of wave propagation in a limited aperture is due to the flow velocity profile. It deforms the wave, and the deformation is determined by the position of the test point on the wave front. Group wave velocity is given by expression $\mathbf{V} = \partial\omega/\partial\mathbf{k}$, where $\mathbf{k}$ is the wave vector, $\omega(\mathbf{k})$ is the dispersion relation

$$\omega = ck + \boldsymbol{Uk}. \tag{4}$$

We introduce the unit vector **s** tangent to the trajectory. If the medium is at rest, $U = 0$, then the test point moves with the velocity $\mathbf{V} = c\mathbf{k}/k$, and since $\mathbf{s} = \mathbf{r}/r$, its trajectory has the form of a straight line. If the velocity is constant, $U = const$, then the wave is carried away by the flow as a whole, and the straight line of the trajectory is preserved, it is only shifted along the flow. However, at the wall $\boldsymbol{U} \neq const.$ a test point trajectory is bent, and the trajectory shape depend not on the coordinates $r_2 - r_1$, but on the distance traveled by the point along the trajectory. $\ell_2 - \ell_1$. When $a \ll \ell$, the element of the trajectory is $d\ell = cdt$, and the density, pressure and velocity u in the packet, $|\ell - c\,t\,| \leq a$, vary according to formula (5):

$$\varrho, \mathbf{p}, u \sim f\left(\frac{\ell - c\,t}{a}\right)\frac{1}{\ell} \tag{5}$$

The trajectory of the test point (wave vector) is described by equation ([2], § 68])

$$\frac{ds}{d\ell} = \frac{1}{c}[\mathrm{rot}\, U, s] \tag{6}$$

In the flat current, it has components:

$$\frac{ds_x}{d\ell} = \frac{s_z}{d\ell}\frac{dU}{dz}, \frac{ds_y}{d\ell} = 0\,, s_z = \sqrt{1 - s_x^2} \tag{7}$$

When a wave occurs on the lower wall of the channel, the boundary conditions take the form

$$x(0) = z(0) = 0, \frac{dx}{d\ell}(0) = \pm\cos\vartheta_0\,,\ \frac{dz}{d\ell}(0) = \pm\sin\vartheta_0\,,$$

Substituting relation $s_z = dz/d\ell$ in (7) and integrating the first equation, we get

$$s_x = s_{x0} + M(z). \tag{8}$$

It follows from this equation that the test point moves away from the wall, but at the same time its trajectory is bent. The direction of the bend depends on the initial angle of inclination ϑ_0 of the wave vector of the test point. When the wave moves against the flow and $s_x < 0$, the trajectory bends away from the wall. When the directions of the motion of the wave and the flow coincide, $s_x > 0$, the trajectory inclines towards the wall. And if the angle ϑ_0 is small enough, then the trajectory moves away from the wall not further than the value $z_{max} = h$.

At maximum h, the tangent to the trajectory is parallel to the x axis. Accordingly, $\vartheta_0 = 0$ and $s_z = 0$. Using the identities $s_{x0}^2 + s_{z0}^2 = 1$ and the inequalities $M^2\,(z) \ll M\,(z) \ll 1$, we replace s_x by s_{x0} and s_{z0}. As a result we obtain the dependence of height h of the trajectory maximum (in scale d) on the initial angle ϑ_0 and the velocity M_0 for its profile (3):

$$h = \frac{1}{2}\left(1 - \sqrt{1 - \left(\frac{1}{M_0}\frac{\sin^2\vartheta_0}{2\cos\vartheta_0}\right)^2}\right) \tag{9}$$

Journal of Physics: Conference Series **1394** (2019) 012007 doi:10.1088/1742-6596/1394/1/012007

Laminar air flow becomes turbulent in the channel d =0.1m at $M_0 \approx 0.001$ (U_0 =0.3m/s). Formula (9) is presented in Fig. 3 as the dependence of the natural logarithm of h on $m = \{1,5\}$, $U_0 = \{0.1,1,2,4,6,8 \text{ m/s}\}$ and $\vartheta_0 \in [0,10°]$.

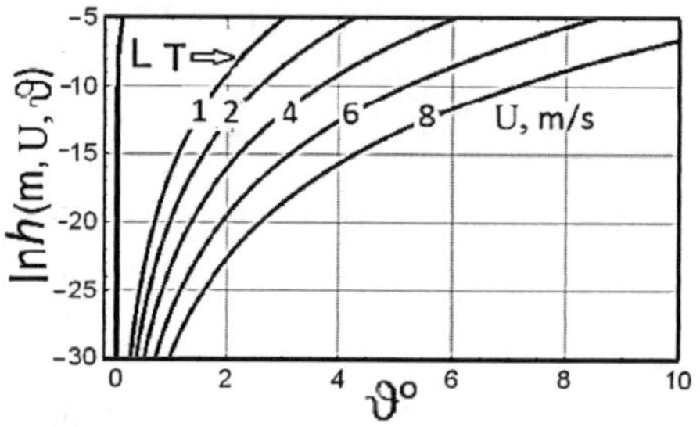

Figure 3. Dependence of parameter h on angle ϑ_0 and velocity M_0.

The h value in the laminar L flow looks like a vertical line at the smallest values of the angle ϑ_0, i.e. as if the boundary layer does not exist. To make sure we build function $h(M_0, \vartheta_0)$ wuthin the range $U_0^{™}[0, 0.3]$ m/s shown in Fig. 4. Surface $h(M_0, \vartheta_0)$ has interesting features. At small ϑ_0, it practically lies on the channel wall, h≈0. Then the trajectory maximum sharply (within interval $\Delta\vartheta_0 < 0.5°$) takes the value h =1. It seems that the boundary layer does not exist. However, by definition, the boundary layer is a flow area thereof properties highly depend on the viscosity. Just in the laminar flow, the effect of viscosity extends to the entire flow. Hence, the boundary layer in the laminar case fills the channel whole cross-section, and it is the Poiseuille formula that outlines the velocity profile, repeat, of the boundary layer.

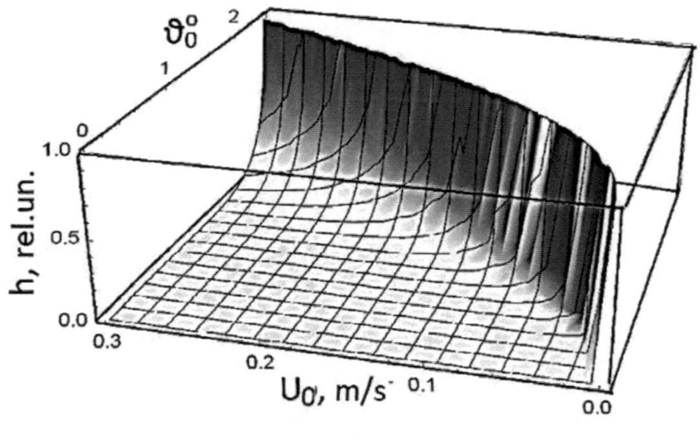

Figure 4. Function h (M_0, ϑ_0) in laminar flow.

Surface $h(M_0, \vartheta_0)$ has interesting features. At small ϑ_0 values, it practically lies on the channel wall, $h \approx 0$. Then the maximum of the trajectory sharply (within interval $\Delta\vartheta_0 < 0.5°$) takes the value $h = 1$. It seems that there is no any boundary layer. However, by definition, the boundary layer is a flow area thereof properties highly depend on the viscosity. Just in the laminar case, the viscosity action extends over the entire flow. Hence, the boundary layer in the laminar case fills the entire cross-section of the channel, and it is the Poiseuille formula that describes the velocity profile, repeat, of the boundary layer.

5. Boundary layer of turbulent flow and its structure

In the turbulent flow, the rapid growth of the velocity gradient at the wall initiates important changes in the propagation of perturbation waves. The maxima appear at the trajectories of the test points of the

Journal of Physics: Conference Series **1394** (2019) 012007 doi:10.1088/1742-6596/1394/1/012007

wave close to the wall. As can be seen from Fig. 3, the value of h is determined by angle ϑ_0, though it also strongly depends on the velocity U_0. Moreover, for each of U_0 values, there exists a trajectory with an angle ϑ_{0cr}, beyond which the trajectories do not extend in accordance with the Thomson theorem. Consider this question in detail.

Parameter h has a real value provided that the difference under the root of formula (9) is not negative. Otherwise, the trajectories have no maximum in velocity field (3), and run over the entire flow, experiencing only some deviation in the direction of the hour hand. Figure 5 illustrates the trajectory for the angles $\vartheta_0 = \{13,14,15,17,20°\}$ in a narrow stripe $z \in [0, 0.1]$.

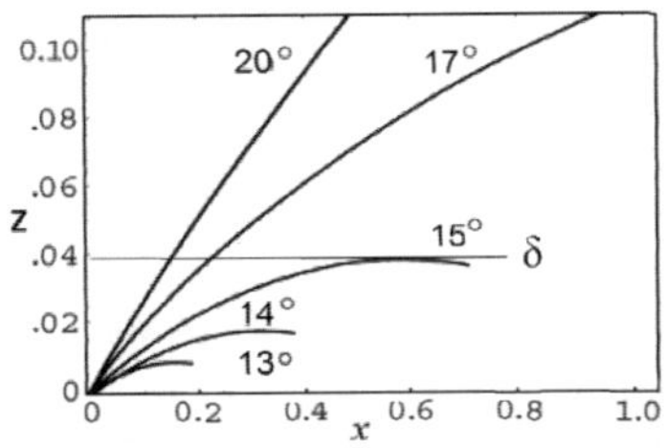

Figure 5. Trajectories for the angles $\vartheta_0 = \{13,14,15,17,20°\}$ in $z\in[0, 0.1]$ stripe

As the angle ϑ_0 decreases, the deviation of the trajectories increases and, starting from the angle ϑ_{0cr}, the trajectories do not go beyond the limits of the δ level, i.e. δ value can be given the status of the outer boundary of the boundary layer. The trajectories with lesser angles ϑ_0 are concentrated within the δ layer, whereto the spherical wave layers reflected from the channel walls arrive as well. The wave propagation is accompanied by fluctuations in the parameters of the medium. They form the boundary layer structure. The value of δ depends on the thickness of the sperical wave layer which is determined by the size of the primary perturbation, and ultimately by the length of the elementary flow waves. Calculations shows that the size of δ is approximately equal to the radius a of the primary perturbation.

The boundary layer structure is being formed by a sequence of flow waves and therein fluctuations, exactly the same as in the core flow. However, a thin layer adjoins the wall, which is often called the "laminar sublayer". The view point on the interpretation of this sublayer structure i is expressed in ([2], §42): "..., there is turbulent movement in it, ..., but it reveals peculiar features that have not yet been adequately interpreted". Therefore, "viscous sublayer" term was suggested instead of the wrong term "laminar sublayer".

The wave model gives this missing interpretation. In fact, a viscous sublayer arises when the perturbation waves fre reflected from the wall. A wave packet bears translational motion energy and energy of pulsations with a specific density of $c\rho u^2$. During reflection, the velocity and its pulsations disappear at the wall, and their energy is transferred respectively to local pressure and to its oscillations. In other words, the energy of these do not disappear, it simply take another form. The viscous sublayer thickness, as well as the boundary layer, depends on the parameters in the primary perturbation. For the parameters of various distribution, the thickness of the sublayer is estimated as $\delta L \sim (0.1-0.2)$a.

The trajectories within the boundary layer are represented in Fig. 5 by ascending branchs. Each of them is limited by a point of maximum $z(x_{max}) = h$. Equation (6) has a singularity in this point. However, it is not a simple singularity. The fact is that approximation of geometrical acoustics is unsuitable for strong interaction, in particular, in the boundary layer. The matter is that the perturbing wave moves in the field of other perturbations, in particular, vorticity formed by curl $\boldsymbol{U}$, and the diffraction divergence. Interaction with these fields leads to phase redistribution and to deformation of the wave flat front, i.e. to aberrations.

Vorticity changes the flow velocity profile. However, the vortex cannot be seen directly, except through the movement of some external body, such as a speck of dust, captured by the vortex motion. The trajectory of a speck of dust coincides with the vortex line, provided that it does not move by itself. If it has its own velocity, then its circular trajectory is deformed as a result of summing of the two velocities.

Journal of Physics: Conference Series **1394** (2019) 012007 doi:10.1088/1742-6596/1394/1/012007

A test point of the wave layer is also an external body, though only in the sense of motion direction and velocity value. "It means the addition of motions in the vortex and in spherical wave layer. Here the difference between the vortex created by the obstacle and the vortex arising at the wall smooth surface should be noted. The obstacle creates some continuous perturbation, while a perturbation on a smooth wall surface occurs from time to time, and the length of its vortex line is limited by the width of the spherical wave layer $2a$. In the latter case, the vortex line breaks, that can be observed in experiments. An intermediate option happens when the channel wall surfaces are rough, and both methods work simultaneously, forming the vorticity of different types.

So, the vortices appear on the descending branches of the test point trajectories of the s. Since the curvature of the trajectories increases in this area, and equation (6) does not describe aberration, the geometrooptical approach with equation (6) must be replaced by the wave method. Searching for the influence of aberrations on the wave propagation, we use an example from optics [10, 11], for this approach has not yet been developted in acoustics.

The aberrations being considered in the classical theory deform the wavefront without changing its topology. Quite a different picture is observed when optical vortices appear in the wave beam: singular points appear on the surface of the wave front, which in many respects are similar to defects in the crystal lattice – screw dislocations. When going around the dislocation, the phase changes exactly by 2π. At the very singular point, the amplitude of light oscillations is zero, and the phase is not determined.

The wavefront of light (acoustic) beams that are close in their properties to a plane wave looks like a family of disjoint surfaces (Fig. 6 *a*) with a distance between them equal to the wavelength λ. The screw dislocation dramatically changes the wavefront topology. The direction of propagation of light energy is given by the Umov-Poynting vector, which is known to be perpendicular to the surface of the wave front at each point. Accordingly, in the vicinity of the screw dislocation, a "swirl" of the energy flow occurs, and an equiphase surface consisting of many sheets shown in Fig. 6 *a* turns into a single surface with a screw structure represented in Fig. 6 *b*.

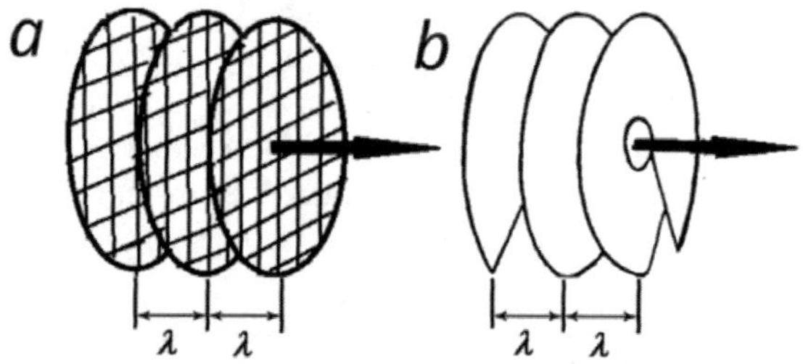

Figure. 6. Wave front structure in the screw dislocation absence (*a*) and if available (*b*)

As you can see, the trajectory takes the form of the vortex. As a matter of fact, the used procedure is ia physical interpretation of the curl operation.

It is worth to add some comment at this place, concerning Thomson theorem which restricts the vortex model application by the boundary layer, while the vortexes ills only some portion of the flow section. whereas the potential part or the flow remains free from vortexes. Meanwhile, the wave model has no such a restriction, for the wave freely crosses the potrntial flow boundary carrying therewith some matter from the boundary layer.

Indeed, the oscillations of hydrodynamic parameters in a spherical wave layer are described by equations ([2], § 71):

$$\left.\begin{aligned} p\sim\rho\sim\tfrac{1}{r}f(r-ct);\\ u\sim\tfrac{1}{r}f(r-ct)-\tfrac{1}{r^2}\int_{r-ct}f(r-ct)d(r-ct). \end{aligned}\right\} \qquad (10)$$

It follows that the velocity expression differs from the pressure and density formula by the convective term. In addition, it changes the sign later, therefore,the wave entrains a part of the medium, and transfers it with frozen vorticity into the potential current. However, the oscillations associated with the convective velocity term decay rapidly, $\sim 1/r^2$, unlike the velocity oscillations in the wave, which decay slowly, $\sim 1/r$.

6. Conclusion

The process of vortex formation was considered in the framework of the concept of a wave model of turbulence. But this process is the starting point of the vortex model. In addition, the wave model made it possible to solve a number of issues that could not be resolved within the framework of the vortex model. These include problems such as limitations associated with Thomson's theorem that action over the entire section is possible only after the boundary layers from all walls fill the section, that is, when the boundary layers merge. The problem of the formation and thickness of the boundary layer, the viscous sublayer, and the structure of the spatiotemporal pulsation field of these layers is solved. In this sense, we can say that the wave concept includes the model proposed by L.F. Richardson [7].

The interpretation of the problems discussed is mainly based on the geometro-optical approximation of the theory. However, the interpretation of the vortex motion in the boundary layer of turbulent flow required the use of ideas about the change in the helical dislocation of the wave structure of the front of the acoustic wave.

References

[1] Cartman Th. The wind and beyond. Autobiograhy. Boston, 1967.
[2] Landau L.D., Lifshits E.M. Hydrodynamics. Course of Theoretical Physics. Vol. VI. PERGAMON PRESS. 1987.
[3] Rayleigh J.W. On the stability, or instability of certain fluid motions. Scientific papers. Cambridge Univ. Press. Vol. 1. Pp. 474, 487 (1878). Part III. Vol. 4, Pp. 203-209 (1895, addendum 1896);
[4] Reynolds O. On the experimental investigation of the circumstances which determine whether the motion of water shall be direct or sinuous, and the law of resistance in parallel channels. Phil. Trans. Roy. Soc. London. 1883. V.174, pp.935-982.
[5] Lorentz H.A. Űber die Entstehung turbulenter Flűssigkeits-bewegungen und űber den Einfluβ dieser Bewegungen bei der Strőmung durch Rőhren. Abh. Theor. Physik., Teubner, Leipzig. 1907, s. 43. (The first publication: Zittingsverlag Akad.Amsterdam. 1897. V.6, p.28).
[6] Landau L.D. To the turbulebce problem. (in Russian) DAN USSR, V.44, № 8, p.339 (1944).
[7] Richardson L.F. Weather prediction by numerical process. Cambrige Univ. Press. 1922.
[8] Pyatnitsky L.N. Ffinite-Aperture Riemann's Wave. J. Experimental and Theoretical Physics. 2018. Vol.127, No.4, pp.778-785.
[9] Pyatnitsky L.N. Turbulence nature and Ithe inverse problem. Springer. 2012. 190c.
[10] J.F Nye. and M. V. "Dislocations in wave trains". Proc. R. Soc. London Ser. A 336, 190–191 (1974).
[11] Allen L., Barnett S. M., Padget M. J./, Optical Angular Momentum (Institute of Physics, 2004).

Air flow control around a cylindrical model induced by a rotating electric arc discharge in an external magnetic field

D. S. Munhoz[1], V. A. Bityurin[1], I. P. Zavershinskii[3], A. I. Klimov[1], I. A. Moralev[1], P. N. Kazanskii[1], N. E. Molevich [2,3], L. B. Polyakov[1], D. P. Porfir'ev [2,3], S. S. Sugak[3], B. N. Tolkunov1

1 Joint Institute for High Temperatures, Russian Academy of Sciences
2 Samara Branch of the Russian Academy of Sciences, Lebedev Physical Institute
3 Korolev State Aerospace University

E-mail: dasmunhoz@yahoo.com.br

Abstract. The structure and dynamics of a near-wall gas flow produced by a rotating electric arc discharge in an external magnetic field around a cylindrical model with and without an incoming flow has been investigated (M <0,2; Re < 9.104). The electric arc on the model has been produced by a combined electric discharge (low-current RF discharge + high-current pulse-periodic discharge). Permanent magnets with induction $B \approx 0.1$ T have been placed inside the cylindrical models. Ring electrodes are arranged on the surface of the model. The structure and dynamics of the near-wall gas flow around the cylindrical model have been investigated using high-speed photography, as well as the shadowgraph and particle image velocimetry (PIV) methods. In the experiment were measured: the flow profile around the cylinder and the aerodynamic forces acting on the cylinder (drag and lift).

Keywords: cylinder flow control, flow circulation, flow separation, flow separation trace past a cylinder, magnetoplasma actuator (MPA).

Intensive research is being conducted nowadays in the field of plasma aerodynamics on the possibilities of the flow control around streamlined bodies by various plasma formations to [1-3]. The main advantage of plasma actuators (active devices for controlling the flow using local plasma formations) is their fast reaction, which is based on the electronic control of the parameters of electric discharges. This property, in turn, allows one to implement feedback control of the development of various gas-dynamic perturbations in the flow near streamlined bodies (for example: to control the moment of transition of the laminar boundary layer to turbulent, flow separation, etc.). An important task in plasma aerodynamics is to control the flow circulation around a blunt body or streamlined body (as a cylinder or a wing). However, there are few experimental works devoted to this issue only [1-3]. The present work fills the gap in this area and it is devoted to study of the flow around a cylindrical model by a magnetoplasma actuator (MPA). To control of the flow circulation around a cylindrical model by a rotating electric arc in an external magnetic field is used in this work. The aim of this work is to study the near-wall gas flow around a cylindrical model with working MPA using high-speed camera, PIV and shadowgraph methods.

1. Experimental setup. Aerodynamic model and MPA

The aerodynamic model used in this work is a cylinder made of a quartz tube with a diameter of 20 mm, length of 95 mm and a wall thickness of 2 mm, Fig. 1-a (1). The measured magnetic field induction on the model surface (between the electrodes) reaches a value of $B \approx 0.1$ T.

Ring electrodes are manufactured from cuprum wire. Spark gap (3) is used for discharge ignition at definit region of an aerodynamic model. Arrangement diagram of the spark gap electrode on the cylindrical models in position $\alpha = 5°$ or 90° relative to the direction of the air flow are shown in the fig 1b. These optimal spark gap angement are used according the recommendation of the works [2, 3].

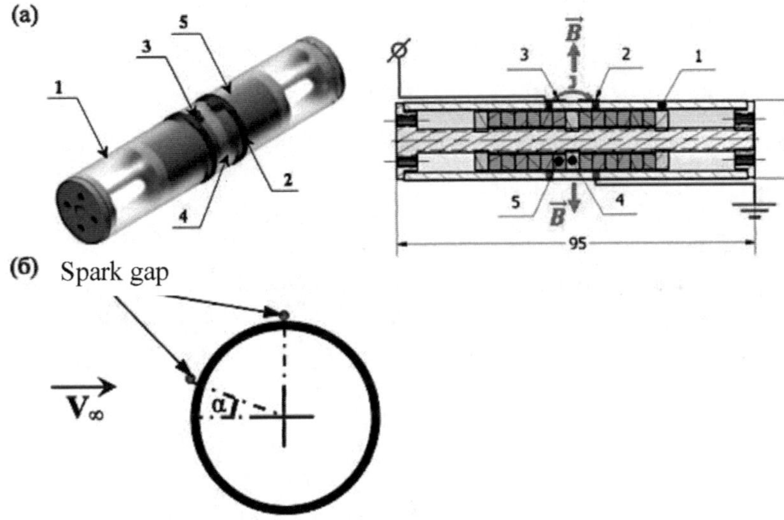

Figure 1. (a) General view of the model and schematic of the MPA on a cylindrical model with a single discharge gap: (1) quartz tube; (2) copper ring; (3) spark gap electrode; (4) polyamide nut and screw; and (5) Nd–Fe–B magnet. (b) Arrangement diagram of the spark gap electrode on the cylindrical models in position α = 5° or 90° relative to the direction of the air flow or to the horizontal (in the absence of flow).

1.1. High-voltage power supply

To initiate an arc discharge, a high-voltage power supply was designed and tested. In our MPA, we used two types of power supplies: a DC source (DC generator) and a power supply for producing a high-power pulse-periodic discharge. For initiating these discharges, we used a high-voltage pulsed discharge produced by a RF Tesla generator. The technical characteristics of the source of a pulse-periodic discharge were as follows: pulse repetition rate $F_i < 3$ kHz, maximal output voltage of the power supply $U_{max} < 20$ kV, maximal discharge current $I_{max} \leq 30$ A, maximal pulse power $P_{max} < 8$ kW, and mean power $P_{mean} \leq 200$ W.

Figure 2 shows typical voltage and current signals measured across the discharge gap of the MPA on the cylindrical model in experiments for flow visualization. It can be seen that the maximal pulsed current on the model surface attains a value of $I_{max} \approx 16$ A; the power pulse duration is $T_i = 12$ ms, and the pulse repetition rate is $F_i = 7$ Hz. The electric voltage amplitude is $U_d = 1.5$ kV. The maximal pulsed power can be as high as $P_{max} \leq 8$ kW; the mean power is $P_{mean} \approx 200$ W.

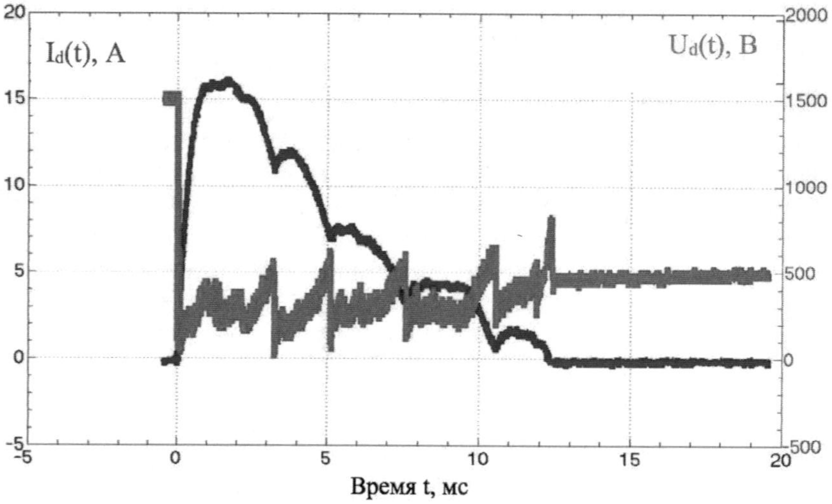

Figure 2. Pulsed current I_d (blue) and voltage U_d (green) in the MPA (pulse-periodic regime).

2. Experimental Results

In order to analyze the effect of the MPA on the flow separation on the model, we used the regime of external synchronization of video recording and the PIV setup. The rotational frequency of the arc discharge measured in this experiment was $F_r \approx 290$ Hz. In this experiment, we used a pulse-periodic discharge with the following characteristics: $F_i = 7$ Hz, $U_{max} < 15$ kV, $I_{max} < 30$ A, $P_{max} < 70$ kW and $P_{med} = 200$ W. Figure 2 shows the typical voltage and current signals measured in the discharge gap of the MPA. The velocity of the air flow corresponded to $Re = 8 \times 10^3$ ($V_\infty = 6.5$ m/s). The spark gap electrode was in the position at $\alpha = 5°$, [2, 3]. The velocity field in the vicinity of the cylindrical model with a working MPA in an external flow in this regime is shown in Fig. 3. The number of measurements for averaging each PIV pattern was 70. Under these conditions, a considerable effect of the rotating arc in the flow past the model was detected. Figure 3 shows the flow field around the model with working MPA after the completion of the second turn of the arc around the model. In this regime, the entire velocity field around the model, including its front and back part, is perturbed. The figure also contains a synchronized video image of the rotating arc discharge at the corresponding instant. An analysis of this figure also shows that the flow circulation around the model changes in the (1) head anode part and (2) tail cathode part of the arc discharge. It should be noted that the region of the stimulated circulation of the near-wall flow also exists in the back part of the model. The flow separation in this region is not observed up to angles of $\theta \approx 260°$. This result makes it possible to determine the averaged field of induced vorticity around the model in this experiment. It is important to note that this result of indused vorticity is absent in the case of MPA operation off. An analysis of the shadowgraph pictures, arc temperature and PIV frames leads to the conclusion that the regime of a weakly compressible flow is realized in the entire layer of the near-wall flow around the cylinder except in the region near the electric arc (small variations in density $\delta\rho/\rho_0 \ll 1$ are observed, where ρ_0 is the density of air, as well as small change in the velocity of the induced flow, $\delta V/V_0 \ll 1$, where V_0 is the flow velocity). Thus, Zhukovsky's theorem is valid for the entire integrated zone with the exception of the electric arc zone (where $\delta\rho/\rho_0 \sim 1$), which, in turn, is too small compared to the entire integrated zone in formulas 2.1 and therefore can be neglected. The value of the evarage stimulated vorticity of the flow around the model turned out to be equal to $\bar{\xi} \approx 1325$ s^{-1}.

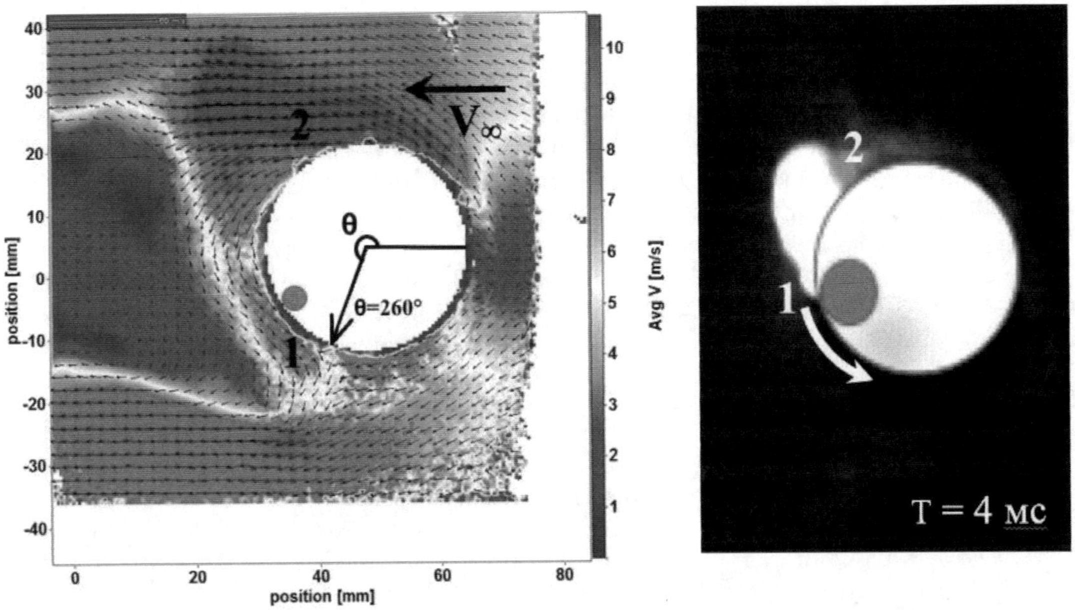

Figure 3. Averaged PIV pattern of velocity field around cylinder with rotating plasma arc. Characteristic Reynolds number of incoming flow is Re = 8 × 10^3 ($V_\infty \approx 6.5$ m/s). Delay time of discharge initiation is $T_2 = 4$ ms.

This high value of everage vorticity ξ (compared to that in the direct-current discharge experiment) was obtained due to the use of discharge with a high pulsed current in the experiment [7].

The magnitude of the stimulated circulation Γ in Fig. 3 was determined using expression 2.1, according to the formula from [4 - 6]:

$$\Gamma = \iint_S (\nabla \times V) dS = \int_0^{2\pi} \int_{R_0}^{(R_0+h)} (\nabla \times V) R_0 d\theta dr \approx 0{,}4 \text{ м}^2/\text{с} \quad (2.1)$$

where R_0 =10 mm and h = 4 mm.

3. Discussions and conclusions

In conclusion, it can be noted that the electric arc has a complex three-dimensional appearance, this result was confirmed using high-speed camera.

A typical width of a perturbed entropic vortex flow or induced flow circulation jet induced by the rotating electric arc around the cylindrical model is about 10 mm or more, which is much more than the luminous electric arc diameter of the order of mm. In Fig. 3, it is clearly seen that there is an asymmetric flow around the cylindrical model caused by an induced flow circulation around the model, which is created by the rotating electric arc. Such an asymmetric flow around the cylindrical model can lead to lateral lift force acting in the model. The practical absence of flow separation on the model in this mode makes it possible to estimate the value of the lift coefficient C_L, due existance of stimulated circulation Γ, according to the formula from [4 - 6]:

$$C_{L,t} = \frac{w\rho_\infty V_\infty \Gamma}{\frac{1}{2}\rho_\infty V_\infty^2 wd} = \frac{2\Gamma}{V_\infty d} \approx 0{,}5 \quad (3.1)$$

where Γ = 0,4 m^2/s and V_∞ = 72 m/s.

It was revealed that the calculated $C_{L,t}$ calculated by formula (3.1) is close to the experimental $C_L, exp \approx 0{,}3$ measured by balance [7]. The lift force in our experiments achieved maximum values of about $L = 0.37\text{-}0.45$ N.

In conclusion, it should be noted that the developed MPA can produce a near-wall vortex flow around a cylindrical model, which affects the external subsonic gas flow past this model ($M < 0.2$, $Re < 9 \times 10^4$). This statement has been confirmed by results of direct measurements, using the PIV, shadowgraph methods and by aerodynamic force measurement by balance [7]. It concluded that the lift force that acts in the cylindrical model emerges due to stimulated circulation of the flow around it.

This study was supported by International foundation CNPq (Conselho Nacional de Desenvolvimento Cienífico e Tecnológico, Brazil) under the program "Science Without Boundaries" for 2012–2015 (CNPq grant no. 200205/2012-9) and in part by the Ministry of education and science of Russia by State assignment to educational and research institutions under project 3.1158.2017/4.6.

References

[1] Biturin V. A., Kazansky P. N., Klimov A. I., Moralev I. A., Study of lift and drag control of circular cylinder by surface HF plasma actuator, Proc. Join EROCOFTAC-PLASMAERO Workshop, Toulouse, France, 10-12 December 2012, P.8

[2] Moralev I. A., Bityurin V. A., Kasansky P. N., Klimov A. I., Chertov D., Flow Separation Control near NACA23012 airfoil by a Capacity Coupled Surface HF Discharge, AIAA Paper 2011-3733, 42nd AIAA Plasmadynamics and Lasers Conference in conjunction with the 18th International Conference on MHD Energy Conversion (ICMHD), Honolulu, Hawaii, June 27-30, 2011

[3] Kazanskyi P. N., Klimov A. I., Moralev I. A., High-frequency actuator control of air flow near a circular cylinder: Impact of the discharge parameters on the cylinder aerodynamic drag, High Temperature (2012), Volume 50, Issue 3, pp 323–330 — https://doi.org/10.1134/S0018151X12030145

[4] Anderson J. D., Fundamentals of Aerodynamics / Editor «Mcgraw-Hill» — Fifth SI unit Edition — 1152 pp. — ISBN 9780073398105 — 2011 —

[5] Wu James C., book «Elements of Vorticity Aerodynamics» / Shanghai Jiao Tong University Press, Shanghai and Springer-Verlag Berlin Heidelberg 2018 — ISBN 978-3-662-44039-1— 140 pp. — 2018— https://doi.org/10.1007/978-3-662-44040-7

[6] Yuan Jiankun, Ph.D. dissertation work: «Circulation methods in unsteady and three-dimensional flows» / Dissertation submitted to the Faculty of the Worcester Polytechnic Institute in partial fulfillment of the requirements for the Degree of Doctor of Philosophy in Mechanical Engineering — 2002 —

[7] V. A. Bityurin, I. P. Zavershinskii, A. I. Klimov, N. E. Molevich, I. A. Moralev, D. Munhoz, L. A. Polyakov, D. P. Porfir'ev, S. S. Sugak, Streamlining of a cylinder with an electric arc rotating in a magnetic field, High Temperature (2016), Volume 54, Issue 4, pp 599–602

Hypersonic plasma flows created by a pulse capillary discharge

A S Pashchina

Joint Institute for High Temperatures, Russian Academy of Sciences, Moscow, Russia

fgrach@mail.ru

Abstract. The possibility of producing by a pulsed capillary discharge the quasi-stationary high-speed plasma flows in a wide range of Mach numbers (M =3-20), electron number densities (n_e~10^{11}-10^{15} cm^{-3}) and temperatures (T_e~3-10 eV), rotational and vibrational temperatures (T_r =3800-6000 K and T_V =9000-15000 K), which can be used for setting up plasma aerodynamic experiments on small laboratory facilities, is shown. The most suitable for this purpose is a mode close to supersonic outflow into a vacuum, the transition to which occurs when the total pressure ratio exceeds $N = p_0/p_\infty$ >5·10^3. This mode permits to obtain wide-aperture flows (D >10 cm) with spatially extended weakly inhomogeneous segments sufficient to accommodate tested bodies of a centimeter scale.

1. Introduction

The creation of hypersonic aircraft (HA) is directly related to the solution of a number of scientific, technical and technological problems, involving experimental studies in conditions close to the real flight of HA. The complexity and high cost of full-scale experiments [1], as well as studies on large-scale test benches [2,3], makes it relevant to search for ways to produce high-speed flows in small laboratory installations, allowing one to solve quickly individual tasks with less stringent requirements for the validity of similarity criteria and flow parameters. In many cases, laboratory experiments are an acceptable alternative to costly bench and flight experiments and, unlike the latter, have important advantages, including efficiency of obtaining the result, as well as more wide options for choosing the diagnostics methods and tools. At the same time, it is highly desirable to be able to produce the flows, whose parameters are close to the real conditions of hypersonic flight, namely: the stagnation enthalpy - h^* >20 MJ/kg, the stagnation pressure - p^*~10^4-10^5 Pa, the Mach number of incident flow - M >6. Such parameters are easily achieved in plasma flows created by a pulsed discharge in a capillary with an evaporating wall. As was shown previously [4,5], this method allows one to produce in an rarefied gas atmosphere the quasi-stationary high-speed plasma flows with a duration of t =1-100 ms in a wide range of Mach numbers M =3-20, gas temperatures T =3000-6000 K and charged particles densities n_e =10^{11}-10^{15} cm^{-3}, that creates wide opportunities for setting up the plasma-aerodynamic experiments. The undoubted advantage of capillary discharge is the ability to select the type and method of supplying the plasma-forming substance that allows one to simulate the conditions of hypersonic flight in the atmospheres with different chemical composition.

The most important requirement for an aerodynamic experiment is to ensure the spatial homogeneity of the flow parameters on the scales exceeding the dimensions of the tested bodies. This requirement can be satisfied simply enough for stationary flows with fully controlled parameters, for example, by installing special supersonic nozzle ($p_e = p_\infty$, p_e – pressure at the nozzle exit, p_∞ - ambient pressure) providing design flow condition. This solution is hardly suitable for pulsed flows with time-varying parameters. In the latter case, a suitable solution is to place the tested bodies in those flow segments, where the flow parameters changes rather slowly. For highly underexpanded supersonic jets, such a segment is located inside a single section of the shock-wave structure that forms in the initial part of the jet. The length of the weakly inhomogeneous segment, as well as the dimensions of the shock-wave section, is determined by the total pressure ratio $N = p_0/p_\infty$ (p_0, p_∞ - pressure inside the capillary and in ambient atmosphere, respectively). It can be expected that for the large values of total pressure ratio $N > 10^3$, the dimensions of the weakly inhomogeneous section will be sufficient to accommodate tested bodies of a centimeter scale. Determining the conditions for obtaining, as well as studying the structure of such flows was one of the objectives of this work.

Another task of the work was to select the appropriate diagnostic methods and to determine the basic electrophysical, gas-dynamic, and thermophysical characteristics of pulsed plasma flows injected into a rarefied atmosphere. This paper focused on the experimental determination of flow parameters such as the Mach number, electron density and temperature, and the temperature of the heavy component. For this purpose, a set of independent diagnostic methods is examined: quantitative spectroscopy, probe diagnostics, visualization of supersonic flows based on gas-discharge and shadow methods.

2. Experimental facility

Principal design of the experimental facility is presented in figure 1. The capillary arrester, whose detailed description is given elsewhere [6], is used for the plasma jet producing. Polymethylmethacrylate (PMMA) is used for capillary fabrication, whose chemical formula $C_5H_8O_2$ determines the initial stoichiometric composition of the plasma. The discharge power source is the energy storage device consisting of a capacitor banks connected to the discharge circuit through a time-delaying inductance. The discharge power source provides following parameters: the storage energy - Q =90–640 J, the amplitude of the discharge current - I_m =100–2500 A, and the duration discharge pulse - τ_d =2–20 ms. The mass-average enthalpy of the obtained plasma at the specified parameters of the discharge power source is h =100-200 MJ/kg. The arrester is placed in a vacuum chamber, the ultimate vacuum is p =13 Pa. The vacuum chamber is equipped with quartz windows, which makes it possible to diagnose the discharge using shadow and spectral methods. Tested bodies of various shape (such as a sphere, a cone, an asymmetrical wedge, a thin plate etc.) were placed in the center of vacuum chamber along the axis of the capillary. The main part of experiments were carried out in an air atmosphere in the pressure range of p_∞ =13-10^5 Pa. In some cases we used argon and helium as tested gases.

To study the flow pattern, we use the registration of the own and shadow images of the selected jet segment using a MotionPro N3 high-speed video camera (exposure duration - 1 μs, frame rate - up to 10 kHz). The emission spectra of the plasma in the wavelength range 280-800 nm were recorded using an AvaSpec UL-2048 spectrometer. The recorded emission spectra were used to estimate the component composition and the main parameters of the plasma — the number density of charged particles, vibrational and rotational temperatures. Also, the Langmuir triple probe method is used to determine the electron number density and temperature [5].

3. Experimental results

One feature of capillary discharge plasma injected into the rarefied atmosphere is a low value of the effective adiabatic index tending to unity $\gamma \to 1$ [4], that is a sign of significant concentrations of polyatomic molecules and condensed particles. The sources of condensed particles are the ablation products of the carbon-containing capillary wall and the electrodes substance. The formation of condensed particles is possible in a limited temperature range T=1500–5000 K [7,8], which, according to the results of spectral diagnostics (see below), are achieved in a plasma jet injected into a rarefied atmosphere. This result is quite expected, since the production of condensed particles was observed earlier for plasma jets injected into a gas at atmospheric pressure [9]. Estimates based on the results of the extinction measurements show that the characteristic size of condensed particles is 2–4 nm, the cluster formation time is 0.2–0.4 μs, and their concentration is 10^{13} cm^{-3}. These estimates are consistent with the results of measurements of the condensed particles parameters by scanning electron microscopy [9].

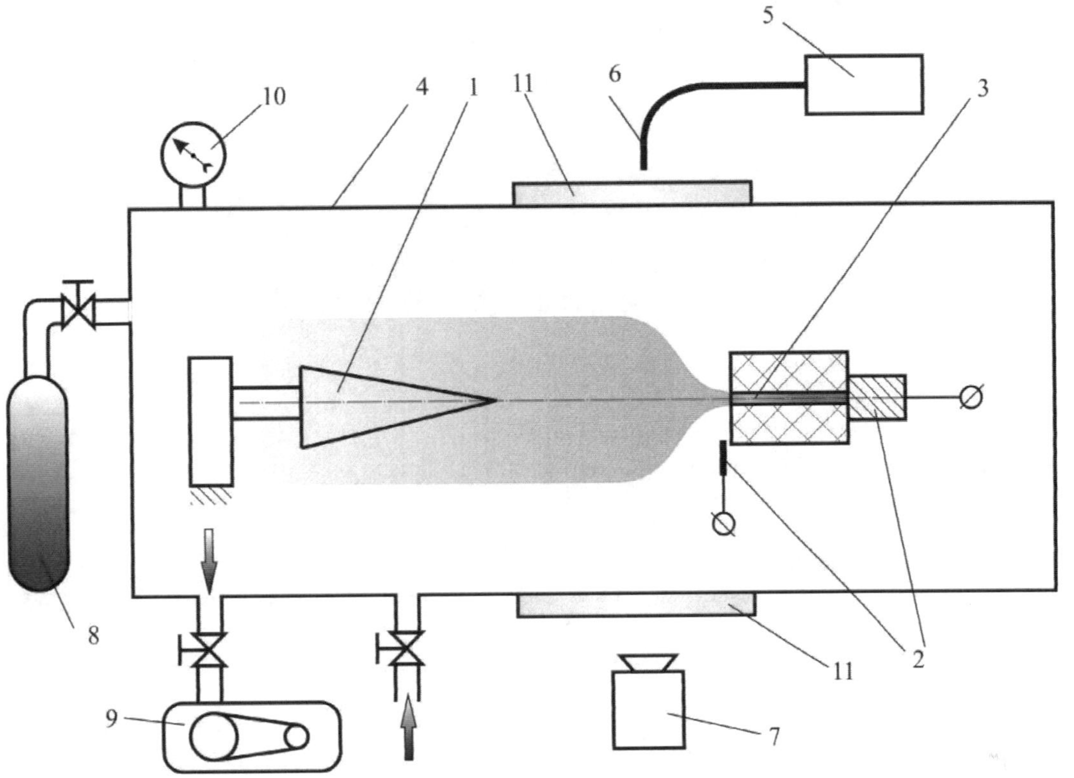

Figure 1. Principal design of experimental facility: (1) the test body, (2) the electrodes of capillary arrester, (3) the capillary, (4) the vacuum chamber, (5) the AvaSpec UL-2048 spectrometer, (6) the spectrometer optical guide, (7) a MotionPro N3 high-speed video camera, (8) the test gas bottle, (9) a vacuum pump, (10) a vacuum manometer, (11) quartz windows.

One of the objectives of the research was to study the conditions for obtaining hypersonic flows with homogeneous segments, the length of which is sufficient to accommodate centimeter-scale tested bodies. For this purpose, a pattern of the flow around a thin plate with a pointed front edge placed at various distances from the capillary outlet was studied. The Mach number of the flow was determined by the inclination angle of the attached shock wave. Two parameters were varied during the research - the distance between the plate edge and the capillary outlet (z =5-100 mm) and the total pressure ratio ($N = p_0/p_\infty$ =2-3·10^5). The total pressure ratio was set by an appropriate choice of pressure in the vacuum chamber (p_∞=13-10^5 Pa) and/or pressure inside the capillary (p_0 =0.2-4 MPa). The value of the last one is determined by the power of the discharge pulse and capillary sizes.

We can distinguish two typical flow modes of a supersonic jet, depending on the total pressure ratio: the mode with the formation of a shock-wave section in the initial part of the jet and the mode close to supersonic outflow into a vacuum. The formation of a shock-wave section is observed up to the total pressure ratios of N~10^3. In this mode, one of the main elements of the shock-wave structure is clearly observed - the central shock wave (Mach disk), which significantly distorts the pattern of the flow around a thin plate, interacting with the attached shock wave (figure 2(a)). The shock wave structure does not form if the total pressure ratio exceed N >5·10^3. In this case, the classical pattern of supersonic flow around a thin plate with the formation of the attached shock wave (figure 2(b)) is observed, and this mode turns out to be close to supersonic outflow into a vacuum.

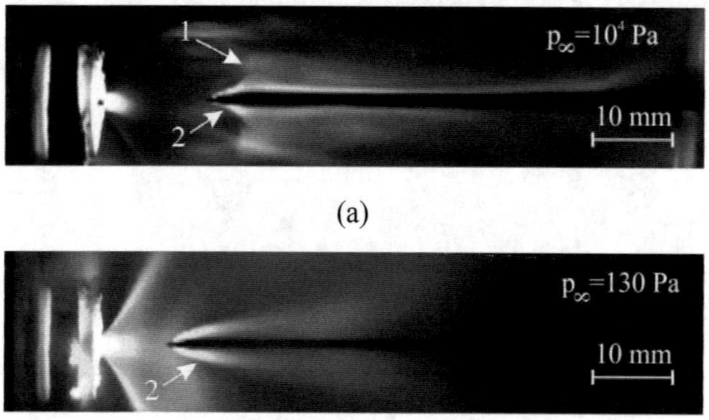

Figure 2. The influence of the total pressure ratio $N=p_0/p_\infty$ on the flow mode of a plasma jet and the pattern of the flow around a thin plate: (a) the mode with the formation of the shock-wave structure ($N\approx100$), (b) the mode close to supersonic outflow into a vacuum ($N\approx7\cdot10^3$), (1) central shock wave (the Mach disc), (2) attached shock wave.

The flow structure in this mode is characterized by the presence of an expansion section, along which the flow accelerates (figure 3), and a stationary section, at which the jet diameter and the Mach number vary slightly. The expansion section length L, as well as the transverse size of the jet D on the

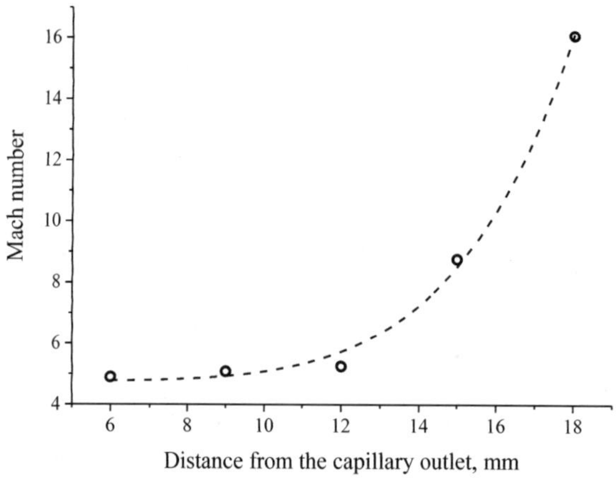

Figure 3. Example of the Mach number distribution along the flow expansion section in the mode close to supersonic outflow into a vacuum for total pressure ratio $N\approx7\cdot10^3$.

stationary section, varies in proportion to the square root of the total pressure ratio, i.e. $\frac{L}{d}\sim\frac{D}{d}\sim\sqrt{N}=\sqrt{\frac{p_0}{p_\infty}}$ (d – diameter of capillary). Thus, when pressure inside the capillary is $p_0=1$ MPa, and the pressure in the vacuum chamber is $p_\infty\sim130$ Pa, the length of the expansion section is approximately $L\approx2$ cm, and the transverse dimension of the jet stationary section exceeds $D>10$ cm. The Mach number at a considerable length of the stationary section varies slightly. For the above mentioned parameters, the length of such segment reaches 8-10 cm at the characteristic values of Mach number of $M\sim 15\text{-}20$.

The emission spectrum of a plasma jet (figure 4) is represented by the intense molecular spectra of the Swan band system of C_2 radical (the bands Δv = -1, 0, +1, +2 of $d^3\Pi_g - a^3\Pi_u$ transition), the violet band system of CN radical (the bands Δv =0, Δv =-1 and Δv =1 of $B^2\Sigma^+ - B^2\Sigma$ transition), and also by weak H_α and H_β lines of the Balmer hydrogen series. Lowering the vacuum chamber pressure below p_∞ <10 Torr leads to the appearance of molecular bands of CH, N_2, NO, CO, O_2 radicals that indicates lowering of the gas temperature – at least below 5000 K. The simultaneous presence in the spectra of the Balmer hydrogen lines and molecular bands indicates high values of the electron temperature (not less than T_e >1 eV) and a significant deviation of the plasma from thermodynamic equilibrium. The latter factor limits the possibility of using the standard methods of quantitative spectroscopy to determine the electron temperature, as well as vibrational and rotational temperatures. In particular, the violation of the Boltzmann equilibrium leads to a variation in the values of the vibrational and rotational temperatures, determined by different sequences of the molecular bands of the C_2 radical. The vibrational and rotational temperatures determined in this way are enclosed in the ranges T_V =9000-15000 K and T_r =3800-6000 K.

Based on the obtained spectra, we estimated the vibrational and rotational temperatures in the regions ahead and behind the shock wave. According to these estimates, higher temperatures correspond to the region behind the shock wave. Based on these data, the estimates of the Mach number ahead the shock wave were made, which are consistent with the results obtained by analyzing the shock-wave structures formed in the vicinity of the streamlined bodies.

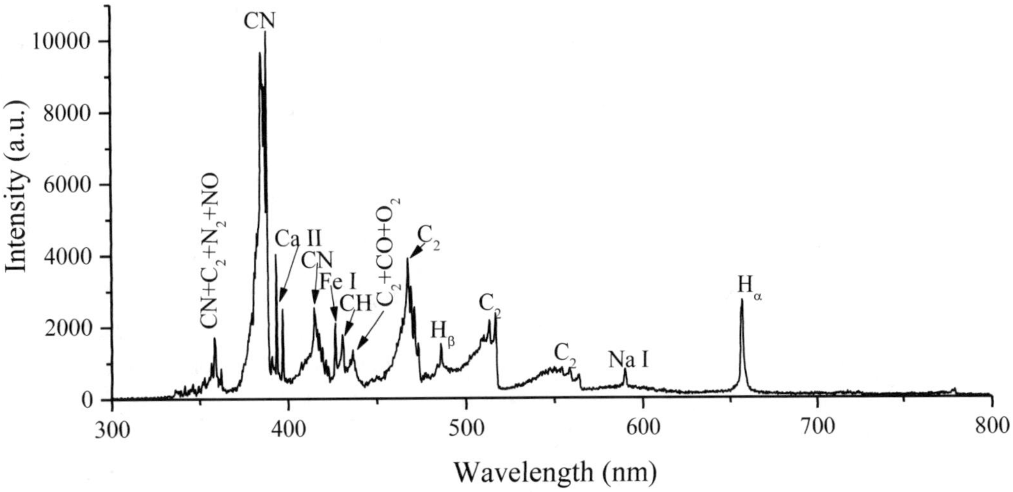

Figure 4. Emission spectrum of a plasma jet. The ambient gas is air, the pressure in the vacuum chamber is p_∞=400 Pa, the total pressure ratio is $N\approx 3\cdot 10^3$.

More suitable methods for non-equilibrium plasma diagnostics are the electric-probe methods, whose advantages are the "insensitivity" to the plasma thermodynamic state and the possibility of local measurements. In this work, we used the triple Langmuir probe method [5], which provides the possibility of simultaneous and continuous measurement of the electron number density and temperature during a discharge pulse. Measurements were performed for a jet flow mode close to supersonic outflow into a vacuum ($N > 5\cdot 10^3$). The probe was placed in the axial zone of the flow at different distances from the capillary outlet - z =5-10 cm. It was found that, at $N > 10^4$, the electron number density and temperature in this section of the flow change only slightly (no more than 10–15%), that further confirms the possibility of obtaining hypersonic flows with spatially extended slightly non-uniform segments sufficient to accommodate centimeter-scale tested bodies. The electron number density depends significantly on the discharge power. Thus, increasing the discharge power by two orders of magnitude (from P =1 kW to P=100 kW) leads to a change in the electron number density by three to four orders of magnitude – from n_e =10^{11}-10^{12} cm^{-3} to n_e =10^{14}-10^{15} cm^{-3} (figure 5(a)). In this case, the electron temperature changes in the range of T_e =3-10 eV and decreases as the discharge power increases (figure 5(b)).

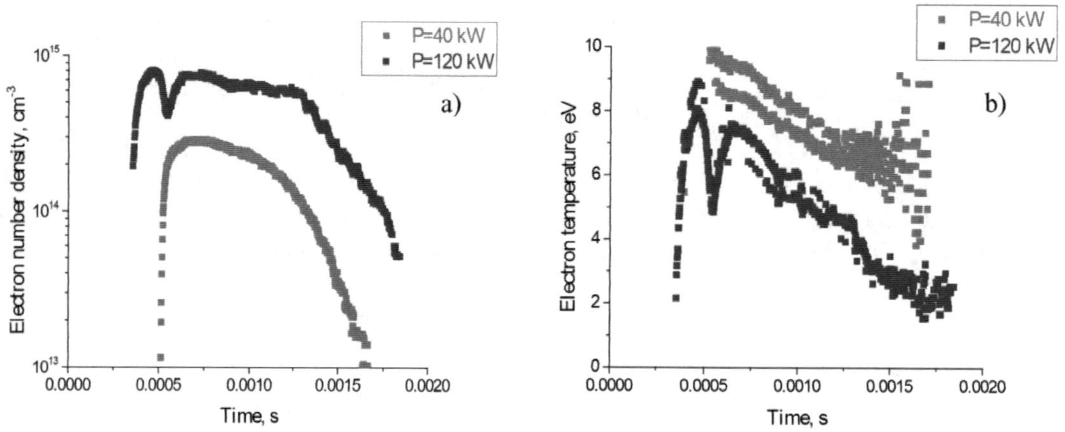

Figure 5. Time dependences of (a) the electron number density and (b) temperature on the weakly nonuniform segment of hypersonic flow, obtained by triple Langmuir probe method for various discharge power. The values of discharge power P are indicated in the plots legend. The probe is placed in the axial flow zone at the distance of z=10 cm from the capillary outlet. The pressure in the vacuum chamber is p_∞=130 Pa, the total pressure ratio is N≈7·10^3.

4. Conclusions

The possibility of producing by a pulsed capillary discharge the high-enthalpy hypersonic gas-plasma flows in a wide range of Mach numbers (M =3-20), electron number densities (n_e~10^{11}-10^{15} cm^{-3}) and temperatures (T_e~3-10 eV), rotational and vibrational temperatures (T_r =3800-6000 K and T_V =9000-15000 K), is shown. Characteristic features of plasma flows injected into a rarefied atmosphere are the strong temperature non-equilibrium (T_e/T_g >10), as well as low values of the effective adiabatic index ($\gamma \rightarrow 1$), indicating intense precipitation of the condensed phase (polyatomic molecules and nanoscale particles). The most suitable for conducting the plasma aerodynamic experiments is the flow mode close to supersonic outflow into a vacuum, the transition to which occurs when the total pressure ratio exceeds $N = p_0/p_\infty$ >5·10^3. This mode permits to obtain wide-aperture flows (D >10 cm) with spatially extended weakly inhomogeneous sections sufficient to accommodate tested bodies of a centimeter scale. An important step in future research should be the preparation and approbation of the quadruple Langmuir probe method [10], which will expand the set of measured parameters by including such important parameter as the plasma flow velocity.

References

[1] Hayes D T and Rotman W 1973 Microwave and Electrostatic Probe Measurements on a Blunt Re-Entry Vehicle *AIAA J.* **11** 675–82
[2] Gulhan A, Esser B and Koch U 2001 Experimental Investigation of Reentry Vehicle Aerothermodynamic Problems in Arc-Heated Facilities *J. Spacecr. Rockets* **38** 199–206
[3] Poniaev S A, Kurakin Y A, Schmidt A A, Bobashev S V, Steffens L, Esser B and Gulhan A 2014 Blackout mitigation in a plasma layer near a high-speed body in ExB fields *J. Phys. Conf. Ser.* **572** 1–5
[4] Pashchina A S, Karmatsky R E and Klimov A I 2017 The creation of hypersonic flows by a

powerful impulse capillary discharge *Tech. Phys. Lett.* **43** 1033–6

[5] Pashchina A S and Karmatsky R E 2018 Using triple Langmuir probe method for study high-speed plasma flow created by a pulsed capillary discharge *J. Phys. Conf. Ser.* **1112** 012023

[6] Pashchina A S, Efimov A V and Chinnov V F 2016 Optical Investigations of the Multicomponent Plasma of the Capillary Discharge: Subsonic Outflow Regime *High Temp.* **54** 488–502

[7] Eremin A V. 2012 Formation of carbon nanoparticles from the gas phase in shock wave pyrolysis processes *Prog. Energy Combust. Sci.* **38** 1–40

[8] Reznikov B I, Alekseev N I, Bobashev S V., Zhukov B G, Ponyaev S A, Kurakin R O and Rozov S I 2011 Equilibrium composition and the effective adiabatic exponent of a clustering carbon plasma *Tech. Phys.* **56** 1106–11

[9] Pashchina A S and Valyano G E 2017 On the influence of the condensed particles on the absorption properties of plasma created by ablation controlled arc in a capillary *J. Phys. Conf. Ser.* **891** 012299

[10] Johnson B H and Murphree D L 1969 Plasma Velocity Determination by Electrostatic Probes *AIAA J.* **7** 2028–31

Pulsed plasma jet interaction with a metal targets

A S Pashchina[1]*, A I Klimov[1], N K Belov[1], B N Tolkunov[1], V Kh Timirbulatov

[1]Joint Institute for High Temperatures, Russian Academy of Sciences, Moscow, Russia

*fgrach@mail.ru

Abstract. The results of experimental studies of the interaction of plasma jets, created by a pulsed capillary discharge with an evaporating wall, with metal targets are presented. It is shown that the main mechanism of targets destruction is the material melting and the subsequent removal of the melt under the dynamic pressure of the plasma jet. The main sources of energy transmitted to the target are: the high-speed flow of high-enthalpy plasma created inside the capillary, and the fluxes of charged particles accelerated in the near-electrode layers. The energy transport provided by charged particles does not exceed ~0.1–0.2 of the total energy input onto the target. The enthalpy transport is capable to provide heat flux densities of ~5-6 kW/mm^2, an order of magnitude exceeding the values achievable in welding arcs.

1. Introduction

The interaction of plasma flows with different materials covers a wide range of scientific and practical problems and underlies numerous processes of plasma technology, metallurgy, plasma chemistry, modelling in experimental technique, etc. Special attention is paid to the study of the processes in the contact area of plasma and target, which are of fundamental importance in plasma physics.

One of the effective means for producing plasma flows with the required chemical composition and high specific parameters is a pulsed discharge in a capillary with an ablating wall [1]. Despite the large amount of experimental and theoretical research in this field, the number of papers devoted to the study of the interaction of plasma flows obtained in this way with various materials is very limited [2–4]. In these papers, the destroying impact of pulsed plasma jets on materials with different thermo- and electrophysical properties was studied. In [4], the heat release in a shock-compressed area, accompanied by melting and subsequent blowing off the melted film, was considered as the main reason of conductors destruction. In [3], the dominant role of this mechanism is questioned. The authors [3] also exclude the possibility of Joule heating of conductors due to closing of the discharge current portion on the target or excessive plasma potential, and note the insignificant influence of radiant energy transport on the target heating. At the same time, Joule heat release, as a possible mechanism for heating the thin metal foils when the discharge current flows through the target, is considered in [2]. It was shown that the discharge current can close through the target if the distance between the target and the capillary edge is not too large. This mechanism, however, cannot cause a noticeable heating of massive conductors, which, just like thin foils, are subjected to intense destruction upon the plasma jet impact.

According to the results obtained in [3], the fraction of energy expended on heating and subsequent destruction of conductors can reach up to 57% of the energy input into discharge. This value is quantitatively consistent with the fraction of energy released on the electrodes of the arc discharge, which value in welding arcs reaches up to 60-80% [5]. Therefore, it seems that the basic physical processes that regulate the energy balance in the contact area of a pulsed plasma jet with a target and in the near-electrode areas of the arc discharge have a common nature. Such processes include: the energy transport by the charged particles fluxes, heat conduction caused by temperature gradients between the plasma and target, forced convection, radiant heat exchange, chemical reactions on the target surface. In this paper, an attempt has been made to estimate the role and contribution of named mechanisms to the energy balance in the contact area of a plasma jet with a target.

2. The object and the methods of research

A capillary arrester, whose detailed description is given elsewhere [1], was used for plasma jets obtaining. A capacitive storage device with series-connected inductor is used as a discharge power source. The algorithm of power supply approximately corresponds to the sine half-wave. The experiments were conducted for two discharge modes, which, for definiteness, will be called as a short-pulse and a long-pulse discharge modes, at a fixed energy of a power source (Q =85 J). The duration of the discharge pulse, the amplitude of the discharge current and the discharge power are the following: t_p =1 ms, I_m =350-450 A, N_m =50-80 kW in the short-pulse mode [6] and t_p =9 ms, I_m =50-60 A, N_m =5-8 kW in the long-pulse discharge mode [1]. The discharge was ignited by applying to the arrester's electrodes of a high voltage pulse (U~60 kV, t~10 μs), formed by an auxiliary power source. Capillaries made of polymethylmethacrylate (PMMA, chemical formula $(C_5H_8O_2)_n$), polytetrafluoroethylene (PTFE, chemical formula $(C_2F_4)_n$) and corundum (Al_2O_3) are used in our experiments, that is allow us to obtain the plasma jets with different chemical composition and thermal properties and permit to estimate the influence of these properties on the energy balance in the contact area.

Experiments were carried out for three operation modes of the target (figure 1): the cathode, anode, and isolated body modes. In the first two modes, the target served as one of the arrester electrodes, and in the third mode it was under floating potential. Targets were placed opposite the capillary, perpendicularly or at different angles of inclination with respect to the jet axis. The typical distance between the target and the capillary outlet was z =(5–10) mm and could vary depending on the problem to be solved. Thin foils, sheet blanks, fine-mesh grids made of metals with different thermophysical properties (nickel, copper, steel, aluminum and lead) were used for targets fabrication.

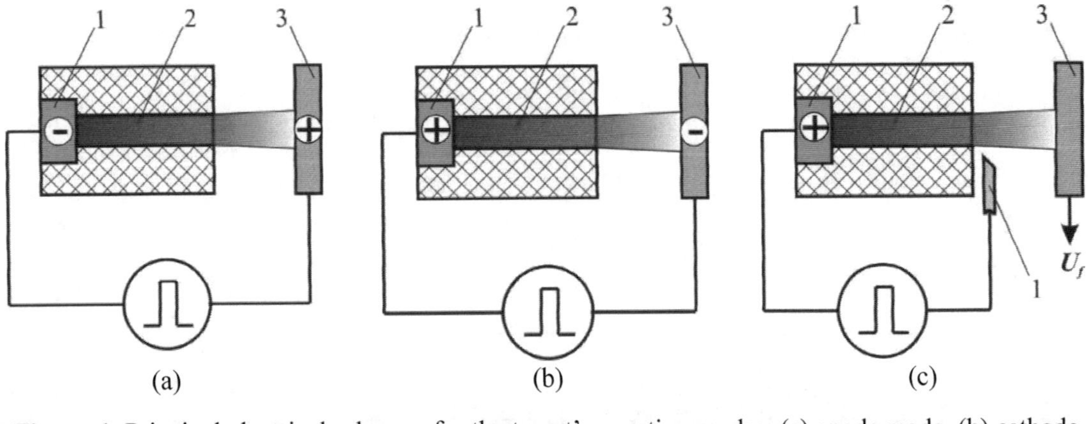

Figure 1. Principal electrical schemes for the target' operation modes: (a) anode mode, (b) cathode mode, (c) isolated body mode, (1) pulsed power source, (2), (4) arrester electrodes, (3) capillary, (5) plasma jet, (6) target.

The following parameters were measured in experiments: the discharge current and voltage drop, the target electrical potential, and the target mass loss. Based on these data, the energy deposited into

Journal of Physics: Conference Series **1394** (2019) 012010 doi:10.1088/1742-6596/1394/1/012010

the discharge was determined, and estimates of the energy release in the contact area were obtained. The interaction process was recorded by MotionPro N3 high-speed video camera.

3. The results of research

3.1. The picture and the main parameters of the target destruction process

The analysis of video recording and of condition of the destroyed targets shows that the substance is carried away from the surface mainly because the target melting and subsequent blowing off of the melt under the plasma jet dynamic pressure. At the first moments of time, the melt appears in the center of the contact area, where the plasma temperature, according to the results of spectral measurements [1,6], reaches maximum values - of the order of $T \sim 1.5$-2 eV. At the further times moments, the melting zone expands, and the melt is pushed out to the periphery of the crater, where it partially solidifies, and partially removed under the dynamic pressure of the plasma jet. Depending on the target thickness and discharge pulse parameters, the plasma jet impact leads to the formation of the through hole or crater (figure 2). The process of destruction and final condition of the affected area do not depend on target polarity.

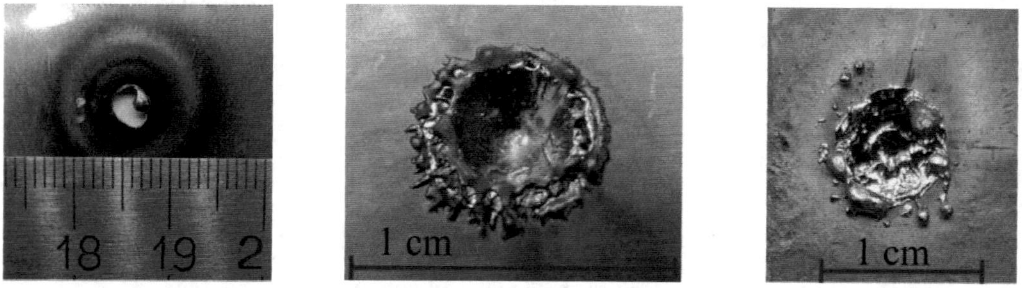

(a) (b) (c)

Figure 2. The results of plasma jet impact on metal targets: (a) nickel foil, δ=0.2 mm, (b) steel sheet, δ=6 mm, (c) lead sheet, δ=3 mm.

The target begin to melt with a certain time delay after contact with the plasma jet, which is necessary for heating the surface to the melting point and depends on the discharge parameters and the thermophysical properties of the target [7]

$$t_{delay} = \frac{\pi}{4} \frac{\lambda_w c_p \rho \left(T_f - T_\infty\right)^2}{q_w^2}, \tag{1}$$

where q_w - heat flux density, λ_w - thermal conductivity coefficient, c_p - specific heat capacity of the target material, ρ - density, T_f- melting point, T_∞ - initial target temperature).

In the range of the studied discharge parameters, the time of heating the target to the melting point is a small fraction (less than 20%) relatively discharge pulse duration. In some cases (for targets with a smooth surface), it was possible to establish the start of melting using the video recording data for this purpose. So, for a target made of nickel foil with a thickness of δ =0.2 mm, the time of heating the contact spot to the melting point is approximately $t_{del}(T_f) \approx 200$ μs for the short-pulse discharge mode and $t_{del}(T_f) \approx 2$ ms for the long-pulse one. Substituting these values into equation (1), one can estimate the heat flux density averaged over the time interval $[t_0, t_{del}(T_f)]$ (t_0 is the moment of contact of the plasma jet with the target), that gives $q_w(t_{del}(T_f)) = 1.7$ kW/mm^2 for the short-pulse discharge mode and $q_w(t_{del}(T_f)) = 0.55$ kW/mm^2 for the long-pulse one. The surface heating time to the boiling point T_b for the indicated values of the heat flux density, in accordance with (1), turns out to be of the order of the discharge pulse duration, $t_{del}(T_b) \leq \tau_p$, for both discharge modes. In fact, since after the beginning of the target melting the heat flux density continues to rise up to the middle of the discharge pulse (according to the power source algorithm), the boiling point in the contact spot is reached only a little later than the melting point due to the inverse proportionality of the heating time to the second degree of heat flux density. For example, in accordance with (1), only a twofold increase in the heat flux density is necessary to reach a boiling point over the same period of time, that is more than ensured by heat fluxes provided by enthalpy transport (see Section 3.2). The heating of the contact area to a boiling point is evidenced by clear signs of evaporation, which are observed over most of the discharge pulse, in particular: intense atomic and ionic lines of the target material recorded in the plasma emission spectra near the contact spot, as well as plasma jets emanating from the contact spot and directed perpendicular to the target surface. Overheating of the surface above boiling point under experimental conditions seems unlikely due to the continuous removal of the melt from the heat supply zone under the dynamic pressure provided by oncoming plasma jet. Therefore, when making estimates, it seems appropriate to take the boiling point as the temperature of the target surface.

During the experiments, weight measurements of the target were carried out before and after interaction with the plasma jet. The accuracy of weight measurements is 0.1 mg, so the error in determining the mass loss for minimum recorded values Δm=1 mg does not exceed 10%. The results of the mass loss measurements were used to estimate the energy spent on the target destruction

$$Q_{destr}(\Delta m) = Q_f + Q_b = \Delta m\left(c(T_s - T_\infty) + L_f\right) + m_b L_b, \tag{2}$$

where $\Delta m = m_f + m_b$ – the total mass loss, including melted mass m_f and vapor-phase mass m_b; c - the heat capacity of the target material (for simplicity, we accept the equality of the heat capacities of the solid and liquid phases); L_b and L_f - the specific heats of evaporation and melting; T_s - the surface temperature at the vapor-liquid interface (it is assumed $T_s = T_b$); T_∞ - the ambient temperature.

The first term in the right-hand side of equation (2) corresponds to the energy Q_f spent on heating the total mass of the removed substance Δm to the boiling point and on its melting, the second term is the energy Q_b spent on evaporation of the mass m_b. To estimate the ratio of these energies, Q_b/Q_f, one can use the continuity equations (in the stationary approximation) for the vapor and liquid phases

$$\frac{m_b}{F_b \tau_d} = \rho_f v_b \tag{3a}$$

$$\frac{m_f}{F_f \tau_d} = \rho_s v_f \tag{3b}$$

from whence

$$\frac{m_b}{\Delta m} = \frac{v_b}{v_b + v_f} \tag{4}$$

where v_b and v_f– the velocities of the phase boundaries "melt-vapor" and "solid-melt"; F_b and F_f – the surface area of the melt zone at the boundary "melt-vapor" and "melt-solid"; ρ_f, ρ_s – the mass density of the substance in the liquid and solid phases, respectively. Hereinafter, we accept for simplicity $F_b = F_f$ and $\rho_f = \rho_s$.

The velocity of melt-vapor boundary can be estimated by the Hertz-Knudsen equation [8]

$$\rho_f v_b(T_s) = \frac{1}{\alpha}\sqrt{\frac{m_p}{2\pi k T_s}}\, p_{sat}(T_s), \tag{5}$$

where $p_{sat}(T_s)$ is the saturated vapor pressure, which is determined from Clapeyron-Clausius equation

$$p_{sat}(T_s) = p_b exp\left[\frac{L_b m}{k}\left(\frac{1}{T_b} - \frac{1}{T_s}\right)\right] \tag{6}$$

Here T_b – boiling temperature at normal pressure T_b =10^5 Pa; α – accomodation factor ($\alpha \approx 1$); m_p – effective mass of the vapor particles.

Journal of Physics: Conference Series **1394** (2019) 012010 doi:10.1088/1742-6596/1394/1/012010

Estimates of the velocity of the "melt-solid" phase boundary were made using the continuity equation (3b) by substituting the measured values of the mass loss Δm (instead of the melted mass m_f) and affected surface area of the target. The estimates obtained in such a way agree satisfactory with the propagation velocity of the heat flux averaged over the discharge pulse duration

$$v_f \approx \sqrt{a/\tau_p}, \tag{7}$$

where a is a thermal diffusivity.

The estimates made on the basis of relations (2) and (4), taking into account (5) and (7), show that for targets made of aluminum, copper, steel, nickel, and lead, which were used in our experiments, the mass fraction of vaporized substance (at $T_s = T_b$) is $m_b/\Delta m$ =0.02-0.07, and the ratio of the energy spent on evaporation to the energy spent on heating and melting the target is Q_b/Q_f =0.09-0.26. In fact, these values are overestimated approximately by an order of magnitude. The reason is that the relation (5) was obtained under assumption that all evaporated atoms are evacuated from the surface into surrounding space and do not return [8]. Equation (5) is applicable for relatively low surface temperatures when the saturated vapor pressure does not exceed p_{sat} <100 Pa [8]. For higher surface temperatures, a significant fraction of atoms flow back to the surface due to collisions, and, therefore, the net rate of substance evaporation under these conditions is approximately an order of magnitude less than that given by (5) [9–12]. Therefore, the second term in equation (2), corresponding to the energy spent on evaporation, can be neglected.

The results of estimates of the target destruction energy calculated by formula (2) (neglecting the second term in the right side) and the ratio of this energy to the energy deposited into the discharge, $Q_{destr}(\Delta m)/Q_{dis}$ ($Q_{dis} = \int_0^{\tau_p} U_d(t)I(t)dt$, where $U_d(t), I(t)$ – voltage drop and discharge current, respectively), are presented in table 1. Here, estimates of the heat flux density to the target surface are also presented

$$q_w(\Delta m) = Q_{destr}(\Delta m)/(F\tau_p) \tag{8}$$

where F – the area of the contact spot, which was determined by the sizes of the hole or crater made in the target. The results of estimations are presented for the three types of targets (the nickel foil with a thickness of δ =0.2 mm, the lead sheet with a thickness of δ =3.2 mm, the three-layers steel mesh with wire diameter of d =0.2 mm and mesh sizes of 1 mmx1 mm), depending on their polarity and capillary discharge mode (the long-pulse (τ_d =9 ms) and short-pulse (τ_d =1 ms) modes). For a target made of a three-layer steel mesh, the results of estimates are also presented for the case when the plasma jet stoichiometry is determined by the PTFE ablation products (chemical formula $(C_2F_4)_n$).

The following specific features attract attention. Firstly, despite the fact that all the experiments, whose results are presented in the table 1, were carried out under the same conditions (fixed discharge parameters, distance to the target and atmospheric pressure), estimates of the energy parameters based on the measured values of the mass loss significantly differ as for various types of targets and discharge modes. Thus, the fraction of the energy spent on heating and melting the target increases from its minimum values of $Q_{destr}(\Delta m)/Q_{dis}$=0.044-0.12, typical for the long-pulse discharge mode, to maximal values of $Q_{destr}(\Delta m)/Q_{dis}$=0.17-0.42, typical for the short-pulse mode. Secondly, the estimates of the heat flux density based on the results of mass loss measurement are noticeably lower than the estimates based on the measurement of the heating time of target to the melting point, $q_w(\Delta m) < q_w(t_{del})$, and cannot explain the heating of the target to the boiling point, whose obvious signs are present in both discharge modes.

The noted features are the consequence of the fact that estimations of the destruction energy based on results of weight measurements do not take into account the mass of the melt remaining on the target surface and the heat losses in the target body. The efficiency of the melt removing to a great extent depends on the dynamic pressure of the incident flow, which differs significantly for supersonic and subsonic plasma jets obtained in the short-pulse [6] and long-pulse [1] discharge modes, respectively. For this reason, a significant fraction of the molten mass, which in some cases reaches more than 90% of the total mass of the melt, remains on the target surface being displaced to the crater periphery under the low dynamic pressure produced by a subsonic plasma jet, while the strong dynamic pressure of a supersonic plasma jet provides removing of more than 60% of the melted mass. Therefore, the estimates of the target destruction energy and the heat flux density, based on the results of mass loss measurements, are much underestimated, especially for the long-pulse discharge mode.

Table 1. The results of evaluation of the energy parameters characterizing the interaction of plasma jet with a target.

Material	Type	Polarity	Q_{dis} (J)	τ_d (ms)	$t_{del}\,(T_f)$ (ms)	Δm (mg)	$Q_{destr}(\Delta m)$ [a] (J)	$Q_{destr}(\Delta m)/Q_{dis}$	$q_w(t_{del})^b$ $\left(\frac{kW}{mm^2}\right)$	$q_w(\Delta m)^c$ $\left(\frac{kW}{mm^2}\right)$
					Discharge in PMMA capillary					
Ni	foil, δ=0.2 mm	cathode	51	1	0.2	1.5	2.25	0.044	0.55	0.062
			58	9	2.0	10	16.5	0.29	1.7	1.5
Pb	plate, δ=3.2 mm	cathode	60	1		44	10.6	0.17		0.88
		anode	57	1		55	13.3	0.23		1.1
Steel	3 layers of mesh	cathode	57	1		13.5	20.6	0.36		1.85
		anode	56	1		15.5	23.6	0.42		2.15
		cathode	50	9		3	4.6	0.09		0.13
		anode	44	9		3.5	5.3	0.12		0.15
					Discharge in PTFE capillary					
Steel	3 layers of mesh	cathode	60	1		8	12.2	0.2		1.1
		anode	57	1		9	13.7	0.24		1.24
		cathode	39	9		1.2	1.8	0.047		0.05
		anode	43	9		1.5	2.3	0.052		0.063

[a] Estimation by the formula (2);
[b] Estimation by the formula (3)
[c] Estimation by the formula (1)

Other discovered features, which will be discussed below, are: an explicit dependence of the energy parameters on the target polarity, as well as on the incoming flow enthalpy depended on the plasma jet chemical composition.

3.2. Main sources and mechanisms of energy transfer

Energy exchange on the target surface, ultimately, determines the amount of heat flux density q_p, which provides heating, melting and evaporation of the substance. The main processes that regulate the energy balance in the contact spot are: energy transfer by charged particle fluxes q_p, radiation fluxes from the plasma onto the target q_{rad}^{pl} and from the target into the plasma q_{rad}^{w}. In the general case, the energy balance in the contact area can be represented as

$$q_p = q_p + q_{rad}^{pl} - q_{rad}^{w} + q_{etc} \tag{9}$$

Here q_{etc} takes into account another possible energy transfer mechanisms, in particular: the heat conduction flux from the plasma to the target, convective heat exchange forced by the plasma jet dynamic pressure, chemical reactions on the target surface, etc.

As a rule, radiation fluxes emitted from the surface and plasma do not influence significantly on the energy balance in the contact region. For example, the radiation flux density emitted from the target surface heated to T =3134 K (boiling point of iron) does not exceed $q_{rad}^{w} < 10$ W/mm^2, which is less than 1% of the total heat flux density. However, in the temperature range T =1-2 eV recorded in the axial zone of the plasma jet [1,6], the fraction of the energy transferred by the radiation flux from plasma to the contact region can be noticeable enough. The radiation flux density in the range of the indicated temperatures can reach up to $q_{rad}^{pl} = \varepsilon\sigma T^4 = 30 - 500$ W/mm^2 (for the grey factor of ε~0.03), which makes it necessary to evaluate this mechanism in each specific case.

The energy transfer by charged particles is determined by the ion and electron fluxes from plasma onto the surface and by the flux of emitted electron into the plasma. The ratio between these fluxes is different for the cathode, anode and isolated body [13–15], that allows us to use this circumstance to estimate the contribution of mentioned mechanism into the energy balance. Experiments have shown that, ceteris paribus, the mass loss is always greater in the case when the target being under anode potential, than when the target being under cathode or floating potential (see figure 1 and table 1). In the latter two cases, the mass loss is about the same.

It should be noted, that since the distance between the target and the capillary outlet in these experiments did not exceed $z < 10$ mm, it was not possible to realize a stable regime of an isolated body (see figure 1). Under these conditions, almost immediately after the plasma jet contacts the target, the discharge switches to a cascade arc mode, consisting of two spatially separated sections, one of which coincides with the plasma jet, and the second section, which provides the discharge current closure, arises between the target and the external electrode (figure 3). In this case, the anode or cathode mode - depending on the polarity of the arrester' electrodes - is established in the contact area of target and plasma jet. In our experiments, the polarity of the arrester' electrodes was such that the cathode mode was established in the contact area. This circumstance, apparently, was one of the reasons that the measured mass loss of targets both being under the floating and cathode potential was approximately the same.

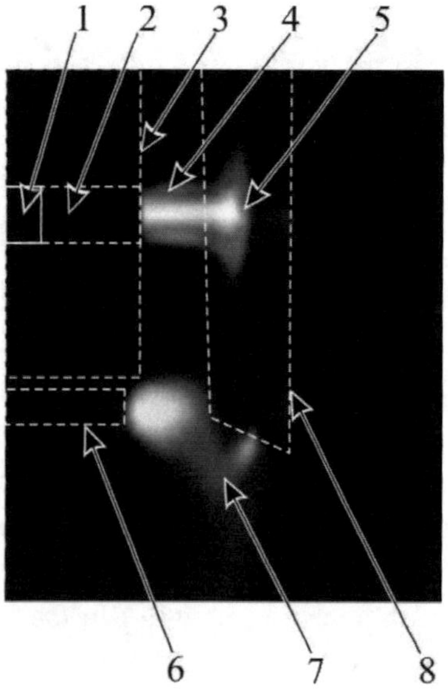

Figure 3. Transition of the discharge to the cascade arc mode during the plasma jet interaction with a metal target initially being under the floating potential: (1) internal electrode,(2) capillary, (3) arrester housing, (4) plasma jet, (5) contact area, (6) external electrode, (7) section of the cascade arc, providing the closure of discharge current, (8) metal target.

Another feature found in the experiments is the different condition of the cathode' and anode' contact areas, indicating the different distribution of the discharge current in these areas (figure 4). A distinctive feature of the cathode contact area is the formation of cathode spots (figure 4(a)), through which, apparently, almost the entire discharge current is closed. Cathode spots arise and move on the periphery of the contact area, and do not penetrate into the zone occupied by the melt. Such pattern is observed on all targets, whose melting points turn out to be insufficient to provide the thermionic emission necessary to maintain the discharge current. This fact suggests that, at least until the surface temperature of the melt reaches the temperature necessary to provide the required emission current, the processes caused by the flow of discharge current do not influence on the energy balance in the melted zone. A different situation occurs when the target being under the anode potential (figure 4(b)). In this case, the discharge current flows through the melt that provides its additional heating.

Estimates of the heat flux densities to the target under the cathode and anode polarity, made on the basis of mass loss data, are presented in figure 5. The difference between the heat flux densities to the anode and cathode in the short-pulse mode is of $\Delta q = q_a - q_c$ =30-200 W/mm^2, which is

Journal of Physics: Conference Series **1394** (2019) 012010 doi:10.1088/1742-6596/1394/1/012010

quantitatively agrees with those that obtained in welding arcs [5], where the charged particles fluxes serves as the main energy source. This value represents a small fraction (approximately $\Delta q/q_a \approx 0.1$-0.2) of the total energy expended to the target destruction. The transfer of the rest energy is obviously provided by other processes, the most probable among which are the following: thermal conduction flux caused by the temperature difference between the plasma and the target, forced convection flux, radiation flux from plasma, chemical reactions on the target surface. The energy source for the first two processes is a high-speed flow of highly ionized plasma created inside the capillary, whose enthalpy, depending on plasma chemical composition and the discharge power, can reach up to h =100-200 MJ/kg [16]. This assumption is confirmed by the experimental results, according to which the heat flux density is twice higher when the PMMA ablation products determine plasma composition, compared to the case when the plasma composition is determined by PTFE or corundum (figure 5).

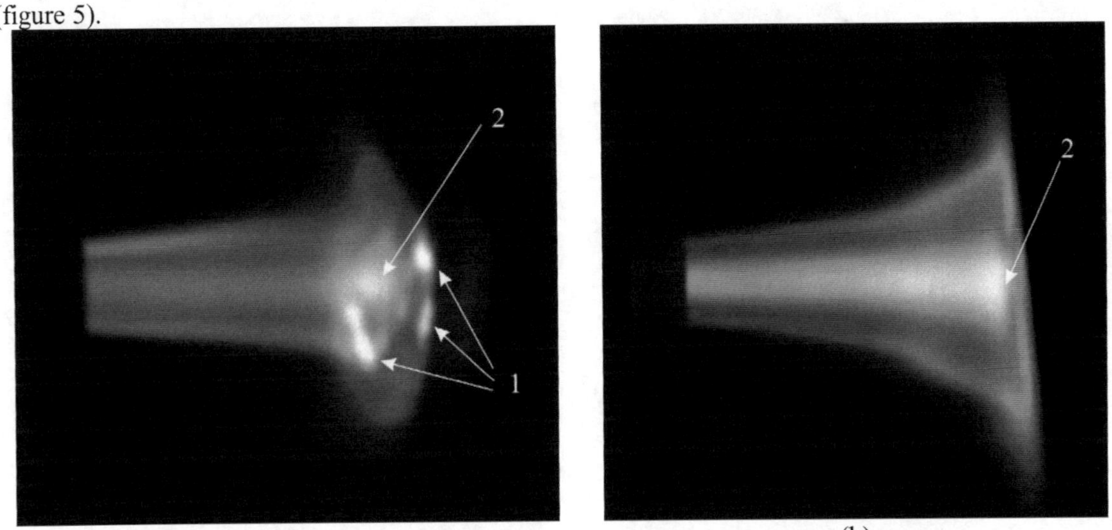

(a) (b)

Figure 4. Images of the contact areas being under the (a) cathode and (b) anode potential: (1) cathode spots, (2) zone, occupied by the melt. The closure of the discharge current to the target being under the cathode potential occurs through cathode spots that arises at the periphery of the contact area and do not penetrate into the zone occupied by the melt.

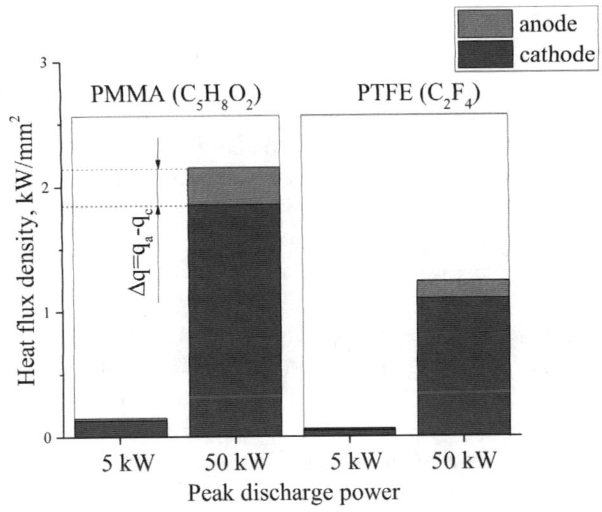

Figure 5. Heat flux densities on the targets being under the cathode and anode potentials depending on the discharge power and chemical composition of plasma-forming substance.

Journal of Physics: Conference Series **1394** (2019) 012010 doi:10.1088/1742-6596/1394/1/012010

The temperature dependences of the heat flux density, which can be provided by enthalpy transport, $q_h = \rho h v$ (ρ and v are the flow mass density and velocity), are presented for some substances in figure 6. These dependences were calculated using the data on thermophysical properties of plasmas of various chemical composition [16], assuming $v = c$ ($c = \sqrt{\frac{\gamma k T}{m}}$ is the speed of sound in a plasma). It can be seen that at atmospheric pressure in the temperature range of T =(15-22) kK recorded in the axial zone of the plasma jet [1,6], the heat flux density reaches up to q_h~6 kW/mm^2 when the PMMA decomposition products determine plasma stoichiometric composition. The heat flux density is about two times lower if the ablation products of PTFE or corundum serve as plasma-forming substance. These estimates are more than an order of magnitude greater than the heat flux densities that are achieved in welding arcs.

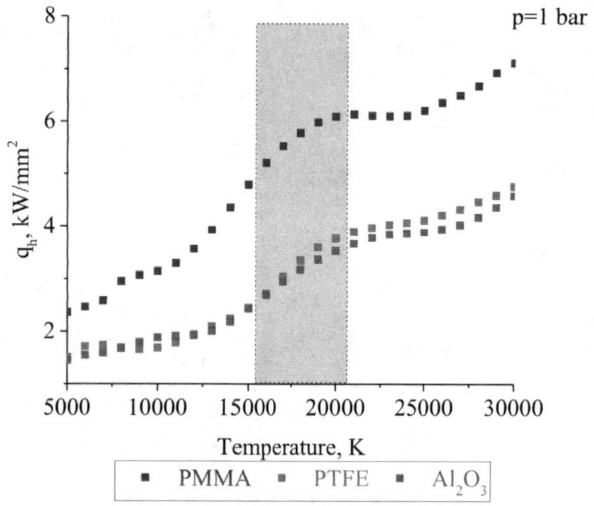

Figure 6. Temperature dependences of heat flux densities, which can be provided by the transfer of enthalpy of plasmas with different chemical composition.

4. Conclusions

The main mechanism of metal targets destruction exposed by a plasma jet is the combined impact of the heat flux and dynamic pressure of the plasma jet. The dynamic pressure plays an important role in the destruction of meltable materials, providing removal of the melt from the contact area.

According to the estimations and obvious signs of target evaporation observed experimentally, the temperature in the contact spot reaches the boiling point. Surface overheating above the boiling point seems unlikely due to the continuous removal of the melt under the dynamic pressure provided by the plasma jet. The mass fraction of the evaporated substance at boiling temperatures does not exceed 7% and does not influence significantly on the total energy balance in the contact spot.

The main sources of energy transmitted to the target are: the high-speed flow of high-enthalpy plasma created inside the capillary, and the fluxes of charged particles accelerated in the near-electrode layers. The energy transport provided by charged particles constitutes a small fraction (~0.1–0.2) of the total energy expended to the target destruction. The enthalpy transport is capable to provide heat flux densities of ~5-6 kW/mm^2, an order of magnitude exceeding the values achievable in welding arcs.

References

[1] Pashchina A S, Efimov A V and Chinnov V F 2016 Optical Investigations of the Multicomponent Plasma of the Capillary Discharge: Subsonic Outflow Regime *High Temp.* **54** 488–502

[2] Ender A Y, Kuznetsov V I, Kolyshkin I N and Shchetinina A N 2011 Features of Jets Formed by Erosive Capillary Discharges *Open Plasma Phys. J.* **4** 40–54

[3] Avramenko R F, Bakhtin B I, Nikolaeva V I, Poskacheeva L P and Shirokov N N 1990 Study

of plasma formations initiated by erosion discharge *Sov Tech Phys* **60** 57–64 [in Russian]
[4] Sultanov M A and Kiselevsky L I 1966 Study of the mechanism of electrodes erosion under the impact of supersonic torches in pulsed discharge *High Temp.* **4** 375–80 [in Russian]
[5] Leskov G I 1970 *Electric welding arc* (Moscow: Mashinostroenie) [in Russian]
[6] Pashchina A S, Efimov A V and Chinnov V F 2017 Optical investigations of multicomponent plasma of capillary discharge. Supersonic outflow regime *High Temp.* **55** 650–64
[7] Carslaw H S and Jaeger J C 1959 *Conduction of heat in solids* (Oxford: Clarendon Press)
[8] Langmuir I 1913 Physical review *Phys. Rev.* **II** 329–42
[9] Li R, Li X, Jia S, Murphy A B and Shi Z 2010 Study of different models of the wall ablation process in capillary discharge *IEEE Trans. Plasma Sci.* **38** 1033–41
[10] Zaghloul M R 2004 On the vaporization of Teflon and heated compound-materials in ablation-controlled arcs *J. Appl. Phys.* **95** 3339–43
[11] Keidar M, Boyd I D and Beilis I I 2001 On the model of Teflon ablation in an ablation-controlled discharge *J. Phys. D. Appl. Phys.* **34** 1675–7
[12] Beilis I I 2007 Laser plasma generation and plasma interaction with ablative target *Laser Part. Beams* **25** 53–63
[13] Raizer Y P 1991 *Gas Discharge Physics* ed J E Allen (Berlin: Springer-Verlag Berlin Heidelberg)
[14] Benilov M S and Cunha M D 2002 Heating of refractory cathodes by high-pressure arc plasmas : I *J. Phys. D. Appl. Phys.* **35** 1736–50
[15] Almeida N A, Benilov M S, Hechtfischer U and Naidis G V 2009 Investigating near-anode plasma layers of very high-pressure arc discharges *J. Phys. D. Appl. Phys.* **42** 045210
[16] Kovitya P 1984 Thermodynamic and Transport Properties of Ablated Vapors of PTFE, Alumina, Perspex, and PVC in the Temperature Range 5000-30 000 K *IEEE Trans. Plasma Sci.* **PS-12** 38–42

Journal of Physics: Conference Series **1394** (2019) 012011 doi:10.1088/1742-6596/1394/1/012011

Determination of the fraction of electrical energy released in the oncoming supersonic airflow using long spark discharge

A S Saveliev[1,2]

[1] Joint Institute for High Temperatures of the Russian Academy of Sciences, Izhorskaya 13 Bldg 2, Moscow, 125412 Russia
[2] Moscow Institute of Physics and Technology, Institutskii lane 9, Dolgoprudny, Moscow Region, 141700 Russia

E-mail: fisteh@mail.ru

Abstract. In present work an indirect measurement of the fraction of stored electrical energy, that by initiating long spark discharge of submicrosecond duration transforms into air heating, that moves near the surface of an aerodynamic model at supersonic speed with Mach number $M = 2$, was carried out. Using the Particle Image Velocimetry method the flow velocity behind the front of an induced quasi-cylindrical shock wave that occurs when heat is released with long spark discharge was measured. This process was modeled in a two-dimensional formulation using a numerical simulation with OpenFOAM package. By comparing the data obtained in the experiment and in the numerical simulation the value of the fraction of stored electrical energy, which goes to the local heating of air, was determined.

1. Introduction

The initiation of an electric discharge near the surface of an aerodynamic body, streamlined by the oncoming airflow, is currently being investigated in many laboratories in the world. The use of energy release in the oncoming airflow is interesting from the point of view of practical application for changing the flow pattern of an aerodynamic body, for example: changing the position of the separation point on the wing model with sub- [1] and supersonic [2, 3] free-stream velocities, increasing lift of the wing at supercritical angles of attack [4], preventing stall flow from the leading edge of the wing [5], controlling the laminar-turbulent transition [6], controlling the ignition of combustion in supersonic flows of combustible mixtures [7], control of the interaction process of a shock wave and a turbulent boundary layer [8]. The main advantages of this method of active action over mechanical or pneumatic are high peak power, low response time, the ability to mount the discharge device flush with a streamlined surface. At the same time, the use of a spark discharge as a instrument for transmitting energy to the incoming airflow has a number of advantages, among which a relatively wide range of energy released in the flow can be noted. First of all, this circumstance is due to the fact that air is predominantly involved in the electrical discharge, and not the surface of the body, which is streamlined by oncoming flow, so that this surface is not subject to the destructive action of high temperature and chemically active components of the discharge plasma. In addition, the initiation of this type of discharge is possible in conditions of high humidity of the incoming air flow [9].

Typically, in experiments, a spark discharge is initiated in a pulsed or pulse-periodic mode, and often the discharge duration is much shorter than the characteristic gasdynamic time. Since the initiation of an electric discharge occurs with usage of a particular high-voltage electrical circuit, the discharge parameters are determined by the parameters of this circuit. In particular, the duration of a spark discharge τ is often determined by a value of order RC, where R is the effective resistance of the discharge circuit during the existence of an electric spark discharge, and C is the total capacitance of the capacitors in the circuit in which the energy is initially stored. On the other hand, the amount of heat energy Q, which is released into the gas in which a spark occurs, is associated with the electrical energy E stored in the capacitor C:

$$Q = \eta E = \eta \frac{CU^2}{2},$$

where U is the value of the voltage on the capacitor at the time of electrical breakdown, η is the fraction of stored electrical energy that goes into heat during electrical discharge. It should be noted that the above formula does not take into account the residual voltage on the capacitor; for simplicity, this value is taken to be zero. From the above relations it follows that the parameters of energy release into the flow can be set by the parameters of the electrical circuit. For example, the flow velocity determines the discharge duration, thus limiting the permissible capacitance of the circuit. In other words, if the energy input should be quasi instantaneous, then the duration of the electric discharge should be set in accordance with the characteristic gasdynamic time. It is possible to formulate a qualitative rule: if the flow velocity is higher, the smaller capacitance of the discharge circuit should be used. Since the value of energy depends on this capacitance, in order to maintain its value, it is necessary to increase the voltage on the capacitor, which discharges to the discharge gap. The value of the fraction η of electric energy E, which is converted into heat of the incoming air flow, remains unknown. The rest of the electrical energy can go to electromagnetic radiation, heat losses in the supply wires and other elements of the discharge circuit (for example, on switching devices), on the excitation of other degrees of freedom of the molecules of air. It should also be noted that if one initiate a spark discharge in a pulse-periodic mode, the discharge frequency will be determined by the value of W/E, where W is the power of the high-voltage source of the discharge circuit.

With the rapid heat release in a gas using a spark discharge, a region with increased temperature and pressure is formed. In this case decay of discontinuity takes place, which leads to the formation of unsteady flow with the presence of a shock wave. If the discharge is pointlike, i.e. the length of the spark channel is comparable to its diameter, the front of the shock wave has a spherical shape. If the length of the spark channel is many times greater than its diameter, then a quasi-cylindrical shock wave is formed, which front shape is a cylindrical surface with an axis of symmetry corresponding to the location of the long spark discharge. It should be noted that a spark discharge can be represented by a set of pointlike discharges which occur simultaneously and are located along a straight line at a distance between adjacent ones comparable to the size of an individual spark. In this case, each discharge is the source of a separate spherical shock wave, and when they interfere one cylindrical shock wave is formed. With initiation of spark discharge on the surface of an aerodynamic model streamlined by airflow, the formation of a shock wave leads to short-term momentum transfer to the flow and to the emergence of local areas of high pressure, which can significantly change the flow pattern. This process can be modeled both experimentally and with the help of numerical simulation. In the numerical simulation it is necessary to know a number of parameters of the physical process of energy release into the incident flow, which will determine the parameters of unsteady flow that occurs, such as the fraction η of electrical energy E converting to gas heat, the duration of pulse heating τ. In the experiment, when using a particular scheme of initiation of the discharge, the discharge duration τ can be measured, and by photographing the discharge the volume of the energy release region can be determined. However, η depends on many process parameters and often cannot be measured directly, but can only be determined indirectly. This paper presents the results of determining the fraction η of stored electric energy, determined indirectly from the results of experiments and numerical simulation of the process described in [3]. In the work, experimentally and with the help of numerical simulation, quasi-instantaneous local heat generation using long spark discharge near the surface of the NACA airfoil streamlined by supersonic airflow with a positive angle of attack and with the presence of a separation region is simulated. The use of long spark discharge allows to assume the two-dimensional nature of the process and to carry out a two-dimensional computer simulation with possibility of fine mesh usage, which allows simulation the viscous flow in the boundary layer and its separation near the trailing edge.

2. Experimental setup

Experiments were carried out on supersonic wind tunnel ST-4 of the atmospheric-vacuum type, and a detailed description of the facility can be found in [2, 3, 10]. The description of the important aspects of the facility and some important parameters are noted below. The description of the aerodynamic model with a discharge device, high-voltage circuit for initiating long spark discharge and systems for diagnosing the flow and measuring the electrical parameters of the process are presented below.

2.1. Supersonic wind tunnel

The facility is intended for the formation of a supersonic jet of air in the working chamber. It consists of the following main parts: a vacuum system with a volume of 240 m^3, a working chamber with a characteristic size of 1 m and equipped with a number of transparent windows for diagnosing the process of flow around a model, de Laval nozzle with an output diameter of 110 mm, and a series of locking devices for starting and stopping supersonic flow. The principle of operation is as follows. After pumping out the gasholders using vacuum pumps to a pressure of less than 10 kPa, they are filled with atmospheric air through de Laval nozzle, which is designed for Mach number $M = 2$, and the jet velocity profile along its diameter has a rectangular shape (under fully expanded regime). The air is taken from the atmosphere, it passes the nozzle, the working chamber, then enters the gasholders and fills them during test run. In the working chamber in a submerged supersonic jet, the jet behind the exit section of the nozzle, an aerodynamic model to be tested in a supersonic airflow is placed. The resulting static pressure in the submerged jet has a value close to $p_{st} = 0.15$ bar, flow velocity $V_0 = 500$ m/s, static temperature $T_{st} = 160$ K. While supersonic flow is maintained behind the output section of de Laval nozzle, the static pressure p_{st} in the jet in this section constant. This jet flows in a space where there is almost stationary air with pressure p_a. This pressure of the ambient air during the test run varies from within 3 kPa to 20 kPa, and the flow regime changes from under-expansion to over-expansion. At the time moment when $p_a = p_{st}$, so-called ideally expanded regime of flow occurs, which is characterized by a rectangular flow velocity profile and the absence of rarefaction waves or shock waves, that start from the nozzle lip. It is important to note that the experimental results presented in this work were obtained near the regime of ideally expanded jet.

2.2. Aerodynamic model with discharger

An aerodynamic model, which was a wing model (model of the NACA64a212 profile) with a chord length of 98 mm and a span of 150 mm, was placed in the submerged supersonic jet. Due to the presence of tilt mechanism inside the working chamber, the model was placed at a certain angle of attack to the oncoming air flow and motionlessly relative to the working chamber, with the leading edge parallel to the nozzle's exit cross-section. At the top surface of the model, at a distance of 38 mm from the trailing edge, a dovetail shaped groove (figure 1a) is milled to install conductive segments of the discharger into this groove. Figure 1b shows a photograph (top view) of an aerodynamic model with a discharger on the surface. It is important to note that the material from which the model is made is fused silica, which is resistant to high temperatures and to chemically active components of discharge plasma. The use of such material allows one to be sure that during the experiments the surface of the aerodynamic model is not subject to destruction (geometry changes), which is important when comparing the results with numerical simulation.

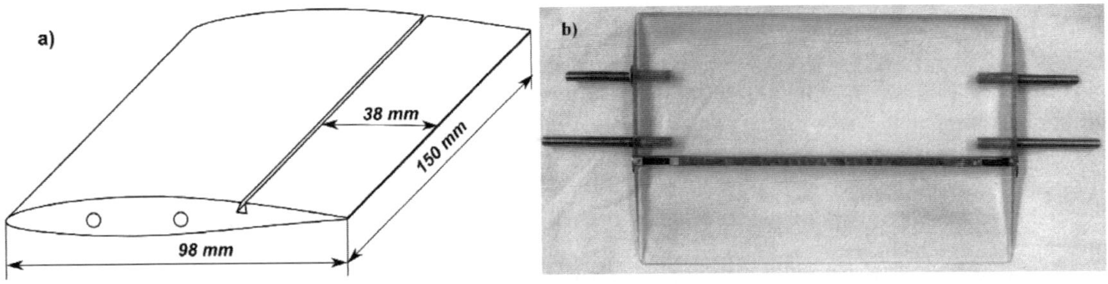

Figure 1. Sketch of wing model (a) and its photo with discharger (b).

An electric spark discharge on the surface of the model was created using segmented discharger. It consists of a set of trapezoidal segments with cross-section corresponding to a slot in the model. Segments have a width of 3.5 mm and are made from a Kovar, which is characterized by a relatively low value of thermal expansion that is close to this value for fused silica. The trapezoid-shaped dielectric layers, made of 0.05 mm thick Teflon film, are placed between the segments. The resistance between the outer segments of the discharger is very large, but when high voltage is applied to them, electrical spark discharge occurs in the air between outer segments. Due to this design of the discharge device, it is possible to significantly reduce the voltage required for gap breakdown. In addition, due to the fact that the discharge occurs in a supersonic airflow, the use of a segmented discharger provides a fixed place for the release of energy, i.e. near the surface of the model above the discharger. The experiments were carried out at angle of attack to the oncoming flow equal to 14 degrees, as in experiment [3].

2.3. Controlled high voltage circuit

To apply high voltage to the discharger gap (i.e. to the outer segments of the discharger), an electrical circuit was used, that is shown on figure 2. The circuit is a Marx generator consisting of two stages of NC_1 capacitance, where N is the number of C_1 capacitors in each stage. The number of capacitors N and the capacitance of each capacitor C_1 can be varied to change the amount of electrical energy E stored in the high-voltage circuit. A controlled switch D_1 (a three-electrode air gap) is connected between the stages, which is designed to start a high-voltage generator through a high-voltage pulse applied to the control electrode of the switch D_1. This pulse is formed using a high-voltage transformer (ignition coil) with an output voltage of about 20 kV, to the primary coil of which a low-voltage pulse is fed, but with large current amplitude. Additionally, the circuit contains an uncontrolled switch D_2 (two-electrode air gap), which is triggered immediately after D_1. Switch D_2 is necessary to prevent current leakage from the circuit while charging its capacitors.

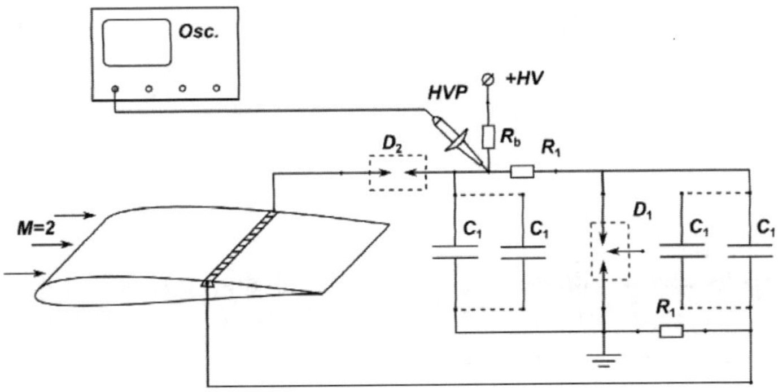

Figure 2. High voltage circuit for long spark discharge initiation on the model surface. HV—high voltage power supply, HVP—high voltage probe, Osc.—oscilloscope.

A distinctive feature of this generator circuit is that none of the two outer segments of the discharger is grounded. If the generator is connected to a high voltage source U, when the switches D_1

and D_2 are triggered, the potential of one of these outer segments becomes close to $-U$ (relative to the ground), and for the other one it becomes $+U$, so the resulting potential difference will be $2U$. In this case, the absolute value of the potential difference between any segments of the discharge device and grounded parts of the working chamber will not exceed the U value. Therefore, the selected high-voltage generator circuit ensures the position of the spark discharge strictly above the discharger surface.

It should be noted that another circuit was used in experiments earlier where the spark discharge occurred spontaneously when the voltage in the outer segments reached the breakdown value [2, 3]. The need to use a controlled high-voltage circuit is due to the use of the Particle Image Velocimetry technique, in which there is a significant value of delay relative to the input trigger that associated with the large value of pumping time of the working body in the laser cavity [11]. In this experiment this delay is greater than characteristic gasdynamic time.

2.4. Diagnostics and equipment

Particle Image Velocimetry (PIV) was used for visualization and to measure velocity of unsteady flow caused by long spark discharge initiation near the trailing edge of the wing model. A more detailed description of the methodology and a description of its application under the experimental conditions are given in [3]. It is important to note that although the discharge must obviously be initiated before the flow pattern was recorder, the PIV shooting was a triggering event for the reason given above. The visible area for measuring the flow velocity was 36x48 mm^2. The PIV video camera was positioned in such a way relative to the wing model, that in the field of its visibility there were the place of initiation of the discharge, the trailing edge and the region of airflow above the model surface. In the experiment, an oil aerosol was used for seeding a supersonic flow, obtained using generator based on Laskin nozzle [11]. The average diameter of a droplet in an aerosol does not exceed 1 μm, and the particle relaxation time in the flow τ_p is near 4 μs [11]. The size of the visible region and the speed of the incoming air flow determined the magnitude of the delay between the laser pulses, which was 0.8 μs. The recording was carried out at times that correspond to the fully expanded regime in the facility ($p_a = p_{st}$) with a maximum possible frequency of 14 Hz and the number of frames equal to 139, i.e. about 10 seconds (the maximum possible operational time of the supersonic wind tunnel is 1 minute).

To register the electrical and temporal properties of each discharge pulse, a high-resolution 12-bit LeCroy WaveRunner 66 Zi digital oscilloscope was used, one of the channels of which received a signal from a high-voltage probe (1:1000 probe LeCroy PPE20KV), connected to high voltage generator. The TTL signal from the PIV system, corresponding to the moment when the flow pattern was recorded, was fed to another channel. The oscilloscope recording was also carried out for 10 seconds with a time discretization of 100 ns. Thus, the voltage at the time of the breakdown U, the discharge duration τ, the residual voltage on the capacitors of the high-voltage generator $U_{\min}$, and the time of recording of each frame of PIV, i.e. the value of the delay Δt between the beginning of the discharge and the moment of recording of each frame. Since in each test run of such frames there were 139, it is possible for each test run to build statistics on the distribution of discharge parameters during the launch time, which will reflect the stability of the parameters of the selected discharge initiation scheme. Value of electrical energy, stored in capacitors of high voltage generator at the moment of discharge, can be calculated as:

$$E = NC_1\left(U^2 - U_{min}^2\right), \tag{1}$$

and the duration of discharge τ was measured by oscillogramms as time period, during which voltage on capacitors falls from $0.9U$ to $0.1U$ ($U_{min} < 0.1U$). It should be noted that in presented work measured value of τ was 0.2–0.6 μs, which is much less than characteristic gasdynamic time, with energy E range from 0.55–6.9 J.

To control the flow regime, synchronously with the PIV-recording, the p_a and p_{st} values were measured in the working chamber. Therefore, after processing the data, it was possible to assign to each frame of the PIV record the flow regime and also to find the frame number that most closely matches the fully expended flow regime. These measurements were carried out using Honeywell vacuum pressure gauges and 16-bit National Instruments NI-6363 ADC.

To determine the place of initiation of discharge on the upper surface of the wing model, synchronous video recording was additionally performed using a RedLake Motion Pro X3 color digital video camera. The frames were taken synchronously with the initiation of the discharge, so that when viewing the resulting recording, it was possible to determine the fact of successful initiation of the discharge and confirm the place of initiation near the discharger surface. The obtained images were used to determine the length of the energy release region and its transverse size.

Synchronization of all recording equipment with discharge initiation was performed using Berkeley Nucleonics BNC-575 pulse generator.

3. Numerical simulation

Computer simulation of the process of rapid heat release on the surface of the wing model was carried out in the OpenFOAM v.5 package, using sonicFoam solver [12]. The calculation was performed on a quad-core processor with a peak computing power of about 100 Gflops. The supersonic flow of a viscous heat-conducting gas (air) was simulated without taking into account turbulence. The computational domain had a characteristic size of 1 m; it contained a wing profile model with geometrical dimensions corresponding to the wing model used in the experiments. In the computational domain a grid was constructed, and it consists of hexagonal computational cells. The grid is thickened when approaching the surface of the wing profile, i.e. the height of the computational cell decreases when approaching a streamlined surface. This is done in order to correctly calculate the flow in the boundary layer near the surface of the model and to obtain the separation flow observed in the experiment. Thus, the computational domain is actually divided into two, one of which surrounds the surface of the profile streamlined by supersonic airflow, and corresponds to the flow in the boundary layer, and inviscid flow is simulated in the rest of the computational domain.

The initial conditions in the whole region were set by the quantities corresponding to the experimental conditions (parameters of the incoming airflow): static pressure p_{st}=0.15 bar, static temperature T_{st}=160 K, speed V_0=500 m/s. Air viscosity was calculated using the Sutherland formula. A zero velocity of gas motion was set on the profile surface, and the surface itself was adiabatic.

As a result of the calculation, the position of the separation point X_{SP} relative to the trailing edge of the wing model, the aerodynamic coefficients of the lift C_l and the drag C_d were determined. It is shown that at angle of attack of 0 and 10 degrees, these parameters X_{SP}, C_l, C_d change by no more than 1% with an increase in the number of computational cells from 230 000 to 300 000. It were considered grids with the number of computational cells from 80 000 to 300 000. Further calculations were carried out with the number of cells 230 000 and with a time step of 10 ns. A comparison was made of the stationary flow patterns obtained in the numerical simulation and measured in the experiment (without initiating heat release) by the position X_{SP} of the separation point and the velocity of the air flow near the heat release place, and this comparing has showed good agreement between experiment and simulation.

Simulation of unsteady process of heat release using long spark discharge was carried out by applying an additional field of increased temperature and pressure to the previously calculated stationary flow pattern. This increase in temperature ΔT and pressure Δp was taken to be equal to [13]:

$$\Delta p = (\gamma - 1)\frac{\eta E}{v}, \Delta T = T_{st}\frac{\Delta p}{p_{st}}, \qquad (2)$$

where v is the volume in which heat is released by the discharge, γ is the heat capacity ratio for air, is assumed to be a constant value of 1.4, E is the electrical energy stored in the capacitor, η is the fraction of this energy that goes into heat. The change in density inside the discharge region was neglected, i.e. a quasi-instantaneous isochoric process of air heating was simulated. The volume of the discharge region v was calculated as the product of the cross section of the discharge channel S by its length L. The length L was specified by the number of discharger segments (in the experiment it was 9 cm), and

S was determined from the discharge photos (the resulting width of the discharge channel was equal to 2 mm).

4. Obtained results

Both in the simulation and in the experiment, the obtained unsteady flow patterns with fast heat release show that the after spark discharge in the oncoming supersonic airflow near the surface of the wing model a cylindrical shock wave forms. Its center of symmetry shifts relative to the surface of the wing model downstream due to the presence of an external supersonic flow. The front of a cylindrical shock wave moves relative to the main supersonic flow, and behind it the air gains some additional impulse due to the expansion of the gas region heated by an electric discharge. The gas flow speed V_2 relative to the speed of the incident flow depends on the time after the discharge and on its energy. On the figure **3** unsteady flow patterns (contours of y-component of velocity vector field), obtained in experiment and in numerical simulation, is presented at site near the trailing edge of wing model. On figure **3**a the averaged PIV-data is presented for angle of attack 14 degrees, stored electrical energy E =3.0 J, delay after discharge Δt =48 µs. On the figure **3**b unsteady flow pattern, obtained in simulation, is presented for the same angle of attack, released energy is 0.2 J, delay after energy release is Δt=50 µs. On both patterns dashed curve is visible part of shock wave front, red circle is place of energy release, open circle is the place, where V_2 is measured. Area near the trailing edge with warm colour is the flow separation region.

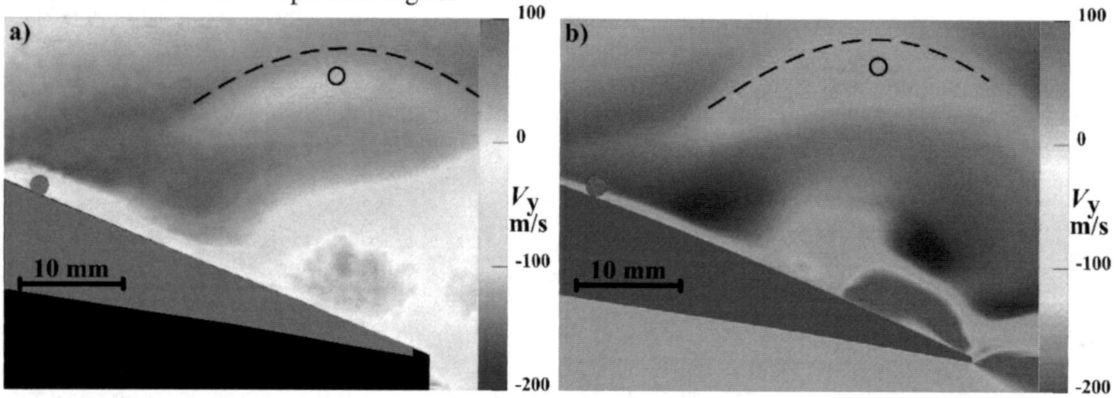

Figure 3. Unsteady flow patterns obtained in experiment (a) and in numerical simulation (b). External airflow with M=2 is from left to right.

The velocity of the front of a cylindrical shock wave decreases with time, and at the same time moment after discharge the magnitude of this velocity and velocity V_2 will be greater if the value of the stored electrical energy E is greater. It is important to note that in the experiment the duration of the electric discharge was measured, and in the energy range E =0.55–6.9 J the duration τ was from 0.2 µs to 0.6 µs, that is much less than characteristic gasdynamic time, and this circumstance makes it possible to consider the process of heat release to be almost instantaneous (as it is in computer simulation). It should be noted that the measured value of V_2 did not exceed 300 m/s in the indicated energy range E =0.55–6.9 J. Considering the value of the tracer relaxation time τ_p =4 µs mentioned in subsection 2.4, gives a distance less than 2 mm. When considering the distribution of the velocity V_2, measured with PIV, from the distance r from the center of symmetry of a cylindrical shock wave, it is clear that the width of the region where the tracers lag behind the flow does not exceed 2 mm (see figure **3**).

The determination of the fraction η of the electric energy E, which goes to gas heating, was carried out by comparing data from flow patterns obtained in experiment and in computer simulation. The values of the velocity V_2 behind the shock wave front (normalized to the magnitude of the velocity of the oncoming flow V_0) were compared at different time moments after discharge initiation and at different values of the energy E. Both experiment and simulation show that over time after the initiation of the discharge in the range of stored electrical energy E =0.1–10 J, the dependence $V_2(t)$, divided by the oncoming flow velocity V_0, has the form:

$$\frac{V_2(t)}{V_0} = kt^n, \tag{3}$$

where k increases with increasing of energy E, and n in the specified range E remains almost constant and equal to −1 in the range of time t from 10 μs to 60 μs. The indicated time range t is limited by the possibility of measuring the velocity V_2 in the experiment: when t <10 μs such measurements cannot be carried out due to the small size of the gasdynamic perturbation region, and at t >60 μs, the shock wave front leaves the region of visualization. Experimental data and computer simulation results with different values of E show that the dependence $k(E)$ can be represented as:

$$k(E) = cE^b, \tag{4}$$

where c and b are constants independent of E, they are determined by approximation of the obtained dependences $k(E)$. Approximation of the data obtained in computer simulation gives the value b =0.26±5%, which corresponds to the results of the analytical model in the theory of a very intense explosion in the two-dimensional case [14]:

$$V_2(t) = \frac{4}{(d+2)(\gamma+1)}\left(\frac{E_1}{\rho_1}\right)^{\frac{1}{2+d}} t^{-\frac{d}{d+2}}, \tag{5}$$

where E_1 is amount of released energy, ρ_1 is density of undisturbed air, d is number of space dimensions (in this case d =2). But this correspondence is only partial, because power n =–1, obtained from experimental data and numerical simulation (see equation (3)), does not correspond to theoretical value (see equation (5)). This is due the fact that in this case rapid heat release cannot be described as very intense explosion, and the pressure behind cylindrical shock wave is not much greater then pressure of undisturbed air (Mach number of front of cylindrical shock wave is less than 2). When comparing the dependencies $k(E)$ obtained in the experiment and in the simulation (see figure **4**), it turns out that the value of η is 0.07±20%, where this error is an approximation error of the experimental data.

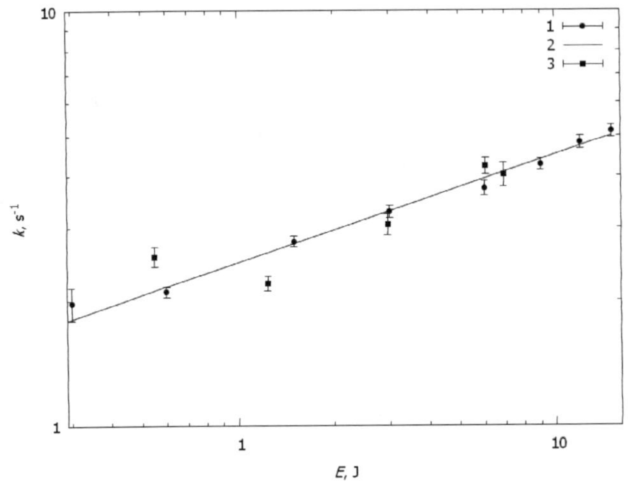

Figure 4. k dependence on E: 1—numerical simulation, 2—approximation by equation (4), 3—experimental data. Plotted vertical error corresponds to error of approximation by equation (3).

5. Conclusions

The experimental study of the process of rapid heat release using long spark discharge on the surface of a wing model streamlined by supersonic airflow with Mach number M =2 and attack angle of 14 degrees and with various discharge energy values was carried out, and a two-dimensional simulation of this process was performed. By comparing the measured and calculated values of the additional velocity that air gains behind the front of a cylindrical shock wave, it was found that the fraction of electrical energy that goes to local gas heating is about 7%. This value turns out to be less than in other works ([13, 15]), and this circumstance suggests that in computer modeling of such phenomena it is necessary to take into account specific conditions and method of discharge initiation. As shown by numerical simulation, additional velocity gained by air after expansion of a small volume of air participating in electric discharge depends on stored electrical energy according to a power law, and under the conditions of described experiment power of this dependence is 1/4, which corresponds to the result in the theory of a very intensive explosion in the two-dimensional case. Obtained result can be used for computer simulation of unsteady process of rapid heat release in supersonic airflow at experimental conditions, described in this paper and in [2, 3].

Acknowledgments
The reported study was funded by RFBR according to the research project № 18-38-00803.

References

[1] Post M L and Corke T C 2004 *AIAA J.* **42**(11) 2177–84
[2] Golub V V, Saveliev A S, Sechenov V A, Son E E and Tereshonok D V 2010 *High Temp.* **48**(6) 903–9
[3] Saveliev A S 2018 *J. Phys.: Conf. Ser.* **1112** 012014
[4] Roupassov D V, Nikipelov A A, Nudnova M M and Starikovskii A Yu 2009 *AIAA J.* **47**(1) 168–85
[5] Correale G, Popov I B, Ratikin A E, Starikovskii A Y, Hulshoff S J and Veldhuis L L M 2011 *Proc. 49th AIAA Aerospace Sciences Meeting including the New Horizons Forum and Aerospace Exposition (Orlando FL)* AIAA 2011-1079
[6] Moralev I, Sherbakova V, Selivonin I, Bityurin V and Ustinov M 2018 Int. J. Heat Mass Transfer 116 1326–40
[7] Leonov S B 2018 *Energies* **11**(7) 1733
[8] Webb N, Clifford C and Samimy M 2013 *Exp Fluids* **54** 1545
[9] Saveliev A, Sechenov V, Golub V, Son E, Aksenov V and Gubin S 2009 *Proc. 47th AIAA Aerospace Sciences Meeting including The New Horizons Forum and Aerospace Exposition Orlando, Florida*. AIAA-2009-696
[10] Son E E 2016 *Physical Mechanics: Laboratory Workshops* (Begell House Publishers)
[11] Raffel M, Willert C E, Scarano F, Kähler C, Wereley S T and Kompenhans J 2018 *Particle Image Velocimetry: A Practical Guide* (Springer International Publishing)
[12] Greenshields C J, Weller H G, Gasparini L and Reese J M 2010 *Int. J. Numer. Methods Fluids* **63** 1–21
[13] Koroteeva E, Znamenskaya I, Orlov D and Sysoev N 2017 *J. Phys.* D: *Appl. Phys.* **50** 085204
[14] Sedov L I 1993 *Similarity and Dimensional Methods in Mechanics* (CRC Press)
[15] Elias P Q and Castera P 2013 *J. Phys. D: Appl. Phys.* **46** 365204

Journal of Physics: Conference Series **1394** (2019) 012012 doi:10.1088/1742-6596/1394/1/012012

Buoyancy driven convection instability and related pulsing of continuous optical discharges

S Yu Lavrentyev, N G Solovyov, A N Shemyakin, M Yu Yakimov

Ishlinsky Institute for Problems in Mechanics RAS
101-1, prospect Vernadskogo, Moscow, 119526, Russia

yakimov@lantanlaser.ru

Abstract. Optical discharges find now ever expanding applications in science, engineering and medicine. In present study authors discuss the phenomenon of periodic pulsing of a continuous optical discharge (COD) on the base of original experimental results. In the literature there is no general agreement regarding the origin of the considered instability. Authors have obtained the dependence of oscillation frequency on the gas pressure and proposed semi-empirical relations to estimate pulse frequency pressure dependence observed. Regular self-sustained oscillations of COD plasma were explained by the pulsations of heated gas bubble surrounding COD followed by buoyancy driven vortex formation that was detected on schlieren images of convection zone. This kind of buoyant convection instability was determined to obey the same similarity law as cycling frequency of puffing or flickering in diffusion and premixed flames. The instability discussed can affect the performance of COD in some important applications such as high brightness broadband light sources, for instance.

1. Introduction

Optical discharges are now widely used for generation and sustaining high temperature plasma of high energy dissipation density in various technical and scientific applications [1-6, 14]. One of the most important features of the optical discharges is outstanding stability of plasma characteristics provided by high stability of supporting laser radiation, together with possibility of sustaining microplasma with high energy dissipation density with size and shape easily controlled by means of focusing optics. All these attractive features may be realized in small scale in portable laboratory equipment. Optical discharges may be initiated and sustained by continuous-wave or periodic-pulse laser radiation of various wavelengths in medium or near infrared bands. Due to plasma transparency for certain spectral bands optical discharge characteristics depending on gas type, pressure and temperature may be easily monitored by spectral optical probes.

Under certain conditions essential for some applications optical discharge plasmas exhibit instability. Just in pioneer experiments on obtaining COD self-excited oscillations were observed related to thermal gravity (buoyancy induced) convection instability [7]. Pulsations of thermally induced convective plume exhibit their selves together with regular oscillations of brightness and spatial location of COD plasmas that make difficult their application as stable high brightness radiation source. Special tricks should be applied to overcome this drawback that may be difficult to attain [5].

There is no general agreement regarding the nature of the considered instability in the literature [8-12] that make further obstacles to suppress it effectively. Authors have obtained the dependence of the oscillation frequency on the gas pressure and proposed general relations to estimate pulse frequency dependence observed given underneath.

The essential feature required of gaining acceptance recently high brightness radiation sources employing COD is exceptional temporal, spatial and spectral stability of their broadband radiation. Thus the problem of COD pulsing revealed in [7] and further studied in [8-12] remains the issue of current interest.

2. Experimental set-up and results

In this work authors have investigated processes manifested their selves first in a form of relatively low amplitude (less than 1% in general) pulsations of plasma brightness of COD sustained in the experimental model of broadband high-brightness light source LPS-50 (figure 1) [14], for instance.

In LPS-50 light emitting plasma is sustained in quartz bulb filled with high pressure xenon due to absorption of the radiation of two diode lasers emitting radiation at $\lambda = 0.97$ μm each at full continuous power of 25-30 W (total rated power $P_l = 2\times26$ W). Two laser beams each focused with f-number $f/3.5$ were arranged to cross near the beam waists at the angle 60^0 forming a location of the discharge zone. Plasma was sustained by both beams in the intersection volume about 170 μm long and 100 μm wide. To eliminate obvious sources of plasma pulsations laser radiation was carefully stabilized, diode operating temperature was fixed by means of Peltier thermo-elements, all possible vibrations of the construction elements and optical fibers induced by cooling fans were excluded.

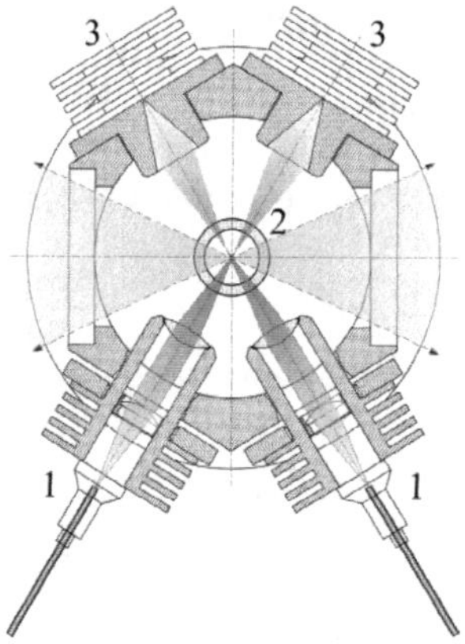

a) *b*)

Figure 1. Photo (*a*) and central cross section scheme (*b*) of the lighthouse of laser plasma broadband light source LPS-50. 1 – fiber laser beam feed coupled to collimating and focusing lenses; 2 – high pressure xenon quartz bulb; 3 – laser beam stop.

In spite of all the precautions pulsations of COD brightness were observed at different xenon pressures (p_{Xe}= 15; 30; 40 bar), set in specially prepared different quartz discharge bulbs. Data were obtained in steady-state conditions when the gas pressure were increased and stabilized after heating by the discharge in a closed volume of the bulb. Pulsations of the brightness in the central region of COD plasma were registered by photomultiplier (PMT) operated in linear response regime. Small portion of plasma radiation was directed to the PMT with achromatic lens that built magnified (×20) image of plasma at the PMT input diaphragm of less than 0.5 mm in diameter. Signal was registered by means of digital oscilloscope and then subjected to Fourier transform to obtain frequency spectrum of the oscillations.

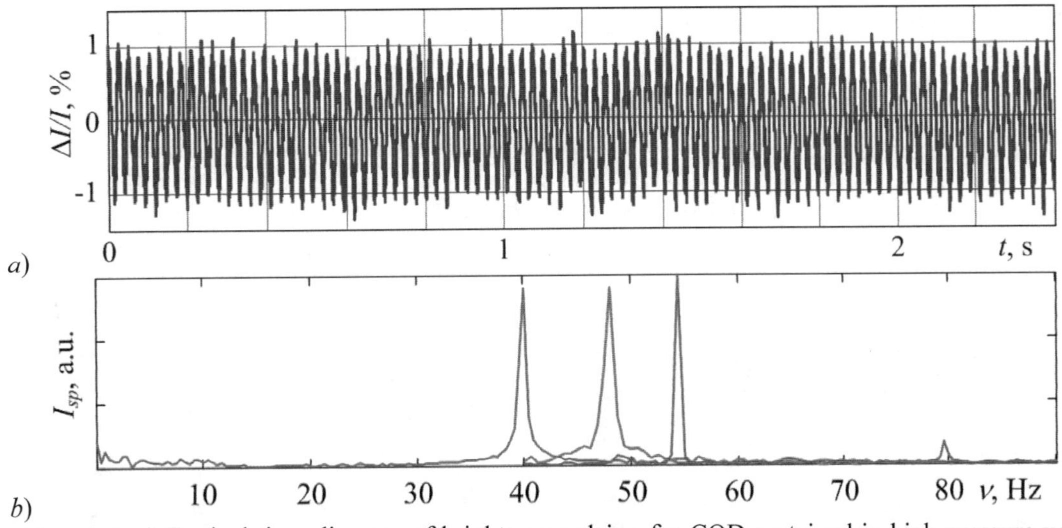

Figure 2. a) Typical time diagram of brightness pulsing for COD sustained in high pressure xenon in two focused laser beams crossing ($\lambda = 0.97$ μm, $P_l = 2\times26$ W, f/3.5, $p_{Xe} = 15$ bar); b) frequency spectra of COD plasma brightness pulsations, obtained at different actual xenon pressures $p_{Xe} = 15$, 30, 40 bar (correspondingly from the left peak to the right: blue, red and green curves).

Figure 2, a) shows sample oscillogram of pulsations of thermal radiation taken for COD in Xe at 15 bar. Figure 2, b) depicts pulsation spectra obtained with fast Fourier transform applied to the oscillograms corresponding to different current gas pressures in the discharge bulbs.

It was found that main pulsation frequency is increased with pressure but weakly depends on laser power dissipated in plasma, whereas laser power absorbed in plasma was declined up to several times while laser power was approaching down to a threshold value of COD sustaining.

The result obtained is mainly in agreement with the data reported in [12] for COD sustained by CO_2 laser in different ranges of gas pressure and plasma characteristics. Discussions on the results and physical model of instability observed presented in [12] cannot be accepted correct, as it was shown in [11]. Possible mechanism of COD plasma oscillations, elaborated in [8], ignores laser beam refraction in plasma and is in contradiction with the results of one-dimensional laser combustion wave stability analysis, consistently performed in [11].

Supposing that real reason of plasma instability is nonstationary convection flow around the plasma, author has employed high-speed filming of the dynamics of gas density gradients around COD visualized by schlieren method (Figure 3).

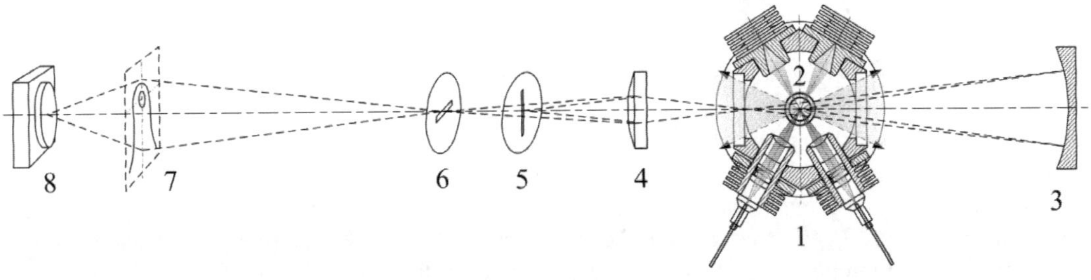

Figure 3. Schlieren imaging scheme. 1 – LPS-50 lighthouse cross-section; 2 – high pressure xenon quartz bulb with plasma; 3 – back reflecting spherical mirror; 4 – achromatic lens objective; 5 – vertical slit diaphragm; 6 – horizontal slit diaphragm; 7 – fine grain rear projection screen; 8 – camera.

Shlieren patterns were obtained using COD plasma itself as a point source. Plasma radiation was partially collected by spherical mirror and reflected back to pass through the discharge zone. The image of the vicinity of the plasma in passed through light was created by lens objective on the fine grain rear projection screen used for filming. Light passed through the discharge zone inside thick wall quartz bulb becomes astigmatic and forms two separate line foci past the objective. Two slits were used to transmit passed through reflected light and suppressing direct light from COD. The same slits were used to obtain dark-field schlieren image. Direct light was also presented on the images by correspondent vertical and horizontal line patches and dimmed plasma image itself.

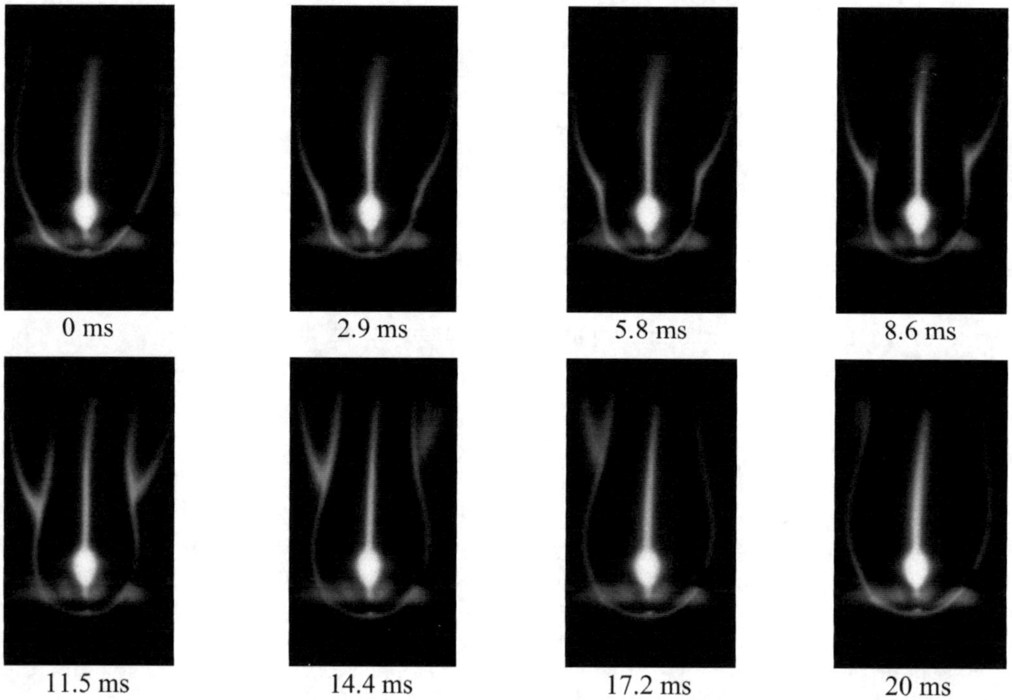

Figure 4. Schlieren patterns of the hot gas bubble surrounding COD plasma showing formation of the bubble, its upward motion and related toroidal vortex formation during one period of pulsation (0-23 ms): 0-6 ms – heated bubble 2.6 mm in diameter just began its upward motion, residual heated bubble 1.3 mm dia. remains around COD; 8-14 ms – floating up bubble rises up at above 150 mm/s, heated zone around COD is expanding from 1.5 to 1.9 mm dia., margins between hot and cold gas make evident vortex motion in the upward bubble; *c*) 15-20 ms – hot gas bubble around COD has expanded up to 2.5 mm dia. just before starting upward motion. Frame size 2.9×5.1 mm^2. The discharge conditions are: $\lambda = 0.97$ μm, $P_l = 2\times26$ W, $f/3.5$, $p_{Xe} = 20$ bar.

Figure 4 shows the frames obtained at xenon pressure 20-25 bar. On the presented frames and high-speed (up to 800 fps) film itself one may observe sequential stages of growing hot gas bubble around COD and strictly periodical buoyancy driven upward motion of the bubbles.

White line indicating maximum refraction index gradient represents the boundary between gas, directly heated from plasma (about 3 kK) and colder (up to 500 K) ambient gas (inside the quartz bulb). Ambient gas temperature inside the bulb is close to the temperature of the built-in electrodes that was measured by IR pyrometer. Current xenon pressure inside the bulb could be derived from known cold bulb pressure being inversely proportional to the temperature. The value of current pressure could be also independently controlled by measuring threshold laser power required to sustain COD [13]. Hot gas temperature on the boundary of the hot gas bubble observed was estimated from the refraction index temperature dependence of high pressure xenon.

Bottom front of a heated gas bubble around COD was found to be semispherical growing from minimal radius $r_0 = 0.65$ mm almost two times up to $2r_0 = 1.3$ mm during about 15 ms. Growing bubble starts floating up due to buoyancy force and acquires the speed of natural convection. All phases repeat with period 23 ms corresponding pulse frequency 44 Hz, equal to oscillation frequency of discharge brightness at the same gas pressure.

3. Discussions on the results

Some qualitative speculations and quantitative estimations should be made to provide theoretical ground for the effect observed. Simple physical model was proposed for rough quantitative estimations.

In LPS-50 light emitting plasma is sustained in quartz bulb filled with high pressure xenon due to absorption of the radiation of two diode lasers in the location of intersecting their focused beams of total laser power $P_l = 2\times26$ W. The expansion of heated gas volume around plasma was obviously produced by thermal component of the energy dissipation in plasma. In typical experimental conditions when incident laser power was 1.5-2 times larger than the plasma sustaining threshold level, plasma absorbed about 50% of the incident laser power. About 50% of the absorbed laser power was reradiated by the plasma in the form of thermal radiation, the rest may be treated as thermal component of plasma energy losses.

At given value of dissipated power thermal component Q the expansion of the heated gas volume will slow down as the front radius r is increased. Simple estimation formula for hot gas bubble expansion velocity $V_t(r)$ could be derived from simplified energy budget. The formula expresses the idea that all thermal energy released is transmitted to the bottom hemisphere front of the bubble (upper front is obviously absent when convective flow is established).

$$V_t(r) = \frac{QM}{2\pi C_p r^2 \rho(T_1)(T-T_1)}, \qquad (1)$$

where M is Xe atomic mass in unified atomic mass units, C_p – Xe heat capacity, $\rho(T_1)$ – Xe volumetric mass density at ambient temperature T_1, T – average hot gas temperature on the bubble front. In this consideration specific physical mechanism of heat transmission – heat conductivity or convection – were not included. In real both mechanisms contribute to the result, but in this simplified consideration we implicitly suppose that the heat is transmitted instantaneously.

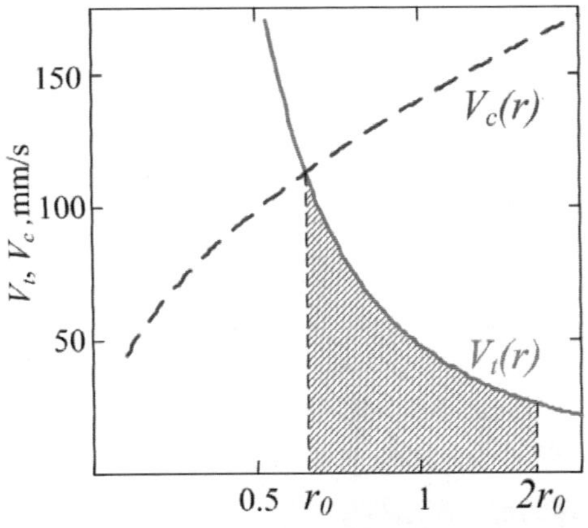

Figure 5. Near-COD heated gas bubble expansion velocity $V_t(r)$ and the bubble floating up velocity $V_c(r)$ plotted against the bubble radius r. Pulsing range (shadowed) from minimum radius r_0 up to maximum radius $2r_0$ of floating up bubble. Calculation performed for $Q = 10$ W, $T_1 = 400$ K, $T = 3$ kK, $p_{Xe} = 20$ bar.

Journal of Physics: Conference Series **1394** (2019) 012012 doi:10.1088/1742-6596/1394/1/012012

Buoyant floating up of the hot gas bubble will oppositely tend to accelerate with radius accordingly to well known formula $V_c(r) = \sqrt{2rg}$, where g is gravity acceleration. The formula may be easily derived from the balance of the drag and buoyant forces of the floating up bubble.

Thermal gravity convection may be treated established when bubble front expansion velocity becomes equal to the floating up velocity. The last condition means that the value of front radius of a steady convection would depend weakly on the other parameters.

Figure 5 shows both dependencies V_t and V_c plotted against the bubble radius r. Parameters used for this calculation correspond to that of used to obtain data presented in figures 2, 4. Intersection point of the curves gives the value r_0 of the expanding front of the bubble stabilized by buoyant convection. As one can see value r_0 corresponds to minimal radius of thermal expansion front presented in figure 4, a). Maximum radius of the floating up heated gas volume roughly equals $2r_0$, as in figure 2, c).

Shadowed range between r_0 and $2r_0$ correspond to variations of thermal front radius, bubble expansion velocity and buoyant convection velocity during the period of pulsation. The bubble radius is increased from r_0 to $2r_0$ during bubble floating up stage. Bubble expanding velocity is decreased as the radius is increased. Estimation formula may be derived for the pulsation frequency ν:

$$\nu = \frac{V_t(2r_0)}{r_0}. \qquad (2)$$

Calculations with the parameters indicated in Figure 5 caption (except for gas pressure) give for p_{Xe}= 15; 30; 40 bar correspondingly ν_{15} = 41 Hz, ν_{30} = 47 Hz, ν_{40} = 50 Hz, being close to experimentally observed pulsation frequencies, correspondingly 40, 47 and 53 Hz.

Thus the emerging picture of the observed plasma brightness oscillations related to the instability of buoyant convection area around COD seems to be as follows. Starting from establishing steady radius of heated gas front stabilized by buoyant convection, the heated gas volume continues growing upward and sideward due to intense vortex convection inside the heated gas bubble from its axis to the periphery. The vorticity of the gas motion manifests itself at a certain stage of the pulsation period when cold ambient gas drawn by the vortex into the convective flow forms additional density gradient surface clearly seen on the schlieren patterns (figure 4, b). As the bubble is growing the floating up velocity is increased and the bubble is periodically risen up out of the COD area inducing buoyant driven pulsations in the ambient gas and plasma. Semi-empirical model based on this physical representation gives oscillation frequency estimation close to the observed ones.

Formula (2) may be reduced to the form well known from theory and practice of diffusion and premixed puffing or flickering flames [15-17]:

$$\nu = \frac{1}{2}\left(\frac{g}{2r_0}\right)^{\frac{1}{2}}. \qquad (3)$$

Replacing $2r_0$ in (3) by D, where D means burner diameter or flame diameter, one will get similarity law for pulsation frequencies of flames widely discussed in [15-17] and many other papers on the base of the results of numerous experiments cowering wide range of flame sizes and pulsation frequencies. Thus COD also follows the same similarity law, but differs from the case of the flames by considerably higher realization pressures and by correspondingly higher pulsation frequencies and smaller diameters of the buoyant convective flow.

4. Summary and conclusion

Based on original experimental results authors of the present study performed the analysis of spatial and temporal effects occurred as manifestation of buoyant convection instability in the unusual case of very concentrated heat source that can be produced by optical discharges in gases under very high pressures. Simple semi-empirical consideration used employs rough physical model that gives some quantitative estimations of the effect of the considered instability on COD plasma characteristics.

In fact buoyant convection instability around concentrated thermal source produces self-sustaining oscillations of the convection flow that also affect brightness and position of high temperature plasma core of COD. Rough physical model permits to derive formula for oscillation frequency dependence on pressure that is in agreement with the experimental results obtained. It was shown that the formula is reduced to the form coinciding with similarity law for puffing and flickering flames, well-known from many experiments and computer simulations. Data derived from experiments with optical discharges will extend the range of workability of the similarity law up to higher pressures and higher oscillation frequencies.

Periodical oscillations of plasma brightness of COD induced by buoyancy driven pulsations play negative role in some important applications, affecting the stability of broadband high brightness light sources, for instance.

Acknowledgements
The study was partially supported by the Government program (contract #AAAA-A17-117021310370-2) and partially supported by RFBR (grant #18–01–00534 A “Experimental study of abnormal quasi-stationary convection effect and resonant gasdynamic phenomena in periodic-pulse optical discharges”).

References

[1] Smith D K, et al. 2008 Patent US #7435982; 2012 Patent US #8309943
[2] Bezel I, Delgado G, Derstine M, Gross K, Solarz R, Shchemelinin A, Shortt D 2015 *CLEO:2015* p ATu4M.2. https://doi.org/10.1364/CLEO_AT.2015.ATu4M.2
[3] Su M N, Lin J J M 2013 *Review of Scientific Instruments* **84** 086106 https://doi.org/10.1063/1.4818977
[4] Kuznetsov V A, Solovyov N G, Shemyakin A N, Shilov A O, Yakimov M Yu, Zimakov V P 2013 *Proc. SPIE* **8600** 860002 doi:10.1117/12.2003658
[5] Antsiferov P S, Koshelev K N, Krivtsun V M, Lash A A 2016 Patent US #9357627 B2
[6] Rudoy I G, Solovyov N G, Soroka A M, Yakimov M Yu 2015 Patent RU #2571433
[7] Generalov N A, Zimakov V P, Kozlov G I, Masyukov V A, Raizer Yu P 1972 *Sov. Phys. JEPT* **34** 763
[8] Gerasimenko M V, Kozlov G I, Kuznetsov V A 1980 *Soviet Technical Physics Letters* **6** 208
[9] Generalov N A, Zakharov A M, Kosynkin V D, Yakimov M Yu 1986 *Combustion Explosion and Shock Waves* **22**(2) 214
[10] Raizer Yu P, Surzhikov S T 1989 *Fluid Dynamics* **24**(4) 593
[11] Makhviladze G M, Selezneva I K 1981 *Journal of Applied Mechanics and Technical Physics* **22**(5) 646 DOI:10.1007/BF00913713
[12] Baranowski A, Mucha Z, Peradzynski Z 1977 *Bulletine de l'Academie Polonaise des Sciences* **25**(4) 361
[13] Zimakov V P, Kuznetsov V A, Rudoy I G, Solovyov N G, Soroka A M, Shemyakin A N, Shilov A O, Yakimov M Yu 2015 *Physical and Chemical Kinetics in Gasdynamics* **16**(2) http://chemphys.edu.ru/issues/2015-16-2/articles/548/
[14] Zimakov V P, Kuznetsov V A, Lavrentyev S Yu, Solovyov N G, Shemyakin A N, Shilov A O, Yakimov M Yu 2016 *Physical and Chemical Kinetics in Gasdynamics* **17**(2) http://chemphys.edu.ru/issues/2016-17-2/articles/653/
[15] Cetegen B. M., Ahmed T. A. 1993 *Combustion and Flame* **93** 157
[16] Xia Xi, Zhang Peng 2018 *Flickering of buoyant diffusion flames: a vortex dynamics revisit* arXiv:1803.10400v1 [physics.flu-dyn]
[17] Boettcher P, Menon S, Ventura B, Blanquart G, Shepherd J 2013 *Journal of Fluid Mechanics* **735** 176 doi:10.1017/jfm.2013.495

Impact of homogeneous and ionization unstable glow discharge plasma on a shock wave in air

A.V. Erofeev[1], T.A. Lapushkina[1]*, O.A. Azarova[2]

[1] Ioffe Institute, Politekhnicheskaya str. 26, 194021, St Petersburg, Russian Federation
[2] Federal Research Center "Computer Science and Control" of RAS, Vavilova st. 40, 119333, Moscow, Russian Federation

*tanyusha@mail.ioffe.ru

Abstract. The work is aimed at exploring the possibility of control of strong shock waves up to their completely destruction. The impact on shock waves by creating a region of gas-discharge plasma in propagation media is proposed. A diffuse or ionization unstable discharge with the ionization strata of different scale on the path of a shock wave propagation is organized. As a result, the shock wave propagates over the region of homogeneous highly non-equilibrium plasma or over the medium of layered structure with higher degree of ionization and electron temperature in the layers. Numerical simulation of the propagation of a shock wave over the region of energy impact was carried out on the basis of the systems of Euler and Navier-Stokes equations with the flow parameters obtained in the experiment. The homogeneous plasma region was simulated by a rarefied gas region with increased temperature and constant pressure. The area formed by the ionization unstable discharge was simulated by a thermally stratified gas area. The strong influence of the ionization of the propagation medium on the transformation of the shock wave with the formation of a new shock wave configuration is shown. Stratification of the medium leads to a distortion of this configuration up to complete destruction of it.

Introduction

The task of controlling supersonic flows by non-mechanical methods with the aim of changing the shape and intensity of shock-wave configurations up to their complete distortion, as well as changing the aerodynamic characteristics of poorly streamlined bodies is widely demanded in modern aerodynamics. Non-mechanical control methods have a significantly better temporal resolution compared to currently used mechanical methods. This problem is solved by a number of authors using gas discharges of various types and structure which are organized in a flow [1-5] or near a streamlined surface [6-7].

This paper proposes to influence the shock waves, their shape, intensity and speed of propagation by creating local ionized zones along the path of their propagation with the help of a gas discharge. The aim of the study is to choose the most optimal configuration of impact. For this purpose, the interaction of a shock wave (SW) with the Mach number 5-6 with the region of a glow gas discharge of different intensity of the discharge current and different types of ionization stability is studied. The impact mechanism can be characterized as thermal - due to heating the heavy component of the medium in

Journal of Physics: Conference Series **1394** (2019) 012013 doi:10.1088/1742-6596/1394/1/012013

these regions during the thermal exchange with electrons. In addition, it can be influenced by plasma effects connected with non-equlibrium processes.

1. Experiment arrangement

The studies are carried out in a straight channel of square section, which is a continuation of a circular shock tube, in which a strong SW is formed, and then enters the working chamber. The scheme of the experimental setup is shown in Fig. 1. The shock tube includes a high pressure chamber (1), a low pressure chamber (2) and a measuring section (3) for flow checking. The cross-section of the shock tube is inscribed in a square section of the working chamber (7), as shown in the diagram (section A-A). The working chamber is separated from the shock tube by a Dacron diaphragm, which allows avoiding transients with the formation of a complex shock-wave structure [8]. It also allows creating a reduced air pressure (7 Torr) in the working chamber, which makes it possible to organize both a uniform diffuse discharge and an ionization unstable one. The working gas is air, the initial pressure in the low-pressure chamber is 20 Torr at room temperature, the Mach number of the shock wave is $M = 6$.

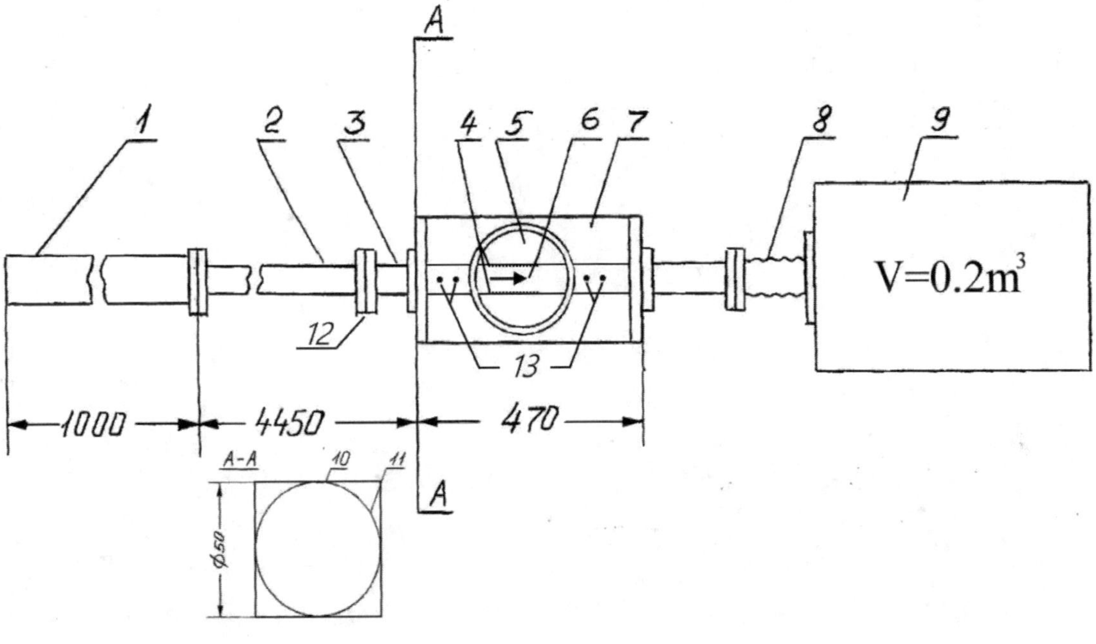

Figure 1. Installation scheme: 1 - high pressure chamber; 2 - low pressure chamber; 3 - measuring section; 4 - electrodes; 5 - optical window; 6 - discharge zone; 7 - working chamber; 8 - bellows; 9 - damper chamber; 10 - a cross section of the working chamber; 11 - a cross section of a shock tube with diameter 50 mm; 12 - Dacron diaphragm; 13 - piezo sensors.

The discharge current is organized across the flow direction. The shape of the discharge was changed by changing the intensity of the gas-discharge current, which was regulated by the value of the load resistance. Depending on the magnitude of the gas-discharge current, it is possible to create a uniformly ionized area of impact or a stratified one with ionization strata of various scales. The width of the discharge zone varied within 4–8 cm. The gas discharge was ignited in advance before the shock wave arrived at the zone of action. The shock wave suppresses the discharge and propagates through the afterglow zone. The time of passage of the discharge zone is 30–40 μs, which is two orders of magnitude shorter than the plasma relaxation time. This suggests that the ionization parameters in the impact zone during the passage of the shock wave are almost the same as in the active discharge zone. The direction of propagation of the shock wave in the impact zone is indicated in Fig. 1 by an arrow.

Journal of Physics: Conference Series **1394** (2019) 012013 doi:10.1088/1742-6596/1394/1/012013

The diagnostics of the flow was carried out by visualizing gas-dynamic discontinuities, namely, density gradients, and their evolution by high-speed multi-frame schlieren system through the optical windows (*5* in Fig. 1). In addition, the time taken by SW for passing between the measuring marks before and after the impact zone (to determine SW propagation velocity) is recorded. To this end, two sensors before and two after of the impact zone were installed in the side walls of the channel at the same distance from each other (13), which also make it possible to measure the static pressure in the working chamber.

Photo of working chamber is shown in Fig. 2. Here one can see a straight channel, a discharge zone with electrodes, optical windows, located in the impact area on the side walls of the chamber, as well as piezo sensors, mounted in front of the discharge zone and after it.

Figure 2. Photograph of the working chamber with the impact zone

2. Interaction of SW with the homogeneous diffuse discharge region

At the gas-discharge current of 100-200 mA, a homogeneous discharge is observed in the impact area (Fig. 3a). When SW enters the impact region, the formation of a new shock-wave configuration (Fig. 3c) is observed; SW is broadened in comparison with the original its form as shown in Fig. 3b. The width of the new formation depends on the magnitude of the gas-discharge current and persists when passing through the impact zone. It was also found an increase in the velocity of SW propagation in the gas-discharge zone with increasing the discharge current intensity.

Numerical simulation of this flow by solving the 2D Riemann problem showed that the observed broadening of SW is associated with the formation of a shock structure consisting of SW and contact discontinuity (CD) moving to the initial SW direction [9]. The numerical density field is shown in Fig. 3d. One can see a good agreement with the experimental data. In the calculated pressure field, this discontinuity is absent, which indicates the formation of a contact surface.

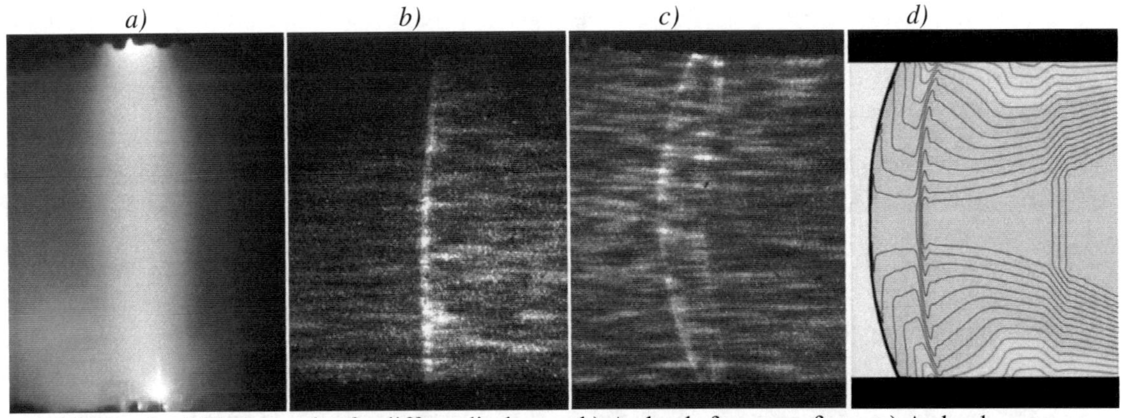

Figure 3. a) Photograph of a diffuse discharge; b) A shock-free waveform; c) A shock-wave configuration in the discharge region; d) Computational flow image with the formation of CD.

3. Interaction of SW with large-scale ionization strata

Unboundedness of the gas discharge in the direction of the flow allows organizing ionization-unstable discharge [10] with ionization strata of different scale, representing the areas of increased ionization rate, temperature and electron concentration [11]. When a gas-discharge current is in the range of 200-300 mA, large-scale strata are formed in the discharge zone, lining up along the current line, 5-7 strata per discharge length. Figure 4a presents a photograph of such discharge. The discharge region in this case is a layered structure with different degrees of ionization and temperature of the electron and gas components. As a result, an impact zone with longitudinal large-scale ionization and temperature inhomogeneities is formed.

When SW passes through the area of such a discharge, its front is distorted and bends, repeating the shape of the discharge area (Fig. 4b). Note that an additional gas-dynamic discontinuity is also formed behind the SW, repeating the waveform. Coming out from the impact zone, SW and CD retain distorted and of unstable shape (Fig. 4c).

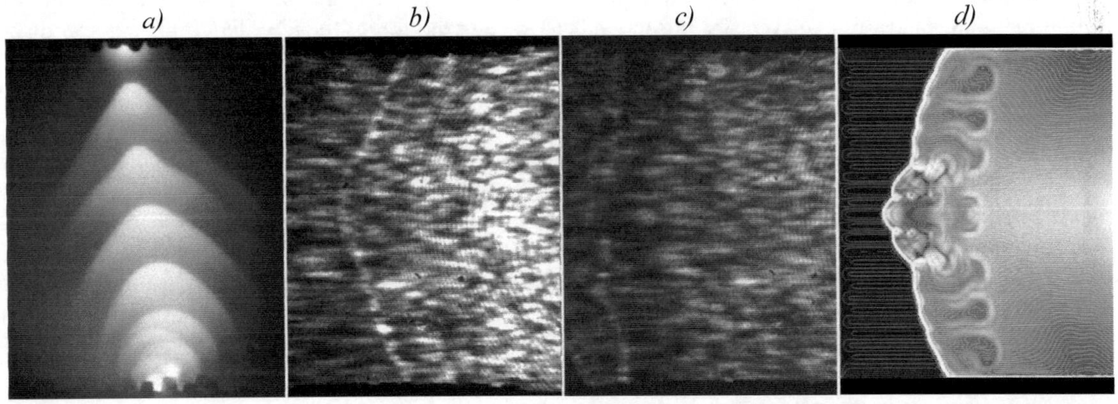

Figure 4. a) Ionization-unstable discharge with large-scale strata; b) Distortion of SW in the discharge zone; c) SW shape after leaving the discharge zone; d) SW interaction with large-scale temperature inhomogeneities, field of density

Simulations of the flow based on numerical solving of the Euler and Navier-Stokes systems of equations assuming the temperature inhomogeneities of such scale in the region of SW propagation are in a good accordance with the experiments (Fig. 4d). It can be seen that the presence of temperature inhomogeneities of the medium causes the development of vortices and instabilities which leads to the distortion of SW and CD.

4. Passage of SW through the region of small-scale ionization strata

Figure 5a presents a photograph of the discharge with small-scale ionization strata up to 20 strata over the discharge length. Such a discharge structure is observed at the gas-discharge current in the range of 300-400 mA and creates a layered structure of the medium with small-scale temperature inhomogeneities in the impact zone. When SW enters the impact zone, it immediately starts to distort (Fig. 5b). In addition to broadening, there is a strong weakening of the density gradients in the discharge zone, as result the gasdynamic discontinuities becomes almost indistinguishable in the Schlieren image (Fig. 3c). Upon leaving the impact zone, SW is not observed, it is completely destroyed [8].

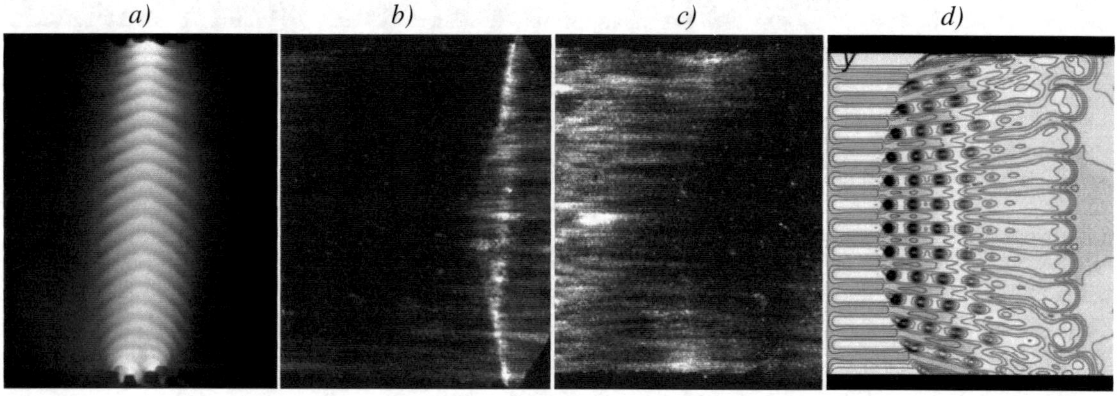

Figure 5. a) Photograph of an ionization-unstable discharge with small-scale strata; b) SW distortion when entering the discharge zone; c) Weakening the shock-wave configuration in the discharge zone; d) Computational density field, interaction of SW with small-scale temperature inhomogeneities

Numerical calculations also showed that the deposition of energy in small-scale temperature inhomogeneities in front of SW propagation leads to the destruction of the SW due to the generation of multipoint shear layer instabilities (Fig. 5d, $M = 2$).

5. Conclusions

1. The interaction of a plane shock wave with a region of stable and ionization unstable discharge characterized by strata of different scales has been investigated.
2. It was shown that the discharge has a strong effect on the shape and stability of a shock wave up to its complete destruction.
3. During the shock wave interaction with an ionized zone a new shock-wave configuration is formed including the shock wave and contact discontinuity which have been obtained experimentally; their formation has been confirmed numerically.
4. It was shown that the most effective impact on shock wave is when energy is dislocated in small-scale temperature inhomogeneities of the medium.
5. Numerical simulation of the flow under the assumption of temperature inhomogeneities corresponding to the experimental scale of strata has shown a good agreement with the experiments.

References

[1] Nemtchinov I. V., Popova M. P., Shubadeeva L. P. Influence of plasma motion on propagation of supersonic radiation waves - *Sov J Quantum Electron*. V.12. No 2. 1982. P. 258–260.

[2] Gridin A.Y., Klimov A.I., Khodataev K.V. Propagation of Shock Waves in a Nonuniform Transverse Pulsed Discharge - *High Temperature*. 1994. V. 32. No 4. 1994. C. 454.

[3] Znamenskaya, I., Kuznetsov, A., Mursenkova, I., Doroschenko, I. The motion of gas-dynamic discontinuities after the interaction of a plane shock wave with pulse discharge- *17th International Workshop on Magneto-Plasma Aerodynamics proceedings*. Joint Institute for High Temperatures of Russian Academy of Sciences. 2018. P. 32-37.

[4] Roupassov, D., Nikipelov, A., Nudnova, M., Starikovskii, A. Flow Separation Control by Plasma Actuator with Nanosecond Pulsed-Periodic Discharge - *AIAA Journal*, 0001-1452 V. 47. No. 1. 2009. P. 168-185.

[5] Khoronzhuk, R.S., Potapeko, D.P., Mashek, I.C., Karpenko, A.G., Lashkov, V.A. Microwave discharge initiated by double laser spark in a supersonic airflow - *Journal of Plasma Physics*. V. 81. No. 3. 2015. P. 905810307.

[6] Houpt, A., Hedlund, B., Leonov, S., Ombrello, T., Carter, C. Dynamic response of an oblique shock wave generated by a Q-DC discharge in pulse-repetitive mode - *48th AIAA Plasmadynamics and Lasers Conference*. 2017.

[7] Lapushkina, T.A., Erofeev, A.V. Supersonic flow control via plasma, electric and magnetic impacts - *Aerospace Science and Technology*. No. 69. 2017. P. 313-320.

[8] Lapushkina, T.A., Erofeev, A.V. Characteristics of the Effect of Low-Current Gas Discharge on a Strong Shock Wave - *Technical Physics Letters*. V. 43. No. 3. 2017. P. 241-243.

[9] Lapushkina, T.A., Erofeev, A.V., Azarova, O.A., Kravchenko, O.V. Passage of a Plane Shock Wave through the Region of a Glow Gas Discharge - *Technical Physics*. V. 64. No. 1. 2019. P. 38–45.

[10] Lapushkina, T.A., Erofeev, A.V., Azarova, O.A., Kravchenko, O.V. 2019. Interaction of a Plane Shock Wave with an Area of Ionization Instability of Discharge Plasma in Air.- *Aerospace Science and Technology*. No. 85. 2019. P. 347-358.

[11] Raizer, Yu.P. Gas discharge physics. *Springer*. 2001.

Journal of Physics: Conference Series **1394** (2019) 012014 doi:10.1088/1742-6596/1394/1/012014

Ignition of gaseous and liquid hydrocarbon fuel in supersonic flow by means of plasma generator

D A Yarantsev[1,*], A A Firsov [1], E V Dolgov[1], K V Savelkin[1], V A Talyzin[2], A A Nikolaev[2] and O V Voloshenko[2]

[1]JIHT RAS, 125412, Izhorskaya str, 13/2, Moscow, Russia,
[2]TsAGI, Zhukovskiy, Russia

*yarantsev.d@gmail.com

Abstract. The paper describes experimental research of the surface electrical discharge influence on the ignition and combustion of the gaseous and liquid hydrocarbon fuel in supersonic airflow without mechanical flameholders. Electrodes of plasma generator are integrated with injectors of gaseous fuel. Tests were carried out on the experimental setup T131B in TsAGI. Diagnostics includes schlieren visualization, measurements of pressure distribution and discharge voltage/current. It is shown that new scheme of plasma generation and pilot fuel injection is rather effective and provides ignition of liquid hydrocarbon fuel in hot supersonic airflow (T_0=750 K) and ethylene ignition in cold (T_0=300 K) supersonic airflow.

1. Introduction

Plasma assisted combustion in supersonic flow is considered as a promising technology for the future scramjet engines. Several labs all over the world carry out the researches in this direction. Detailed review of the electrically driven combustion (not only supersonic) was made in [1]. In particular, paper [2] describes experimental research and flow modelling of ethylene and JP-7 (liquid fuel) combustion in supersonic flow by means of direct-current (DC) constricted arc. Flames for both fuels were observed by means of planar laser-induced fluorescence. Ignition of hydrocarbon fuels by means of plasma torch in supersonic flow was studied in [3]. The recent study in ONERA [4] demonstrated the ignition of the hydrogen by means of multi-electrode quasi-DC discharge in cold (T_0=270 K) supersonic flow. Experiments with multichannel spark igniter are described in [5]. Igniter was mounted on the bottom wall of cavity in supersonic channel and demonstrated the capability of ethylene ignition.

Previous research showed [6-11] that surface electrical discharge can initiate ignition of ethylene and provide flameholding on the plane wall of supersonic duct T131B without mechanical flameholders. At the conditions of T131B experimental setup ethylene, as well as ethylene+liquid fuel ignition required simultaneous operation of the two plasma generators arranged along the flow. Preliminary experiments with new scheme of plasma generation, which includes anodes integrated with fuel injectors, on the experimental setup PWT-50 of JIHT RAS in the cold flow (T_0=300K) demonstrated higher efficiency of this method [10-11]. This study describes results of experiments and computational modelling for the same plasma generator on experimental setup T131B in TsAGI at higher flow parameters (pressure and temperature). Testing at the conditions of setup T131B is important, because this allows us to perform experiments in a hot flow and with liquid fuel.

Journal of Physics: Conference Series **1394** (2019) 012014 doi:10.1088/1742-6596/1394/1/012014

Parameters of the experiments on T131B of TsAGI are listed in the Table 1:

Table 1. Experimental parameters.

Parameter	Value
Mach number at the entrance	M =2.5
Static pressure inflow	Pst = 0.4-0.5 atm
Stagnation temperature	T_0 = 300 - 900 K
Duration of steady state mode	ΔT= 10-30 s
Duration of the fuel injection	ΔT_{fyel} = 0.2-4 s
Duration of the discharge generation	ΔT_{dis} = 0.1-0.4 s; single and pulsed periodic
Fuel type	hydrogen, ethylene, liquid hydrocarbons (LH)
Mass air to fuel ratio	λ = 12-18 for ethylene λ = 1-4 for LH
Electric power	W = 8-22 kW
Voltage of power source	up to 5 kW
Type of power supply	quasi-DC
Cross-section (h:w)	3:10, width=d

Arrangement of plasma generators and pilot fuel injection is presented in the Figure 1. Two plasma generators are mounted on the plain wall of the supersonic duct. The downstream generator consists of single ceramic insertion that forms the wedge and includes the raw of electrodes (cathode-anode-cathode…) that is transversal to the flow direction. The breakdown occurs between the neighbor electrodes but after that, discharge may also connect to the metal wall downstream the ceramic insertion. Operation of this generator is described in more detail in [7]. The upstream plasma generator has electrodes integrated with fuel injectors [10]. Each electrode is an anode and it is mounted in separate ceramic insertion with the duct wall being a cathode.

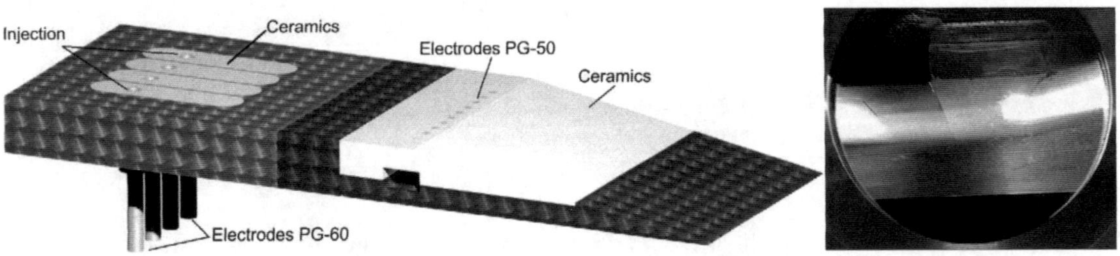

Figure 1. Scheme of plasma generation and discharge photo.

Liquid hydrocarbon fuel was injected using pylons installed on the distance = 3d upstream the first plasma generator integrated with gaseous fuel injectors. Such arrangement of pylons provides a good mixing of hydrocarbon fuel with airflow before its interaction with ethylene ignited by electrical discharge.

2. Numerical simulation

Numerical simulation of the flow was performed by means of FlowVision 3.10.02 CFD software [12] for the aerodynamic duct T131B. Flow simulation was made by solving 3D unsteady Navier-Stokes equations with modified k-ε turbulence model of FlowVision (URANS approach). Rectangular grid with local adaptation to the flow structure was used in simulation. The simulation project consists of the half of the aerodynamic duct. Boundary condition of zero gradient of the physical parameters was applied on the symmetry plane. Incoming flow conditions correspond to the T131B incoming flow with stagnation temperature 700K. In the first part of this work the influence of pylon for the liquid fuel injection on the supersonic flow structure was investigated. Flow around the pylon results in complex oblique shock wave structure, as it is shown in Figure 2. It provides an uneven pressure distribution with local maxima and minima of pressure along the test section. But such pressure distribution is undesirable for discharge operation because the local pressure maximum in the discharge breakdown area could prevent the normal discharge activation. Therefore, based on the calculation, the position of the discharge for future experiments was corrected so that the discharge initiation was organized in the region between two shock waves and in the absence of strong pressure gradients.

At the next step of simulation the variation of the energy release simulating fuel combustion was performed for obtaining the information about pressure distribution in the case of high completeness of combustion and further comparison with experimental one. Plasma-assisted combustion for cases of ethylene injection was modelled by means of volumetric heat sources. The geometry of such volumetric heat source is shown in the Figure 3. Calorific power of the ethylene and liquid fuel is about 48 MJ/kg; ethylene air to fuel ratio in experiments was about λ =16-12. It was assumed in modelling that completeness of combustion of the ethylene is 100% so the power applied to the volumetric heat source was about 330 kW. Mach number distribution for described case is presented in Figure 3. It is clearly seen that high energy release caused by combustion should result in formation of subsonic separation regions in the duct.

Figure 2. Grid and shock wave structure in the test section behind the pylon

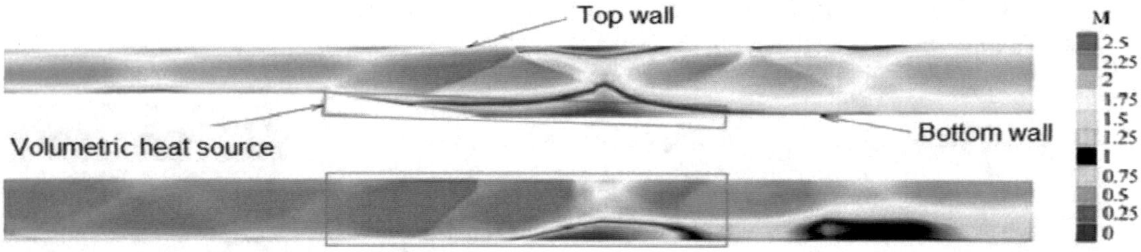

Figure 3. Mach number distribution at the ethylene combustion (front and top view).

Simultaneous combustion of liquid fuel and ethylene was simulated by the volumetric heat source with a higher volume. In this case the heat source was the same length as at ethylene only combustion simulation but occupied the entire cross section. Equivalent total air to fuel ratio was in the range of λ =5-1.7. Based on the simulation results the following conclusions can be made: separation propagation during combustion occurs from the sidewalls due to the rectangular shape of the channel. Analysis of Mach number distributions for different energy releases that are corresponding to the different fuel flow rates show that the flow has a supersonic or transonic core down to λ =2.5. At lower air to fuel ratio the flow become subsonic close to location of the plasma generator. And thermal choking of the duct occurs if the air to fuel ratio is less than λ=1.7.

3. Experimental data

3.1. Experiments in the cold flow

Ignition of the ethylene was achieved by means of upstream plasma generator at the air to fuel ratio λ less than 14 and discharge power about 15 kW for the conditions of the cold flow (T_0 =300 K). Dynamics of the static pressure downstream the wedge is presented in the Figure 4. Ignition causes the pressure rise within the discharge region. As a result, discharge becomes unstable, several plasma filaments (of four) stop operating (see Fig.4, U_2 voltage) in this configuration. The flame is gradually blown away by a flow and combustion does not reach the stationary level.

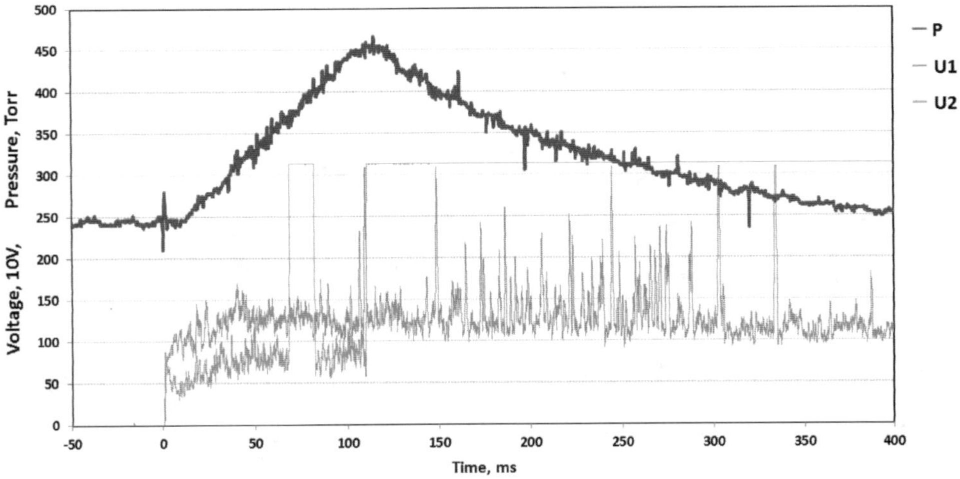

Figure 4. Dynamics of static pressure downstream the wedge at ethylene injection λ_{C2H4} = 12 and voltage of discharge filaments. Saturation of discharge voltage corresponds to the shutdown of discharge filament operation.

Figure 5 presents schlieren visualization of the flow structure without discharge, with discharge and during the combustion. Regions disturbed by the discharge and combustion are emphasized on the photos by the magenta line

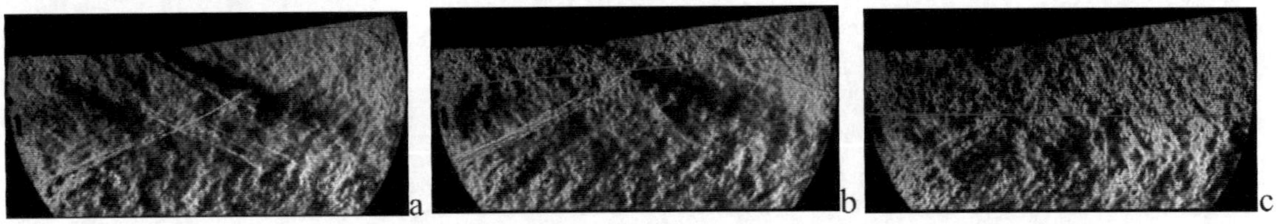

Figure 5. Schlieren photos of the flow structure: a – supersonic flow M=2.5; b – discharge; c – ethylene combustion. λ_{C2H4} = 12.

3.2. Experiments in the hot flow

Hot flow experiments were made for the T_0 = 750 K. Ethylene was injected through the upstream plasma generator and liquid fuel was injected through the one or two pylons that were mounted upstream the plasma generator.

Single plasma generator occurs to be insufficient to ignite the liquid fuel, therefore experiments with liquid fuel were carried out with two plasma generators. Two central anodes of the each generator were used during simultaneous operation of two generators. Total discharge power remains unchanged but the length of plasma area along the flow is increased that leads to the increasing of the plasma-fuel interaction.

At the ethylene injection the air to fuel ratio was varied within the range of λ =18÷12, and for liquid hydrocarbon λ was from 4 to 1. If liquid hydrocarbon air to fuel ratio was less than 2.5 two pylons for injection were used. Liquid fuel ignition was obtained for the whole range of flow rates for the discharge power about 15 kW and T_0 = 750 K. Reducing the liquid fuel and ethylene flow rate caused just to the reducing of the pressure rise due to the combustion, as it is seen in the Figure 6.

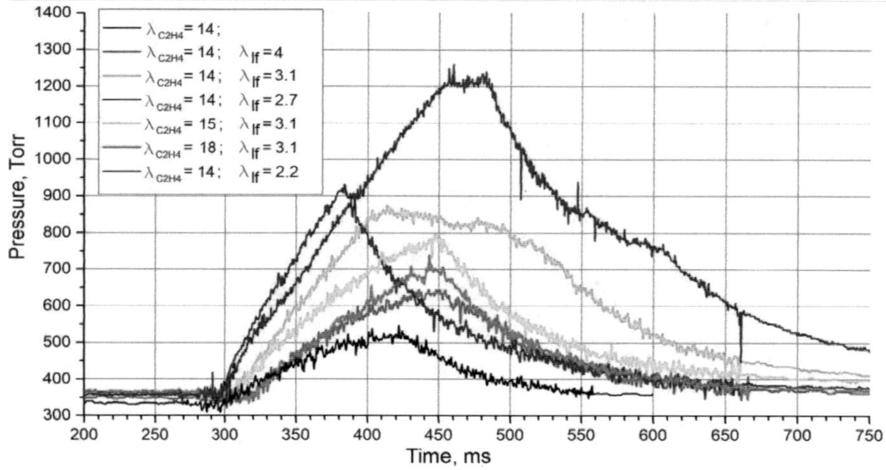

Figure 6. Static pressure dynamics for different fuel flow rates of C_2H_4 and liquid fuel.

Liquid fuel air to fuel ratio less than λ =1.7 caused to the massive flow separation within the discharge area according to the static pressure distribution along the channel. Thermal choking of the aerodynamic duct occurs at the λ =1. If liquid fuel injection was less than λ =2.2 than combustion becomes unstable with pressure pulsations that influence on the discharge operation, Figure 7. Detailed analysis of voltage measurements for all plasma channels allows concluding that downstream plasma actuator is shutting down first and ~2ms later the upstream actuator is shutting down too. After ~2ms of downtime the upstream actuator resume the operation, and ~3ms later the downstream actuator resume the operation too and then normal operation of all plasma filaments takes place about 5ms. The frequency of oscillations was about 120Hz. Such dynamics of discharge operation corresponds to the pressure wave motion in combustion chamber.

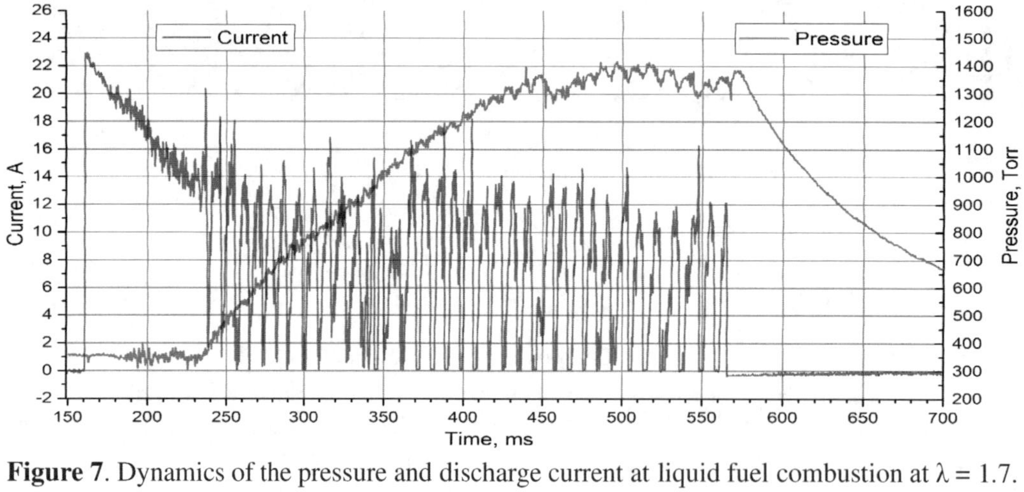

Figure 7. Dynamics of the pressure and discharge current at liquid fuel combustion at λ = 1.7.

Comparison of modelled and experimental pressure distributions in Figure 8 shows that completeness of combustion in experiments may be estimated as 25-30 %. Nevethereless, it should be noted that position and volume of the volumetric heat source in modelling was not changed depending on fuel mass flow rate.

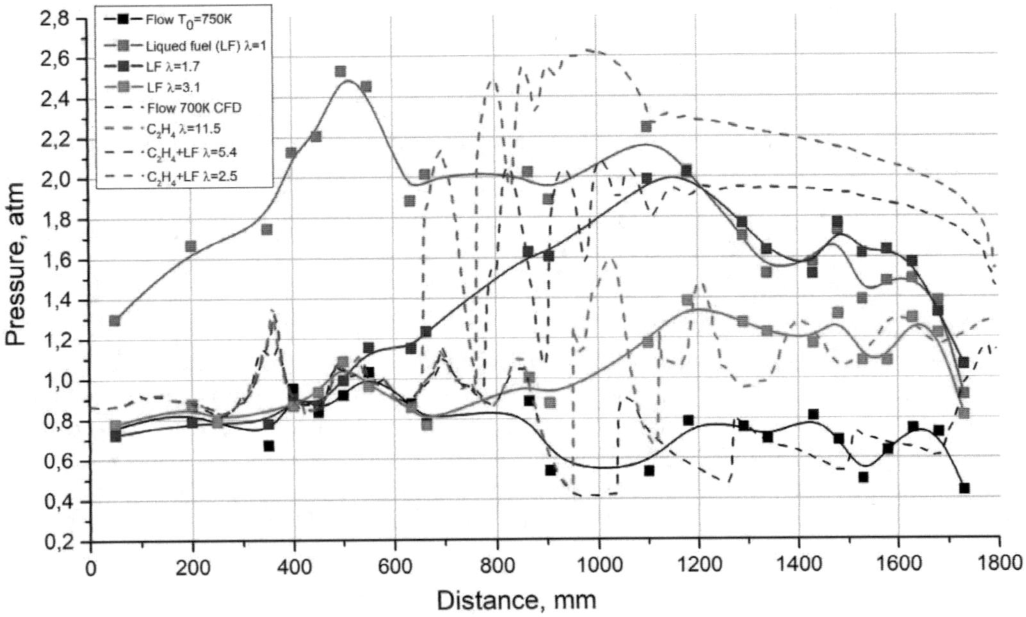

Figure 8. Static pressure distributions: dash lines – modelled; cont. lines – experimental.

4. Conclusion

The conclusion can be made that for the first time the new scheme of the discharge-fuel interaction based on integration of the electrodes and fuel injectors allowed to ignite the ethylene without using of additional plasma actuator on the T131B testbed at low temperature conditions (T_0 =300 K, P_{st}=0.4 atm). Future optimization of this approach may allow reducing the length of the supersonic combustion chamber. Nevertheless, in the hot flow (T_0 =750 K) ignition of the liquid hydrocarbons causes the significant pressure increase and unstable discharge operation, as a result, an additional plasma generator is needed in order to sustain the combustion. Mechanisms of the interaction between the discharge and gas jet within the supersonic flow are a subject of further research.

Acknowledgments
The work was supported by TsAGI-RAS program. The authors are grateful to Prof. S.B. Leonov for a useful constructive discussion of the results presented.

References

[1] Leonov S B, Electrically driven supersonic combustion. *Energies* **2018**, *11*, 1733.
[2] Jacobsen L, Carter C, Baurle R, Jackson T, Williams S, Bivolaru D, Kuo S, Barnett J and Tam C-J, Plasma-Assisted Ignition in Scramjets, *J. Propuls. Power* **2008**, *24*, 641–654.
[3] Takita A, Shishido K and Kurumada K, Ignition in a supersonic flow by a plasma jet of mixed feedstock including CH4. *Proc. Combust. Inst.* **2011**, *33*, 2383–2389
[4] Vincent-Randonnier A, Packan D, Sabelnikov V, LeJouan F, Rouxel B, Roux P and Leonov S, First Experiments on Plasma Assisted Supersonic Combustion at LAERTE Facility; AIAA Paper 2017-1975; AIAA: Reston, VA, USA, **2017**.

[5] Shengfang Huanga, Yun Wub, Huimin Songa, Jiajian Zhuc, Zhibo Zhanga, Xiliang Songc, Yinghong Lib, Experimental investigation of multichannel plasma igniter in a supersonic model combustor, *Experimental Thermal and Fluid Science*, **2018**, *99*, 315-323.
[6] Leonov S; Carter C and Yarantsev D, Experiments on Electrically Controlled Flameholding on a Plane Wall in Supersonic Airflow. *J. Propuls. Power* **2009**, *25*, 289–298.
[7] Yarantsev D A, Firsov A A, Chernyshev S L, Nikolaev A A and Talyzin V A, Ignition and flameholding of hydrocarbon fuel in supersonic flow by means of surface electrical discharge, *AIAA SciTech Forum*, **2019**, San Diego, California
[8] Yarantsev D A, Leonov S B, Firsov A A and Savelkin K V, Experiments on Hydrocarbon Fuel Ignition by Means of Electrode Discharge in Supersonic Combustor without Mechanical Flameholders, *16th International Workshop on Magneto-Plasma Aerodynamics*, April 5-7, **2017**, Moscow
[9] Savelkin K V, Yarantsev D A and Leonov S B, Experiments on Plasma-Assisted Combustion in a Supersonic Flow: Optimization of Plasma Position in Relation to the Fuel Injector, *Journal AerospaceLab*, **2015**
[10] Firsov A A, Savelkin K V, Yarantsev D A and Leonov S B, Plasma-enhanced mixing and flameholding in supersonic flow. *Phil. Trans. of the Royal Society A*, **2015**, 373.
[11] Savelkin K V, Yarantsev D A, Adamovich I V and Leonov S B, Ignition and flameholding in a supersonic combustor by an electrical discharge combined with a fuel injector. *Combustion and Flame,* **2015***, 162*, 3, 825-835.
[12] Zhluktov S V, Aksenov A A, Karasev P I, Simulation of separation flow using two-parametric turbulence model] // Computer Research and Modeling, **2016**, V. 8, N.1, P. 79–88

Journal of Physics: Conference Series **1394** (2019) 012015 doi:10.1088/1742-6596/1394/1/012015

Experimental study of high-speed ionized gas flow around plate and wedge model

V P Fomichev, M A Yadrenkin, E K Shipko

Khristianovich Institute of Theoretical and Applied Mechanics, 630090 Russia, Novosibirsk, Institutskaya str., 4/1

fomichev@itam.nsc.ru

Abstract. In the development of magneto-plasma methods for controlling the aerodynamics of promising high-speed aircraft, an important place is occupied by studies of the interaction of ionized gas flows with a shock-wave flow structure. In this work, the flow near a model (a wedge and a plate) by a pulsed stream of air plasma is investigated. As a result of the work, it was shown that a change in the local Mach number in the stream ionized by an electric discharge leads to a change in the shape of the shock wave generated by the leading edge of the model. The temperature of the discharge plasma was estimated, which showed the predominance of the thermal effect on the flow structure.

Experimental techniques

An experimental study of the flow near a plate and a wedge with a high-speed ionized air flow was carried out on the magnetohydrodynamic (MHD) test rig of the ITAM SB RAS [1]. This test rig based on a shock tube allows to obtain a high-speed gas flow using various gases, in this work – air. Figure 1 shows the installation diagram. The low pressure channel 3 (LPC) is filled with the work gas. The high-pressure chamber 1 (HPC) is filled with a high-pressure pushing gas — helium under the test conditions. Quasi-stationary conditions are formed behind the reflected shock wave at the entrance to the nozzle 5 to obtain a flow of working gas at the nozzle exit with the necessary quasi-stationary parameters for 1–2 ms. A nozzle on the Mach number M = 6 was used in the work.

Visualization of the ionized flow pattern near the models was carried out by an optical schlieren-system with a high-speed camera. The optical scheme with an adaptive visualizing transparency as a Foucault's knife [2] made it possible to register both the wave pattern of the flow of models in the green light of a laser and the radiation from the plasma part of the stream.

The flow was ionized using a pulsed high-voltage (3 kV) discharge initiated between the electrodes installed in front of the model. The discharge pulse duration was 20 and 120 μs, depending on the experimental conditions. Measurement of discharge parameters was carried out using current transformers without galvanic coupling of the object being measured with oscilloscopes. Measurements were made of all electrical quantities that determine the required mode of operation of the installation.

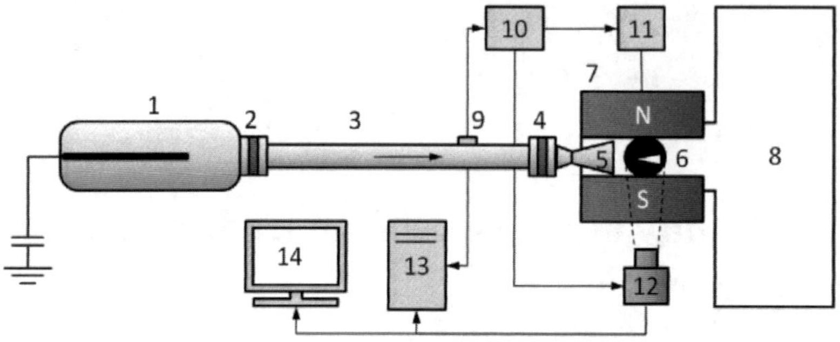

Figure 1. Scheme of experimental test rig. 1 - HPC with power supply, 2 - diaphragm, 3 - LPC, 4 - diaphragm; 5 - nozzle, 6 - working chamber, 7 - magnetic system, 8 - receiver, 9 - piezoelectric pressure sensors, 10 - pulse generator, 11 - power of ionizing devices, 12 - optical system, 13 - diagnostic complex, 14 - monitor.

Experiment statement

Case of air flow with Mach number M = 6 were considered. The distance from ionizing flow electrodes to the leading edge of the models varied from 5 to 20 mm depending on the experimental conditions. Fig. 2 shows the experimental scheme with a flow around a plate and a wedge.

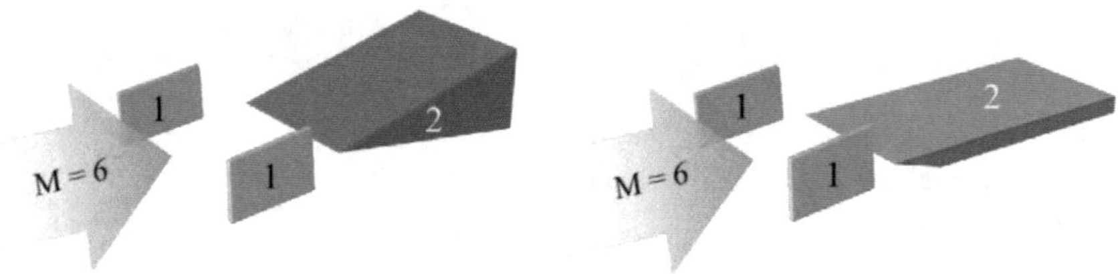

(a) (b)

Figure 2. Scheme of MHD interaction in flow near wedge and plate model: 1 - electrodes, 2 - model.

Experimental results

1.1. Ionization of a flow before a wedge by a short pulse discharge

To clarify the effect of heating the flow during gas ionization, an experiment was conducted with a short discharge time according to the scheme shown in Figure 2,a. When the flow was ionized by a pulse discharge with a duration of 20 μs, the distance from the electrodes of the ionizing discharge to the leading edge of the model was 5 mm. The width of the electrodes is $b = 15$ mm, length is $l = 25$ mm. The flow passed the electrodes in 12.5 μs with rate of about 2000 m/s, almost ending at the time of passing the discharge through the electrodes space. At the same time, the heated region continued to move at an average flow rate. Figure 3 shows three schlieren-photographs of a flow around a wedge of 50 mm length. A flow was taken at a frequency of 330000 frames per second. The frame exposure time was 1.2 μs. Fig. 3 shows that as soon as the gas cools between frames *b* and *c* in 3 μs, the flow pattern of the upper edge of the wedge with a continuous shock wave corresponding to the change in the Mach number of the flow is restored.

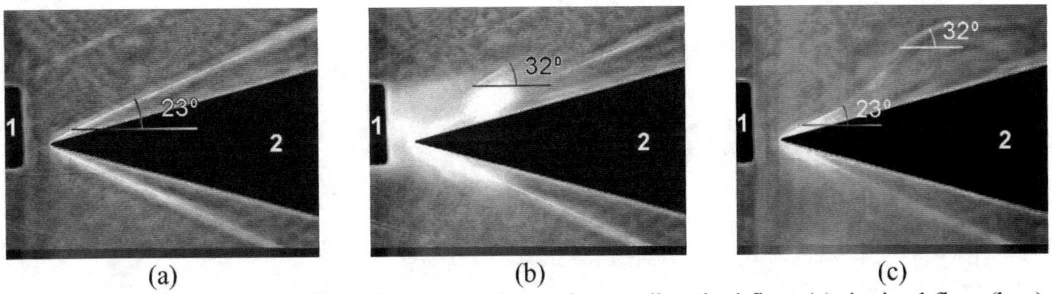

Figure 3. Schlieren-photos of the flow around a wedge: undisturbed flow (a), ionized flow (b, c). 1- electrodes, 2- wedge.

The slope angle the oblique shock wave in the undisturbed flow, Figure 2,a, is 23 degrees, which corresponds to the flow around the wedge with Mach number M = 5.8. The slope of the shock in the central part of the MHD-interaction region, Figure 2,b and Figure 2,c, is 32 degrees, which corresponds to the Mach number M = 3 in this part of the flow. You can see that the slope angle of the shock in the middle part of the flow persists when the disturbed area passes over the wedge surface. In this case, in the flow region near the leading edge of the wedge, after passing through the disturbed zone, the angle of inclination of the shock is restored to the undisturbed value. Comparison of two consecutive frames, the time between which is 3 μs, allows us to conclude that the flow recovery time under the experimental conditions does not exceed 3 μs. These experiments showed that the plasma flow acts as a well-heated gas. The gas heated by an electric discharge leads to the modulation of the local Mach number under the test conditions.

1.2. Ionization of a flow before a wedge by a long pulse discharge

With a longer discharge time (120 μs) the temperature and conductivity of the flow were maintained during the ionized gas flows around the wedge. This led to a change in the shock wave structure of the flow. Figures 4 and 5 show several frames (frame rate of 160 000 fps) of plasma flow over the wedge and the plate. From the pictures in Fig. 4 and 5 it can be seen that the plasma flow destruct the shock wave generated by the leading edge of the models. The shock is restored after passing through the plasma region (Fig. 4,d-f). When the discharge plasma is localized under the shock wave formed in the cold flow, a change in the shock wave angle is observed, Fig. 4,e and Fig. 4,f. This suggests that in the observed flow patterns, the plasma flow acts as the gas heated up to a high temperature.

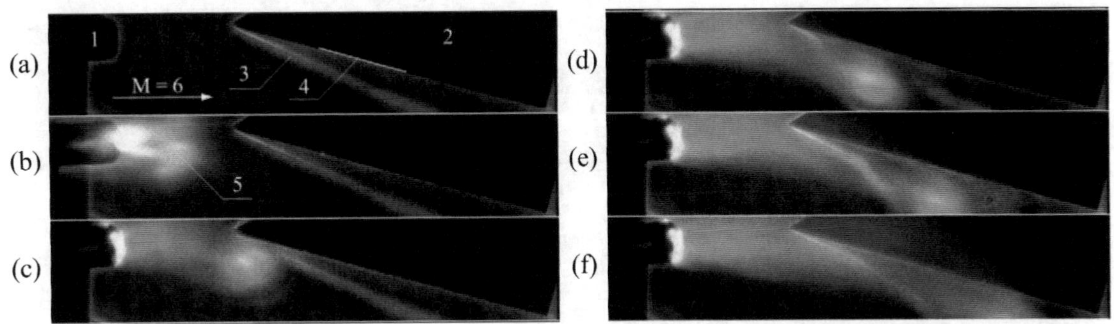

Figure 4. Hypersonic air flow ionized by pulsed electrical discharge around wedge model: 1 – electrodes, 2 – model, 3 – shock, 4 – measuring electrodes area, 5 – discharge plasma.

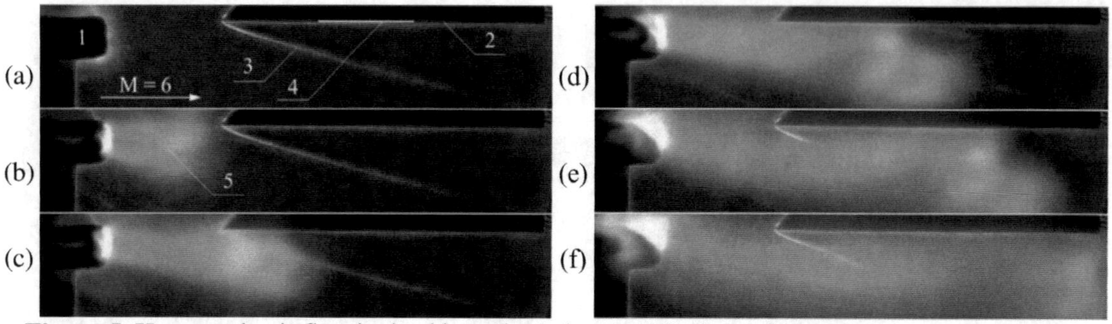

Figure 5. Hypersonic air flow ionized by pulsed electrical discharge in accelerating magnetic field around plate model: 1 – electrodes, 2 – model, 3 – shock, 4 – measuring electrodes area, 5 – discharge plasma.

Fig. 5 shows hypersonic flow with electrical discharge near the model in magnetic field ($B = 0.17$ T). Direction of discharge current and B-field induction leads to accelerating the gas downstream. One can see that the shock wave on the leading edge of the plate is destructed in the plasma region, Fig. 5c, and is not generated again when it moves further along the plate, Fig. 5,d. With further advancement of the plasma part above the plate surface in the cold part of the flow, the shock wave is restored, Fig. 5,e, Fig. 5,f.

Discussion and conclusions

To clarify the thermal effect on the flow by an electric discharge, the determination of the gas temperature in the plasma part of the flow was performed. The temperature was determined from the plasma radiation power using the “gray body” radiation model, with the emissivity taken from table [3]. To determine the plasma emissivity, we used the static pressure in the flow and the thickness of the radiating plasma layer, which was assumed to be equal to the distance between the electrodes, Fig. 2. After exiting the area between the electrodes, this part of the plasma flow was connected to the electrodes with relatively thin, in the direction of the flow, parts of the electric discharge arc directed along the flow, Fig. 5,c-f. Therefore, the acting plasma region of the flow was only that volume in which the current flowed crosswise the gas flow. Long parts of the arc discharge in flow direction are visible, but between them there was a cold gas with the temperature of the incoming flow, Fig. 5,d-f and Fig. 6,e,f.

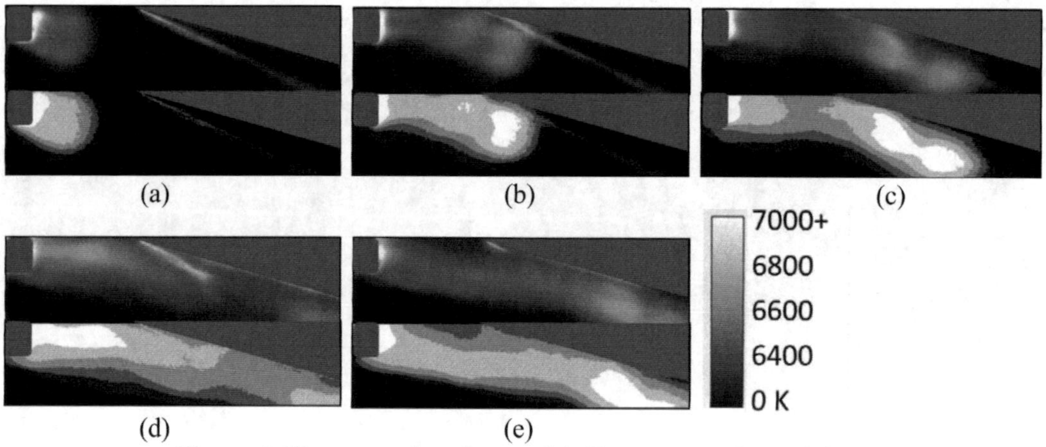

Figure 6. Flow around wedge model. Temperature determining.

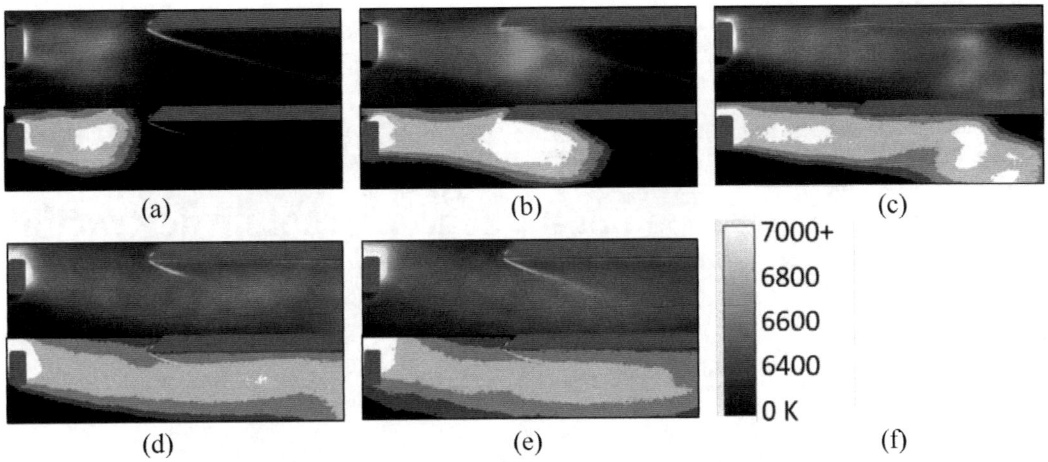

Figure 7. Flow around plate model. Temperature determining.

Fig. 6 and Fig. 7 shows the temperature of the plasma region of the flow near the wedge and the plate. The plasma temperature can be estimated as 6500-7000 K, which is corresponds to the temperature for low-pressure arc discharge [4]. According to the temperature of the plasma flow, it was obtained the sonic rate of about 2100 m/s. The flow velocity determined by the photography was about 1900 m/s. The Mach number of this part of the flow was about 0.8 – 0.9. Since the flow was subsonic, a shock did not form. It can be assumed that this was the main effect of the ionization of the flow of electric discharges under the conditions of this experiment.

Acknowledgments
The research has been carried out at the expense of the grant from the Russian Science Foundation (project No. 17-19-01375).

References

[1] Korotaeva, T.A., Fomichev, V.P., Shashkin, A.P. et al. Tech. Phys. (2011) 56: 327. https://doi.org/10.1134/S1063784211030108
[2] Shevchenko, A.M., Golubev, M.P., Pavlov, A.A. et al. Tech. Phys. Lett. (2017) 43: 473. https://doi.org/10.1134/S106378501705025X More references
[3] Biberman L.M. *Opticheskiye svoystva goryachego vozdukha (Optical properties of hot air).* Nauka Publishing House, Moscow, 1970.
[4] Fridman A. A., Kennedy L. A. *Plasma Physics and Engineering.* Taylor & Francis Routledge. 2004. p. 500.

Journal of Physics: Conference Series **1394** (2019) 012016 doi:10.1088/1742-6596/1394/1/012016

Temperature distribution in the zone of graphite destruction by a nitrogen plasma jet

Kavyrshin D.I. [1]*, Korshunov O.V. 1, Sargsyan M.A. 1, Chinnov V.F. [1], Scherbakov V.V. [1]

1 Joint Institute for High Temperatures of the Russian Academy of Sciences (JIHT RAS) 125412, Izhorskaya st. 13 Bd.2,Moscow, Russia

*dimakav@rambler.ru

Abstract. Using the automated system of high-speed visualization and spatial-temporal spectral diagnostics, the destructive effect of an atmospheric pressured nitrogen plasma jet with a temperature of 0.7 - 1.0 eV on samples of isotropic and anisotropic graphite was studied. Measurements of the electron, vibrational, and rotational temperatures of the plasma in the zone of destructive interaction were performed, and the pattern of the spatial-temporal distribution of the products of graphite destruction (carbon atoms and cyanogen molecules) was revealed. The efficiency of local spectral analysis of near-surface plasma is shown using a combination of longitudinal and transverse periodic scanning of the emission spectra and recording of instantaneous 2D spectra of the selected spectral region containing the CN and N_2^+ bands. The spectral estimate of the concentration of the graphite destruction main product (carbon atoms) is in quantitative agreement with the results of measuring the mass loss rate of the sample material, made by two-positioned visualization of its surface. In the zone of realizable heat loads of 0.5-1.5 kW/cm^2, heating of the graphite surface to temperatures of 2800 ÷ 3600 K was achieved and the rate of loss of its material was 3-15 mg/cm^{2}s, which is consistent with the literature data.

1. Introduction.

Presently, modelling of spacecraft atmospheric re-entry conditions is performed in Russia and in the world on large-scale and mock-up installations in close cooperation with developers of new heat-resistant coating materials with required properties [1, 2]. Work [3] studies interaction between high-velocity nitrogen plasma jet with samples of composite carbide-silicon material with heat-resistant coating. Studies of carbon material ablation in chemically active environments shows the principal role of graphite oxidation processes in border layers and on the surface of the heat-resistant material [1, 3, 4] as well as that of carbon nitriding process [5]. The works of particular interest are experimental works dedicated to the roles of nitrogen plasmachemistry processes in decomposition of heat-resistant materials which have been published in recent years [2, 3, 5].

This paper presents the results on spatial and temporal optical emission spectra diagnostics of interaction region of air pressure nitrogen plasma jet impacting graphite surface. The plasma is generated by a DC electric arc plasmatron with nozzle exit temperature T_∞= 10000 ÷ 12000K interacts with a graphite surface heated by the jet to $T_W \leq 3600$ K.

2. Measuring system.

A slightly divergent ($2\alpha = 12^0$) plasma jet with a diameter of 5 to 10 mm with an enthalpy of 10–100 kJ/g is formed at the output section of the plasma torch with a plasma-forming gas flow rate of 1–3 g/s and an average mass plasma temperature at the outlet of 10000– 20000 K. From the expanding anode channel of the plasma torch a high-enthalpy plasma jet flows into the free space of air at atmospheric pressure. Samples of graphite are mounted on supporting tungsten rods and are located at the front of the jet at a distance of 15-25 mm from the outlet of the plasma torch.

Automated measuring complex includes (see Fig.1) the following measuring systems.

Synchronized video recording system for the interaction of incident plasma flow with sublimating heat-resistant sample, consists of high-speed cameras Motion Pro X3 (USA) and VS-FAST (Russia) and ultra-high-speed (with a frame rate of up to 100000 s-1) camera Phantom (USA). In combination with long-focus lenses with a system of camera extension rings, the image of the "jet-sample" interaction zone is recorded at a selected scale during the entire observation period (usually 60–200 s).

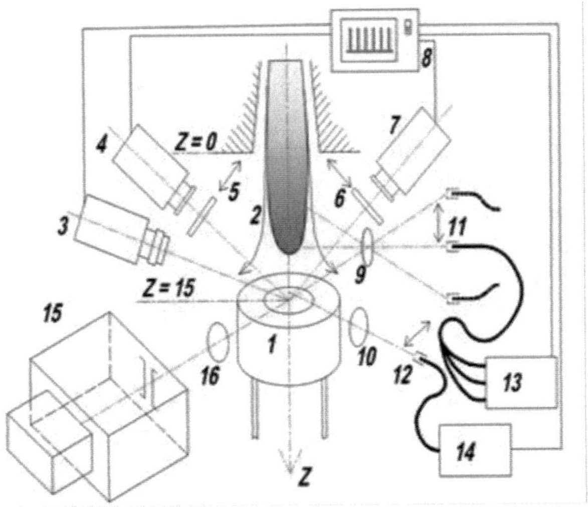

Figure 1. Schematics of the diagnostics set-up. Designations: 1 - sample, 2 - plasma jet, 3.4 - high-speed cameras, 5.6 - interference filters, 7 - ultra-high-speed camera, 8 – control system, 9,10,16 - condensers, 11, 12 – optical fibers, 13 - AvaSpec 3648 spectrometer, 14 - AvaSpec 2048 spectrometer, 15 - MS-257 spectrometer with Andor camera.

To use these cameras as a high-speed micro-micrometer, in front of the lens, interference filters with a bandwidth $\delta\lambda_{1/2}$ = 10–12 nm are introduced before a lens.

Simultaneous video recording of the interaction region by two high-speed cameras provide the means to determine in real time the loss rate of the sample material and the temperature of its surface [6].

Spectral measurements are performed using a three-channel and a single-channel fiber-optic spectrometers AvaSpec2048 and AvaSpec 3648. A single-channel fiber-optic spectrometer AvaSpec 3648 with a spectral range of 220–1100 nm and spectral resolution of about 1 nm monitors the radial distribution of plasma jet radiation from the near-surface plasma, during the continuous transverse optical fiber guide displacement. Measurements provide control over the temporal change (with a frequency of 2 spectrum/s) of the plasma chemical composition and plasma parameters in the interaction zone ($Z \approx$ 1–5 mm above the sample surface). AvaSpec 2048 three-channel fiber optic spectrometer (spectral range 220–1000 nm) with a spectral resolution of 0.2–0.5 nm records plasma emission along the longitudinal coordinate of the plasma flow incident on the sample. Spatial-temporal changes in plasma emission spectra are monitored by scanning the plane of the intermediate sharp image of a plasma jet formed by condensers 9, 10 (Fig. 1) with spectrometers optical fiber guides.

Journal of Physics: Conference Series **1394** (2019) 012016 doi:10.1088/1742-6596/1394/1/012016

3. Differences in plasma jet flow patterns around isotropic and pyrographite samples

The investigated samples of isotropic graphite MPG-6 had a cylindrical shape, a diameter of 20 mm and a length of 15–30 mm and were turned to a free-flowing jet with a diameter of 6-10 mm either with its front or side surface. Samples of pyrographite (Scientific Research Institute Grafit) - the parallelepiped 20x12x4 mm^3 are facing the jet with a plane 20x12 mm^2. The systems of longitudinal and transverse scanning of the emission spectra of a near-surface plasma record the radiation from extended plasma regions that have a complex movement pattern in the stagnation zone. As shown by the measurements, the possibilities of recording spatial and temporal changes in the radiation intensity of the main products directly in the zone of destruction are different for these types of graphite.

In isotropic samples, the formation of a crater in the stagnation zone and an increase in its depth leads to the appearance of a “stagnant” flow region with a large residence time of plasma particles and degradation products in the crater, which increases the efficiency of plasma processes involving C and CN. In this case, the registration of the spectral composition of the near-surface plasma itself is possible only in the initial period of exposure, while the crater depth is small $h \leq 1$ mm. After that, the near-surface plasma region is blocked by the crater edge, and the observed spectrum characterizes the state of the plasma above the edge, relative to which the bottom of the crater is diminishing with time. For this reason, the registered $2D$ spectra can be attributed to the near-surface plasma only during the first 10–15 seconds of plasma exposure to an isotropic sample.

In experiments with pyrographite, the screening effect of the edge is practically not manifested. Plasma deceleration occurs on a flat surface of 20x12 mm^2, the role of a shallow crater is small, and the recorded spectra carry information about the near-surface plasma. And although this spectral intensity is integrated along the line of sight of the optical recording system, a simultaneous longitudinal scanning of the spectra across the plasma jet allows us to correlate the longitudinal scanning spectra with the parameters of the axial plasma region in the corresponding z coordinate.

4. Spatial-temporal scanning of the emission spectra of the plasma and the destroyed sample

Spatial-temporal changes in the intensity of the radiation of the main radiating components of the incident plasma flow (NI, N_2, N_2^+) and the products of its effect on the samples (CI, CN*) were obtained (Fig. 2). The plasma emission spectra in the interaction zone in the local thermodynamic

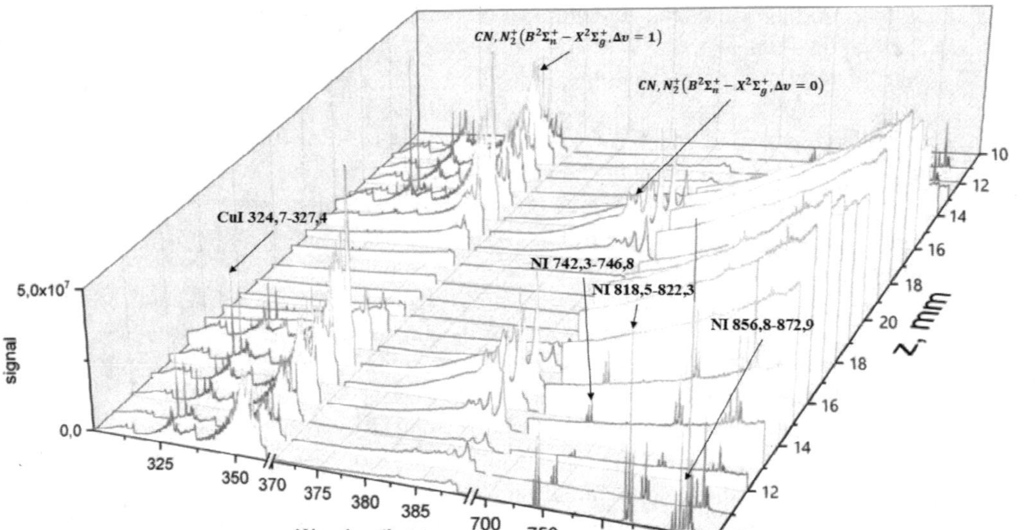

Figure 2. 3D - spectra of the nitrogen plasma jet radiation at different distances from the sample surface (mm) in the quasi-stationary period of its heating. Arc current 300 A.

equilibrium approximation (which is valid under the experimental conditions) established the pattern of spatial-temporal temperature variations of the nitrogen plasma electrons, as well as the nature of changes in the vibrational and rotational temperatures of the CN, N_2 molecules and the N_2^+ ion during heating and destruction of the samples.Fig. 3 (isotropic graphite) and Fig. 4 (pyrographite) show the dynamics of changes in the radiation intensity of the most important radiating components of plasma: spectral lines of atomic nitrogen NI (measurement of electron plasma temperature), atomic line of carbon CI 247.8 nm (indicator of "blowing" of the samples destruction material into the stream), the molecular spectrum of the radical CN (the main product of interaction of C with nitrogen plasma), as well as continuous thermal radiation of the sample (with a temperature at its surface T_W~3000 K) in the region with a wavelength of 775 nm free of the spectrum lines.

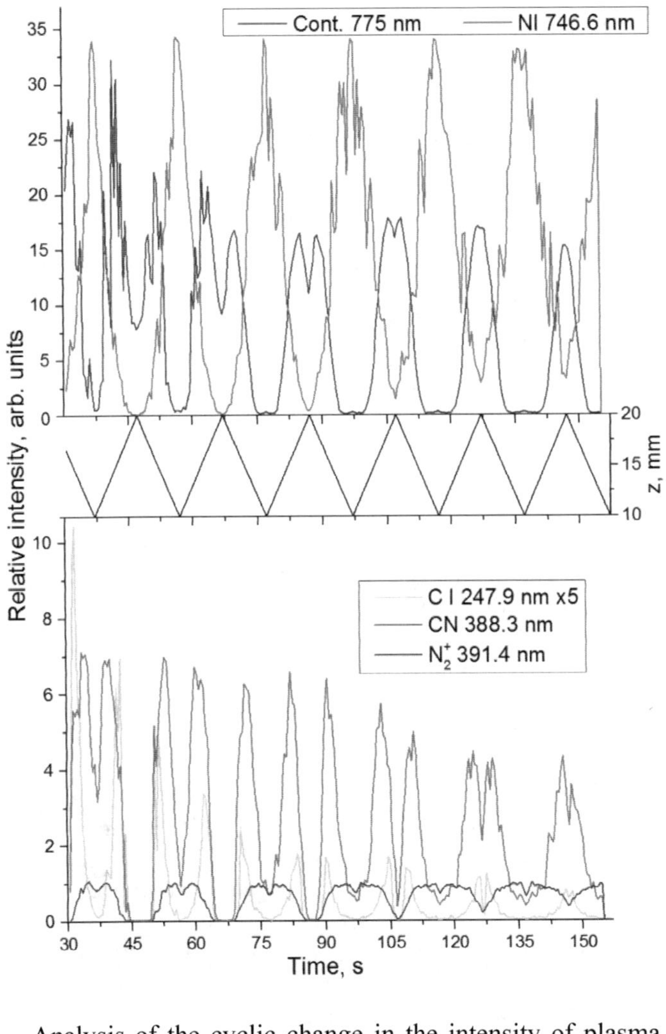

Figure 3. Cyclograms of changes in the intensity of the atomic nitrogen line, the thermal continuum (above), and the intensity of the atomic carbon line, the molecular bands of CN and N_2^+ (below) in the interaction zone. The sample is MPG-6 graphite, with a diameter of 20 mm and a height of 30 mm. Arc current 300A. A sawtooth insert between cyclograms establishes a connection between the longitudinal coordinate z and the moment of impact.

Analysis of the cyclic change in the intensity of plasma radiation acting on an isotropic sample (Fig. 3) shows that the atomic line NI reaches its maximum values near the plasma nozzle, at the greatest distance from the target, at z = 10 mm, and minimum - at z = 20 mm. At the same time, the intensity of the samples surface emission reaches its maximum at its upper edge with the initial coordinate z ≈ 15 mm. As the sample warms up and its mass decreases, this maximum (corresponding to the coordinate of the most heated sample surface at the stagnation point) slowly shifts down from cycle to cycle along the z axis, up to z ≈ 17-18 mm.

Journal of Physics: Conference Series **1394** (2019) 012016 doi:10.1088/1742-6596/1394/1/012016

The spectral lines of atomic carbon served as an indicator of the main primary product of the destruction of graphite in our study. Other carbon-containing particles in the gas phase (C_2, C_3 ... C_8) were not detected in the spectra. The cyclograms show the change in the region of the interaction of the intensity of the CI line at 247.8 nm with the excitation energy E* = 7.68 eV and the transition probability $3.4 \cdot 10^7$ s^{-1} [7]. The maximum of its radiation, which monotonously decreases with exposure time due to the gradual lowering of the crater bottom surface (where the actual concentration of atomic carbon is maximum) from the upper edge of the sample, is near this edge. The secondary product graphite destruction is the radical CN, the change in the radiation intensity of which is also shown in the cyclograms.

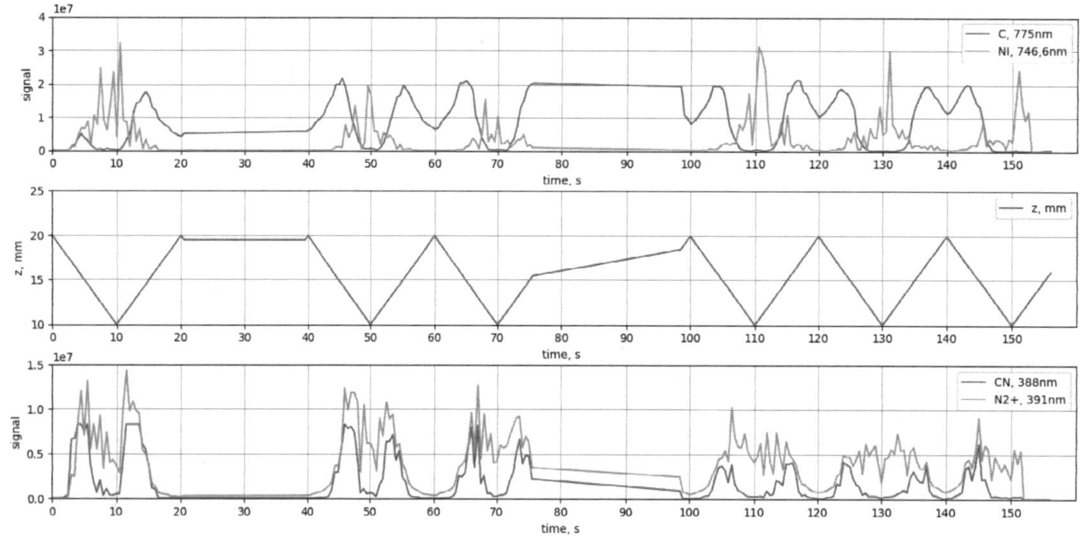

Figure 4. Cyclograms of changes in the intensity of the atomic nitrogen line, the thermal continuum (top), and the intensity of the atomic carbon line and the molecular bands of CN and N_2^+ (bottom) in the interaction zone. The sample is pyrographite. Arc current 150A.

It is noteworthy that a quasi-stationary mode of samples heating is established after 20-30 s of continuous plasma effect on samples, during which the temperature of the sample surface facing the jet for a long time (up to 150 s) remains unchanged.

5. Results of 2D diagnostics of the boundary layer.

The best possibilities of recording the intensity of radiation with high spatial resolution are provided by system 15 (Fig. 1), consisting of an MS-257 spectrometer and Andor matrix camera. To record the 2D spectra of the near-surface interaction region, a sharp image of the stagnation region with a vertical length of about 6 mm is projected onto the entrance slit of the spectrometer, capturing a plasma jet (~ 4 mm) and 1-2 mm of the samples side surface. When using a diffraction grating of 1600 lines/mm a matrix with a length of 25 mm (1200 pixels), accommodates a portion of the spectrum 40–45 nm in length, which is recorded with a resolution of better than 0.1 nm (see Fig. 5).

New possibilities for analyzing the state of the plasma in the interaction zone are provided by the electron-vibrational-rotational structure of the CN spectrum, whose simulation allows us to estimate both the vibrational T_v and the rotational T_r temperatures of the CN molecules. This is even more important because in the near-surface region, where a plasma jet with $T_e \approx 7000$ K runs onto the surface with $T_w \approx 3000$ K, T_v and T_r may not coincide.

As a result of processing of numerous spectra, an averaged pattern of changes in the vibrational T_v and rotational T_r temperatures in the near-surface region of pyrographite interacting with a nitrogen plasma jet was obtained, as shown in Fig.6.

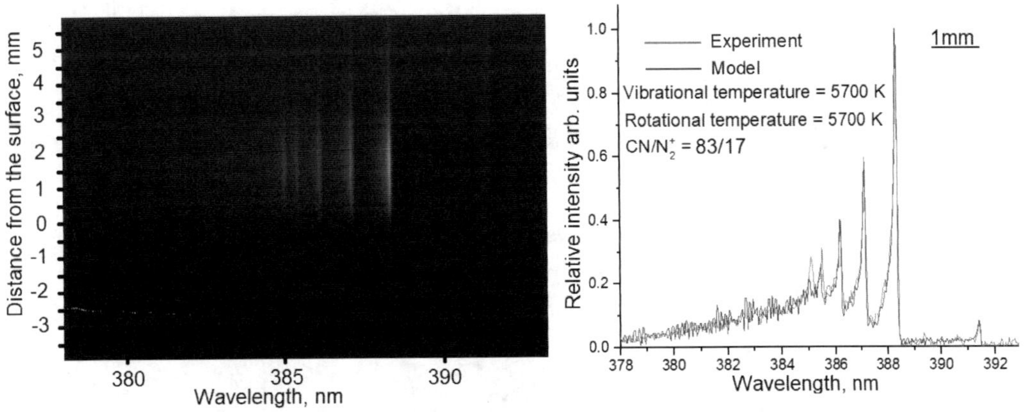

Figure 5. The matrix spectrum of the nitrogen plasma incident on the graphite sample, and the result of its comparison with the model spectrum in order to determine T_v and T_r. The 0 coordinate in the left figure corresponds to the sample surface.

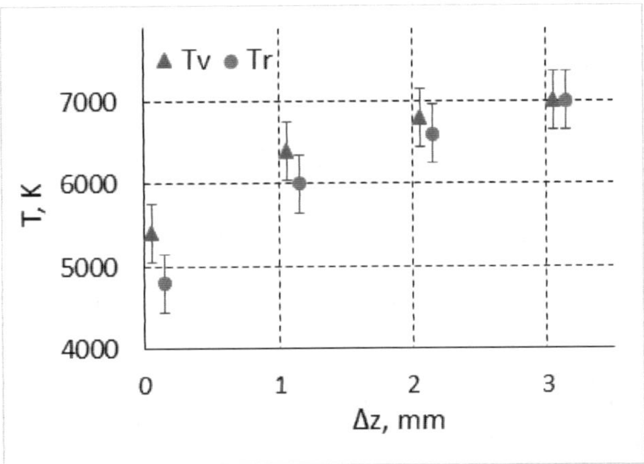

Figure 6. Vibrational and rotational temperatures of the CN radical according to the results of molecular diagnostics of the near-surface layer. Δz - distance from the surface.

The spectrum bands of the 1^- system of the N_2^+ ion (transition $B^2\Sigma_u^+$ - $X_2\Sigma_g^+$, isoelectronic with the transition CN) are close in their spectral position to the corresponding bands of the CN radical and can be used to estimate the concentration [CN]. At the same time, the almost identical excitation energies of the $B_2\Sigma^+$ system of the N_2^+ and CN molecules, as well as an important feature of the composition of the equilibrium nitrogen plasma in the temperature range 5-10kK, consisting in the presence of a maximum concentration of $[N_2^+] \approx 10^{14} cm^{-3}$ at $T \approx 7000K$ [8], make this estimate of the concentration of [CN] fairly accurate.

An important analysis tool is the comparison of the observed molecular spectra of N_2 (2^+ system), N_2^+ (1^- system) and CN (violet system) in the 300-430 nm wavelength range, containing several sequences of these molecules, with their model representation by varying T_v and T_r values for these molecules, which provide the best agreement between the experimental plasma radiation intensities and the sum of their model values. In fig. 7 such a comparison was made for the plasma region at the distance of 3 mm from the pyrographite surface. Within the error range of 15–20%, all considered molecular components of the plasma in the interaction region are characterized by a single temperature $T = T_v = T_r = 7000 \pm 1000$ K.

Journal of Physics: Conference Series **1394** (2019) 012016 doi:10.1088/1742-6596/1394/1/012016

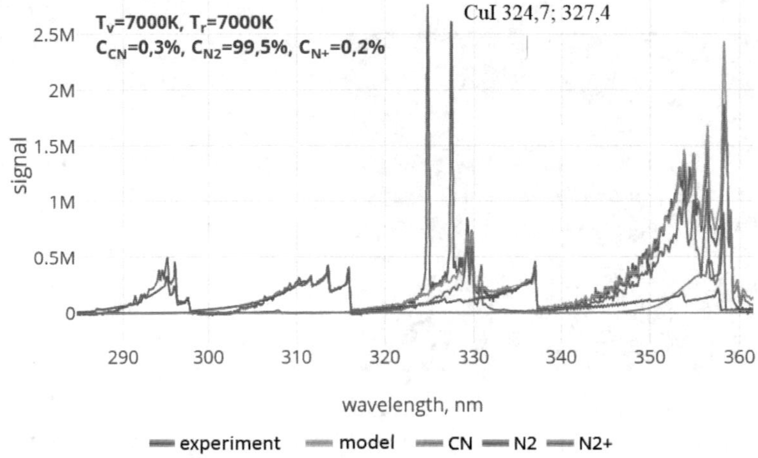

Figure 7. Experimental and model plasma spectra 3 mm from the surface of graphite. Along with the molecules, a resonant doublet of CuI 324.7, 327.4 nm is observed.

6. The rate of loss of material.

Experimental estimation of the mass loss rate of the sample material in real time is performed by the method of two-positioned synchronous visualization of the “jet-sample” system. The results of measuring the rate of destruction of isotropic and pyrolytic graphite under the influence of a nitrogen plasma jet are shown in Fig.8.

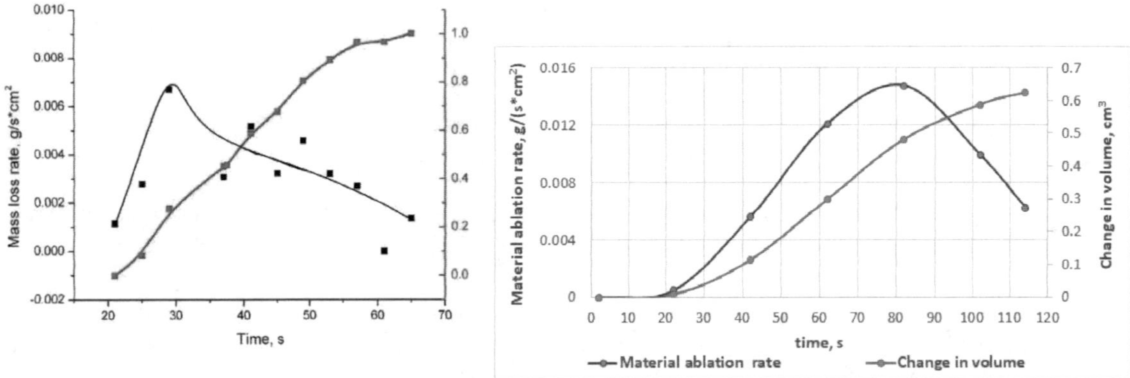

Figure 8. The change in time of the mass loss rate of isotropic (black curve on the left) and pyrolytic (blue line on the right) graphite under the influence of a nitrogen plasma jet: 150 A, $q_0 \approx 1.0$ kW/cm^2

Two estimates of the flux of atomic carbon particles with thermal velocity u_C (T_W) leaving the graphite surface

$$J_C = u_C \cdot [C] \approx \dot{\mu}_C \cdot (N_0/M_C) = 10^{20} \text{ particles/cm}^2\text{s},$$

in which data is used on the spectral estimate of the maximum carbon concentration $[C] \approx (1 \div 2) \cdot 10^{15}$ cm^{-3} (second term of equality) and the results of measuring the mass loss rate of the material n&(third member of equality), they give the same results.

7. Conclusion.

We have measured spatial and temporal variations of the following plasma parameters in the region of graphite sample decomposition by the high enthalpy nitrogen plasma jet: electron temperature, vibrational and rotational temperature of the CN molecule. In the area of real heat loads of 0.5-1.5 kW/cm^2, heating of the graphite surface to temperatures of 2800÷3600 K was achieved. At the same time, the mass loss rate of the material was 3-15 mg/cm^{2}s, which is in good agreement with the previously performed measurements [1].

In the zone of graphite destruction, the main process regulating the relationship between [C] and [CN], apparently, is the substitution reaction [9]

$$C + N2\ (v > 1) \leftrightarrow CN + N\ (\geq -2.01\ eV),$$

since the main components of plasma N_2 and N are involved in it on both sides. In more distant (from the samples surface) regions of the plasma jet, where $[N] > [N_2]$ and $n_e \geq 10^{15}$ cm^{-3}, the recharge process can become significant

$$N^+ + C \rightarrow N + C^+.$$

Analysis of the cyclic change in the intensity of the nitrogen atomic lines and nitrogen ions molecular bands during the quasi-stationary heating of an isotropic sample (see Fig. 4) indicates that when there is a crater and a “stagnant” near-surface zone monotonous heating is observed, eventually leading to a small “ablative” cooling of the sample.

Acknowledgments
The study was carried out with the state support to young Russian scientists, PhD, the Grant of the President of the Russian Federation No. MK-2460.2019.8.

References

[1] *Mihatulin D.S., Polezhaev Yu.V., Reviznikov D.L.* Heat transfer, thermochemical and thermal-erosive destruction of thermal protection. Moscow, Janus-K, 2011. 520 p.

[2] *Bernd Helber, Alessandro Turchi, James B. Scoggins, Annick Hubin, Thierry E. Magin* “Experimental investigation of ablation and pyrolysis processes of carbon-phenolic ablators in atmospheric entry plasmas” International Journal of Heat and Mass Transfer 100 (2016) 810–824.

[3] *Zhestkov, B.E., Vaganova, M.L., Lebedeva, Y.E. et al.*, “ Influence of Nitrogen High-Speed Flow Impact on the Structure and Chemical Composition of High-Temperature Coating on SiC Composite”, High Temp (2018) 56: 378. https://doi.org/10.1134/S0018151X18030215

[4] *Francesco Panerai, Bernd Helber, Olivier Chazot, Marianne Balat-Pichelin.* “Surface temperature 2 jump beyond active oxidation of carbon/silicon carbide composites in ectreme aerothemal conditions” Carbon journal, vol 71. 2014. Pp.102-119.

[5] *Lei Luo, Yiguang Wang, Liping Liu, Liuyang Duan, Guolin Wang, Yonghong Lu.* “Ablation behavior of C/Sic composites in plasma wind tunnel” Carbon journal, vol 103. 2016. Pp.73-83.

[6] *V. F. Chinnov, A. S. Tyuftyaev, D. I. Kavyrshin, A. G. Ageev, M. A. Sargsyan and M. Kh. Gadzhiev.* Comprehensive Study of the Effect of Plasma Stream on Heat-Resistant Materials// High Temp. 2018. V. 56. N 1. P. 25–32. https://doi.org/10.1134/S0018151X18010224

[7] *Kramida, A., Ralchenko, Yu., Reader, J., and NIST ASD Team* (2018). NIST Atomic Spectra Database (ver. 5.6.1), [Online]. Available: https://physics.nist.gov/asd [2019, May 22]. National Institute of Standards and Technology, Gaithersburg, MD. DOI: https://doi.org/10.18434/T4W30F

[8] *Sokolova I.A., Vasilevskij S.A., Andriatis A.V* Description of the Sova software package designed to calculate the equilibrium composition and transport coefficients of low-temperature plasma in the higher approximations of the Chapman-Enskog method. // Moscow, vasil@ipmnet.ru.

[9] NIST Chemical Kinetics Database Standard Reference Database 17, Version 7.0 (Web Version), Release 1.6.8, Data Version 2017.07 (http://kinetics.nist.gov/kinetics)

Nanosecond volume discharge in the flow with diffracting shock wave in the rectangular channel

I Znamenskaya[1], D Tatarenkova[1], T Kuli-zade[1]

[1] Faculty of Physics, M.V. Lomonosov Moscow State University, Moscow 119991, Russia

tatarenkova.darya@yandex.ru

Abstract. New effects of plasma redistribution shock wave (M = 2.8–3.4) diffraction on an obstacle at flow ionization with pulse volume discharge are presented. The obstacle has size 48.0 × 6.2x1.9 mm^3 and consist of forward facing and backward facing steps on the wall. Localization of the nanosecond combined discharge plasma (with pre-ionization in the volume) in non-stationary gas flow was visualized with plasma glow imaging. It is shown that the discharge images visualize flow evolution structure, mainly - zones of reduced density. Image are compared with 2D CFD simulation.

1. Introduction

Nanosecond discharges proved to be rather effective for active gas flow control [1-4] Pulse discharge is accompanied by a rapid temperature increase, which is caused by the high energy deposition in local area. The gas may be heated in time interval less than 1 μs, plasma zone generates a shock (blast) waves. Their configuration and intensity depends on discharge localization, gas flow structure and density distribution. Surface discharges may be used to control boundary layer [3–7]. Pulse volume discharge with pre-ionization changes dramatically the flow in the supersonic channel [8–10].

The objective of the work is: to analyze the instant redistribution of the pulsed volume discharge plasma at shock wave moving in the channel including in a density field being well-studied on shock diffraction on step, and its further movement in the channel. This redistribution (energy localization) may change the post-discharge flow due to local energy input (see for example [9]).

Shock diffraction on obstacle is one of the most basic phenomena in high speed gas flow dynamics. It has been often adopted as a numerical test for the simulation of unsteady compressible flows [11]. Experimental results were obtained for shock wave diffraction on upward facing and backward facing steps in shock tubes [12–14]. Most of them were obtained using shadow methods and interferometry [11, 12]. Here we recorded the glow of unsteady flow fields at pulsed volume ionization by the discharge initiated at various stages of the gasdynamic process of shock diffraction. Flow in rectangular channel 24x48 mm is close to two-dimensional. Localizing the energy input into specified flow segments and control of the discharge and flow structure are of current interest. We compare the gasdynamic fields experimentally visualized by plasma glow with the numerically simulated density fields of the corresponding complex unsteady flows (with obstacles in the flow). We recorded images of the volume discharge glow distribution in a known field of parameters. Thus the gasdynamic flows can be adjusted via the influence of the localized energy contribution on flow elements.

Journal of Physics: Conference Series **1394** (2019) 012017 doi:10.1088/1742-6596/1394/1/012017

2. Experimental Setup

Experiments were conducted using a diaphragm shock tube with a special discharge chamber [8]. The energy redistribution of a pulsed volume discharge with preionization from two plasma sheets (electrodes) during its initiation in a rectangular shock tube channel with an oblique shock wave is investigated. The sidewalls of the chamber were transparent windows through which registration was possible. The bottom and top walls were flat plasma electrodes (fig. 1). The initial plane shock wave with homogeneous flow behind it passed over the obstacle in the form of a parallelepiped placed on the lower wall of the shock tube chamber (Fig.1). The instant plasma glow configuration occurring at the initiation of volume discharge was studied. The objective was: to investigate the patterns of nanosecond volume discharge energy redistribution when it is initiated in a non-stationary gas dynamic flow close to $2D$ in the channel. The discharge was initiated at various stages of the plane shock wave passage through the channel with an obstacle on the wall and after the formation of a stationary flow. By means of a low-inductance shunt we measured the discharge current. The pulsed volume ionization of the gas-dynamic channel section with a length of 10 cm was investigated. Rectangular parallelepiped with dimensions of 48 mm x 6.2 mm x 1.9 mm was fixed at a distance of 1.5 cm from the leading edge of the plasma electrode (downstream). The duration of the discharge current was 150-300 ns, and the duration of the glow was up to 1 μs [8]. The discharge was initiated by 25 kV voltage pulses. The interaction of a plane shock wave with Mach number M =2.8–3.4 and the flow behind it with Mach number 1.2 – 1.4 with this obstacle is considered. Flow visualization is available through the side walls which are quarts glasses. Integrated images of the discharge glow in the flow field were recorded from two sides of the test (discharge) chamber. Oscilloscope curves were measured of combined discharge electric current in flow with moving shock.

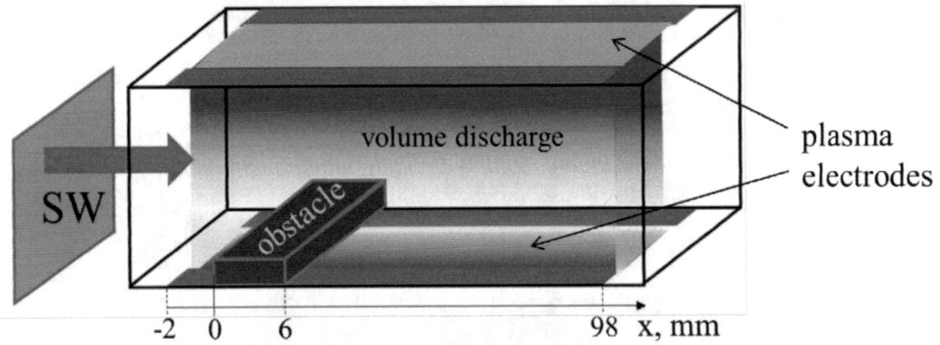

Figure 1. Scheme of discharge chamber channel with a shock wave.

Objects of particular interest are 1) the region of unsteady interaction of the flow with an obstacle on the lower wall (on shock-wave moving past a upward-facing step and backward-facing step; 2) the configuration of oblique shock wave and the flow after initial shock wave had left the discharge area; 3) the interaction of the oblique shock wave with the upper wall and the boundary layer after initial shock wave had left the discharge area.

3. Results and discussion.

In this paper we investigate the first problem - the redistribution of the discharge energy in the flow when moving shock wave is inside the discharge gap. Every glow image is instant visualization of flow structure as far as its exposition is less than 1 μs (gas charge glow time). It is correlated with the distribution of the density in flow due to the effect that it depends on value of an ionization coefficient. Local ionization coefficient is a non-linear function of E/N, (here E is instant electric field, N is the concentration of gas molecules).

Four images of flow at different stages of shock wave propagation are on Fig. 2. Grey rectangular is obstacle position. On the first one can see shock wave (movement direction marked with arrow) which has passed by the downward facing step. Combined discharge glow is separated by shock front – it is instantly localized in low density area [9], the same is on fig. 2b, c. Time Td is 28 and 47 μs – it is time from front edge of the obstacle to moment of discharge initiation. Behind the diffracted shock the vortex area [11,12] is visualized with intense glow. On fig. 2b (Td = 28 μs) shock front bottom part is curved after the diffraction; all the flow is in front of it. On fig. 2c (Td = 47 μs) shock front is already quite straight. On Fig. 2c we can see the flow configuration after initial shock wave had left the discharge area (Td=173 μs): the oblique shock wave; its interaction with boundary layer at the upper wall, separation zone at the backward facing step. This is stationary flow slightly changing in time.

It is important to mark that the gas flow almost does not change in image exposition time and it changes dramatically after the discharge, supplying locally electric current energy in structural flow [9].

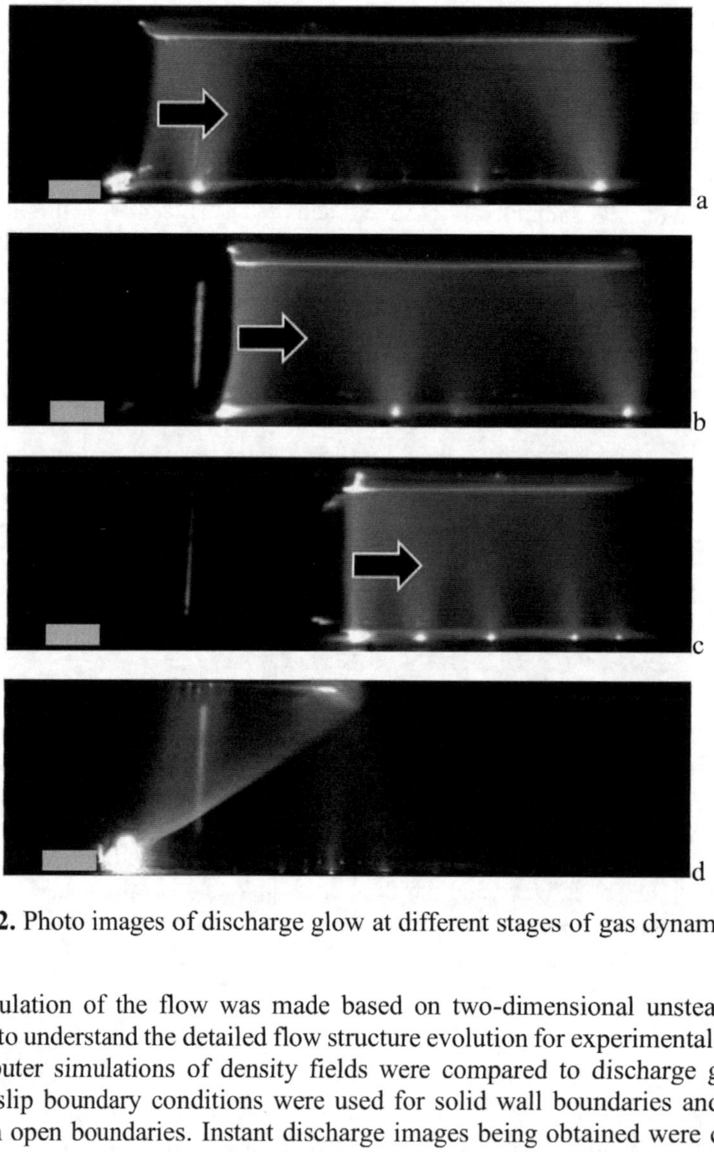

Figure 2. Photo images of discharge glow at different stages of gas dynamic process.

Numerical simulation of the flow was made based on two-dimensional unsteady Navier-Stocks equations in order to understand the detailed flow structure evolution for experimental parameters. Two-dimensional computer simulations of density fields were compared to discharge glow distributions [10,15]. The non-slip boundary conditions were used for solid wall boundaries and "non-reflecting" ones were used on open boundaries. Instant discharge images being obtained were compared to CFD calculations of $2D$ non-stable flow, visualized in gray scale. Image of discharge glow at shock diffraction and CFD simulation of the same stage at T_d 17 μs are on fig.3.Vortex behind the step (lowest density area) is the brightest zone, the darkest is area behind bow shock formed in supersonic flow in front of upward facing step.

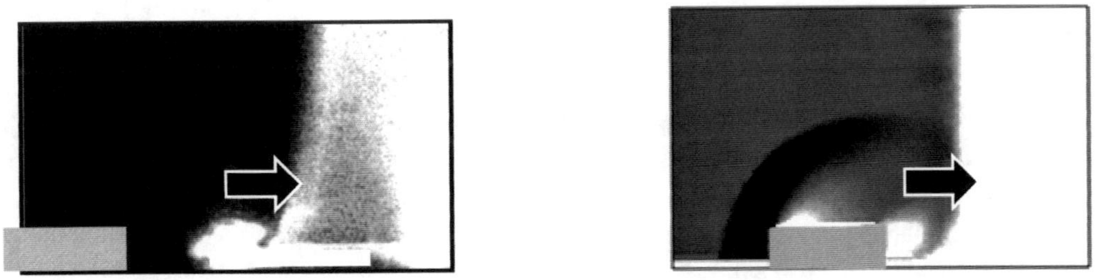

Figure 3. Instant photo image of discharge glow at shock diffraction (left) and CFD simulation of the same stage (right).

The analysis of oscilloscope curves of discharge current in flow with moving shock allowed us to investigate the dependence of the time and amplitude characteristics of the discharge current on the instantaneous structure of the gas-dynamic flow. On Fig. 4: Mach number $M = 3.23+-0.05$, SW position was: $x = -1$ mm; 9 mm; 14 mm; $x = 26$ mm in 1-4 subsequently. It was shown that they do not differ greatly while shock wave is inside gap. When part of energy is directed into vortex low density channel

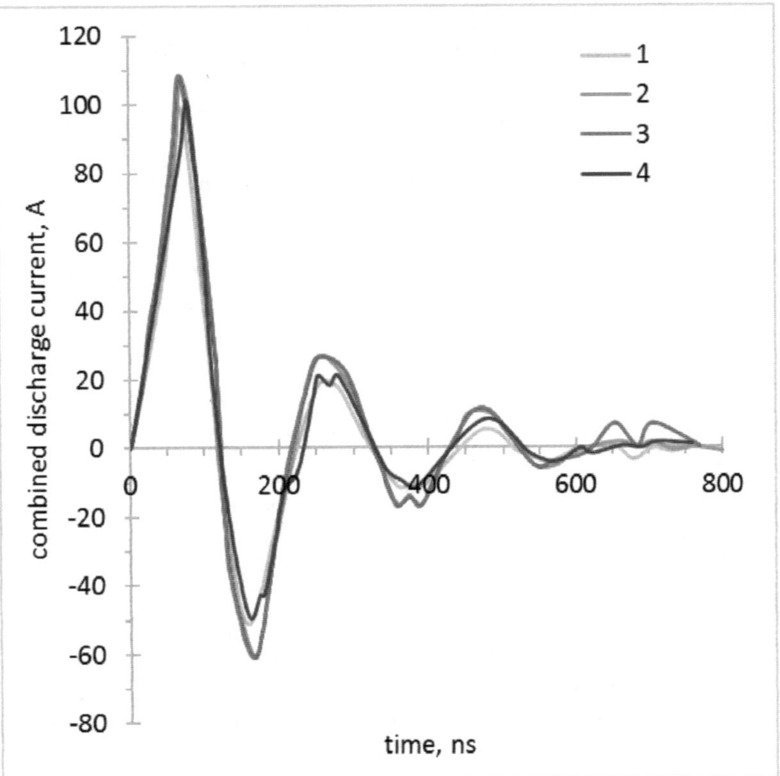

we have a small changing on the curve (3 on Fig.4).

Figure 4. Oscilloscope measured combined discharge current curves in flow with moving shock

When the shock wave passes between the plasma electrodes (on the upper and lower walls of the channel), the initiation of the discharge leads to the localization of almost the entire discharge current in channel area separated by the shock front [1]. At the same time, the front curved when the shock passed over the obstacle and part of the energy was localized to the vortex area behind the obstacle.

4. Conclusions

This work is focused on details of shock wave interaction at pulse discharge initiation in channel with step, which was not previously explored in available publications. We investigated the patterns of nanosecond volume discharge energy redistribution when it was initiated in a non-stationary gas dynamic flow (close to 2D in the rectangular channel).

We recorded the instant glow images of unsteady flow fields at nanosecond combined discharge ionization being initiated at various stages of the gasdynamic process. Shock wave ($M = 2.8 - 3.4$) diffraction on rectangular parallelepiped on the wall and transition to stationary flow ($M_f = 1.2\text{-}1.5$) with oblique shock wave was visualized with volume discharge at submicrosecond exposition (discharge glow time). Volume discharge energy is instantly redistributed in non-homogeneous density field.

When shock is moving inside gap (10cm long) at pulse discharge initiation moment T_{dis}, discharge glow is localized in low density zones: in front of shock and in vortex at downward step wall.

When initial shock wave had left the discharge area, oblique shock wave, formed downstream the obstacle, separation zone at the backward facing step and the shock-boundary layer interaction on the opposite was are visualized in channel flow. Upward facing step has a dark zone in front of it - due to bow shock formed in supersonic flow (with high density area behind it). The calculated flow density distribution patterns at different time moments had been obtained by solving the Navier-Stokes equations [10]. We compared the gasdynamic fields experimentally visualized by plasma glow with the numerically simulated corresponding grey scale density fields of the $2D$ flow. Oscilloscope measured current curves in flow with moving shock were analyzed. Thus we can predict the plasma energy pulse localization in non-stationary flow. Sharp density gradients attract energy fluxes and this is followed by flow alternation. A predictable volume-discharge current redistribution seems to be promising in flow control in channel flows.

Acknowledgment. The work was conducted under the support of RSF Grant № 18-19-00672.
The authors would also like to thank Dr Irina Mursenkova, Dr Fyodor Glazyrin for their assistance in preparation of the experiments

References

[1] Ukai T., Russell A., Zare-Behtash H., Kontis K. – Temporal variation of the spatial density distribution above a nanosecond pulsed dielectric barrier discharge plasma actuator in quiescent air. Physics of Fluids 30, (2018)

[2] Shkurenkov I., Burnette D., Lempert W.R., Adamovich I.V. – Kinetics of excited states and radicals in a nanosecond pulse discharge and afterglow in nitrogen and air. Plasma Sources Science and Technology Volume 23, Issue 6, (2014),

[3] Bayoda K. D., Benard N., Moreau E.– Nanosecond pulsed sliding dielectric barrier discharge plasma actuator for airflow control: Electrical, optical, and mechanical characteristics.. J. Appl. Phys., vol. 118, no. 6 (2015).

[4] Kinefuchi K., Starikovskiy A. Y., Miles R. B.– Numerical investigation of nanosecond pulsed plasma actuators for control of shock-wave/boundary-layer separation. Phys. Fluids 30 (2018).

[5] Zheng J. G., Cui Y. D., Li J., Khoo B. C. - A note on supersonic flow control with nanosecond plasma actuator. Physics of Fluids 30 (2018)

[6] Znamenskaya I. A., Latfullin D. F., Lutskii A. E., Mursenkova I. V. – Energy deposition in boundary gas layer during initiation of nanosecond sliding surface discharge. Tech. Phys. Lett. (2010)

[7] Mursenkova I. V., Znamenskaya I. A., Lutsky A. E. – Influence of shock waves from plasma actuators on transonic and supersonic airflow. J. Phys. D: Appl. Phys. 51 (5) (2018)

[8] Arkhipov N. O., Znamenskaya I. A., Mursenkova I. V., Ostapenko I. Yu., Sysoev N. N. - Development of nanosecond combined volume discharge with plasma electrodes in an air flow. Moscow Univ. Phys. Bull. 69 – 96 (2014)

[9] Doroshchebko I., Znamenskaya I.,Koroteev D., Kuli-zade T.- When shock is shocked: Riemann

Journal of Physics: Conference Series **1394** (2019) 012017 doi:10.1088/1742-6596/1394/1/012017

problem dynamics at pulse ionization of a shock wave. Physics of Fluids. Vol. 29, no. 10 (2017)

[10] Znamenskaya I. A., Ivanov I. É., Kryukov I. A., and Kuli-Zade T. A.– Pulsed Volume Discharge with Preionization in Two-Dimensional Gasdynamic Flow. Journal of Experimental and Theoretical Physics. V. 122, No 6, pp. 1198–1206. (2002)

[11] M. Sun, K. Takayama: A note on numerical simulation of vortical structures in shock diffraction Shock Waves. Fluid Dynamics Volume 35, Issue 1, 2000, Pages 117-122 (2003)

[12] Skews BW – The perturbed region behind a diffracting shock wave. J. Fluid Mech. 29: 705 (1967)

[13] Bazhenova, T.V., Bormotova, T.A. , Golub, V.V., Shul'meister, A.M., Shcherbak, C.B. Type of reflection of a near-wall shock wave in the process of diffraction from a square channel. Fluid Dynamics V. 35 (1) pp 117-122 (2000)

[14] Zheng J. G., Cui Y. D., Li J., Khoo B. C., Skews B.W., The perturbed region behind a diffracting shock wave. J. Fluid Mech. 29: 705 (1967)

[15] Znamenskaya I. A. , Kulkin M. K. , Ivanov I. E. , Doroshchenko I. A. - Visualization of shock wave interaction with wedge flow. Journal of Flow Visualization and Image Processing. V.25 (2). Pp. 77–89. (2018).

The operating mode of the water-cooled MHD accelerator of SMGDU with a magnetic field that changed along the channel

D S Baranov[1], V A Bityurin[1], A N Bocharov[1], S S Bychkov[1]*, V A Grushin[1], E V Kusmartseva[1], N V Tretyakova[1], N I Batura[2], N G Zhurkin[2] and N M Kolushov[2]

[1]JIHT RAS, Moscow, Russia
[2]TsAGI, Zhukovsky, Russia

*serg-bychkov@yandex.ru

Abstract. The work is aimed at solving the problem of creating in laboratory conditions hypersonic high-enthalpy gas flows that parameters are kept constant for 5-10 seconds. For this purpose the test section of the MHD channel with water-cooled electrodes and side walls was under investigation at the set up with a magnetohydrodynamic accelerator in TsAGI. A new series of runs was performed in which the MHD-acceleration stage was successively increased to 5 seconds. It was determined that the electrical and gas-dynamic characteristics of the MHD accelerator did not depend on the duration of the run. It was first demonstrated that the use of a magnetic field special shaped along the axis of the channel made it possible to avoid catastrophic damage to the side walls of the channel at the entrance and exit of the MHD accelerator, which increased the service life of the installation.

The experiments were carried out at TsAGI on a unique in the world wind tunnel with a magnetogasdynamic accelerator (SMGDU). The feature of the test stand is that the acceleration of the conducting gas to hypersonic speeds is provided using the MHD accelerator that converts the main part of the electrical energy supplied from outside directly into the kinetic energy of the flow. The schematic diagram of such a tunnel with a heat-capacitive MHD channel and its characteristics are described in detail in paper [1]. The same review presents the experimentally obtained maximum values of the velocity of the hypersonic flow formed in the Eiffel chamber which reached 8 km/s with a Mach number of 15. Numerous researches have been performed at SMGDU over the years of its employment. The results of experimental investigation of airflow past simplest model bodies at different velocities are presented in [2], as example.

The use of a channel that accumulates heat in structural elements limits the time for the existence of a high-speed gas stream up to 1 second. The time-limit is due to the melting of copper electrodes and the rapid erosion of insulators exposed to high thermal and electrical loads. This circumstance significantly reduces the range of application of the facility for the study of promising technologies and materials. The possibility of increasing the duration of the acceleration stage by an order up to 10 seconds while maintaining the level of total pressure was demonstrated in [3, 4 and 5]. To achieve this goal the standard channel of the MHD accelerator was replaced by a new channel in which the electrodes and side walls were cooled with running water. However the destruction of the side walls at the ends of the channel was observed in separate runs when the energy input was increased. It is

Journal of Physics: Conference Series **1394** (2019) 012018 doi:10.1088/1742-6596/1394/1/012018

known that end effects manifest themselves due to the inhomogeneity of the boundary conditions that have effect on the distribution of currents and potentials at the input and output of the MHD channel [6]. In our case the melting of the insulators and after them the metal walls up to the water circuit was caused apparently by electric currents arising as a result of shorting the induced Hall potential through an external circuit. This phenomenon was recorded only in the edge zones, where there were no longer loaded electrodes, but there was still a strong magnetic field.

The task of this work was to search for modes of trouble-free operation of the MHD accelerator equipped with a water-cooled channel. Measures were taken to weaken the magnetic field at the entrance and exit of the MHD accelerator not much changing it in the zone of the central loaded electrodes since the efficiency of acceleration increases rapidly with increasing magnetic flux density.

The magnetic field induction on the axis of the MHD accelerator was precomputed by the FEMM 4.2 software. Since the program considers only plane and axisymmetric tasks the cross section of the electromagnet was considered in which the dimensions of the core, the position of the coils, of the steel magnetic insertions and the MHD channel were specified. An example of the calculation for ten coils connected in series with a total number of turns 420 and a direct current of 200 A is demonstrated in Figure 1, in which the field distribution is shown by magnetic induction lines.

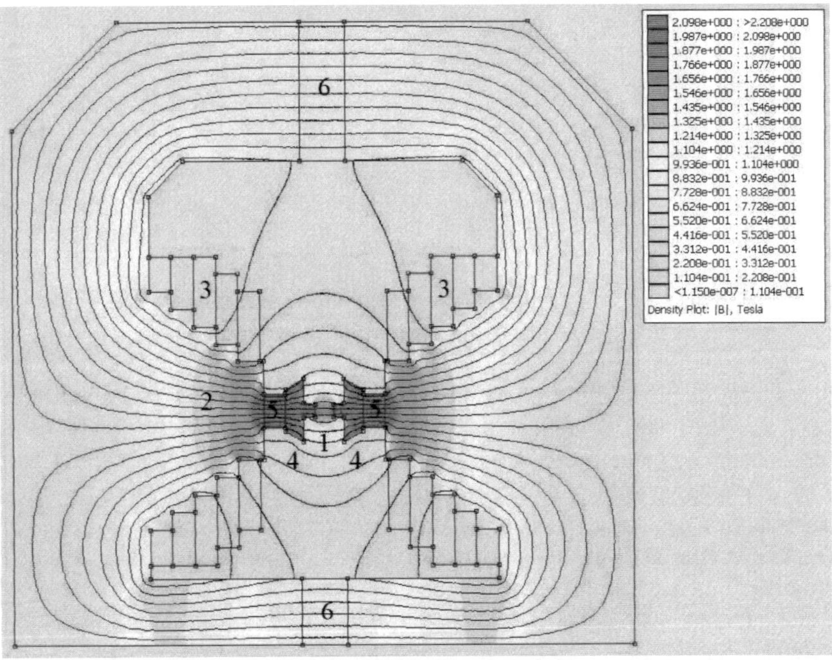

Figure 1 A vertical section of the electromagnet with the results of the calculation of the magnetic field distribution.

The electromagnet of the installation was divided into two equal halves. Each of them was an E-type core assembled from electrical steel sheets. The MHD accelerator was located in the gap 1 between the poles of the central core 2 on which the coils of the current windings 3 were fixed. The MHD channel was bounded on sides by steel walls 4 insulated from a flow of conducting gas, and from above and below by sectioned electrode walls consisting of electrode-insulator pairs (are absent in Figure 1). The lateral walls were adjacent directly to the pole tips of the electromagnet. The gas flow was directed perpendicular to the plane of the drawing. Thus, the magnetic induction vector was oriented across the velocity vector of the working medium and the electric field intensity vector.

The magnetic field inside the channel was uniform in usual mode. The resistance of the magnetic circuit that is closed through the side parts of the magnetic core was increased due to the organization of the air gap to weak the field at the edges. This gap was filled with inserts of magnetically soft steel

5 and 6 in the middle part of the electromagnet. Thus, the magnetic field had maximum value in the central zone of the MHD channel that covering approximately 11 electrode pairs.

The magnetic flux density distributions along the MHD channel axis were measured with a teslameter at two current values in the electromagnet winding: 117 and 200 A. The data obtained and their approximations by polynomials (solid lines) are shown in Figure 2.

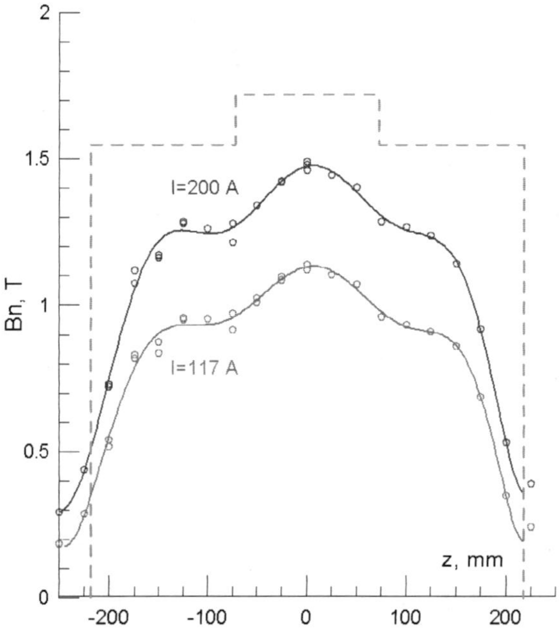

Figure 2 The distribution of the magnetic field along the axis of the MHD channel: dotted line - calculation; circles - measurements.

Zero on the *z* axis matches with the middle of the MHD channel. The dashed line is constructed according to the results of calculating the magnitude of the induction for two vertical cross-sections of the magnet core when 200 A electric current flowed in the winding. The central part of the graph corresponds to a cross-section in which all the gaps that appeared were filled with magnetic conductive material and the parts of the graph with lower values of induction that located to the left and to the right of it fits cross-sections where air gaps were originated. The magnetic field values calculated had exceeding slightly the data measured. The fact is that it was assumed in the calculation that the core and the inserts were made from pure iron that is the properties of real materials were not taken into account.

As a result of the preparatory work a case of a magnetic field changing along the gas path was realized. It was not constant and reached a maximum value only in the center of the channel. Nevertheless it is possible to distinguish three regions with different average values of induction: the central region where the field had the highest values and two regions symmetrically located along the edges of the channel in which the field was weaker by about 15%. Further the field had weakening quickly to the ends of the channel.

New series of runs were carried out at SMGDU wind tunnel with an MHD channel cooled by water and a magnetic field that varied along the axis of the MHD accelerator. The test unit was a quadrangular channel with two side insulating walls and with two electrode walls with sectioned electrodes. The number of electrode sections was 21. The sectioning step (total length of the electrode and insulator) was 13 mm. The width of the cooled copper electrode was 10 mm. The insulators were made of boron nitride. Their thickness was 3 mm. Steel side walls were also insulated from the flow with boron nitride plates. The 555 mm long channel expanded in the vertical direction from 15x15

mm^2 inlet cross-section to 15x25 mm^2 outlet cross-section. Each electrode pair of the MHD accelerator was connected to an independent power source according to the Faraday scheme.

Fifteen pairs of electrodes from 4 to 18 were involved during the runs. Extreme electrodes 1 through 3 and 19 through 21 were not connected to the power supplies. The maximum magnitude of the magnetic field at a current of 117 A in the electromagnet winding was about 1 T. All runs were performed under the same conditions and differed only in the duration of the gas flow acceleration stage. In each case the distributions of the main electrophysical parameters along the channel were recorded: currents and voltages between the electrodes, as well as Hall (longitudinal) voltage. The data obtained are presented in Figure 3 (currents) and Figure 4 (voltages) respectively as dependencies of the measured characteristics on the sequence number of loaded electrodes.

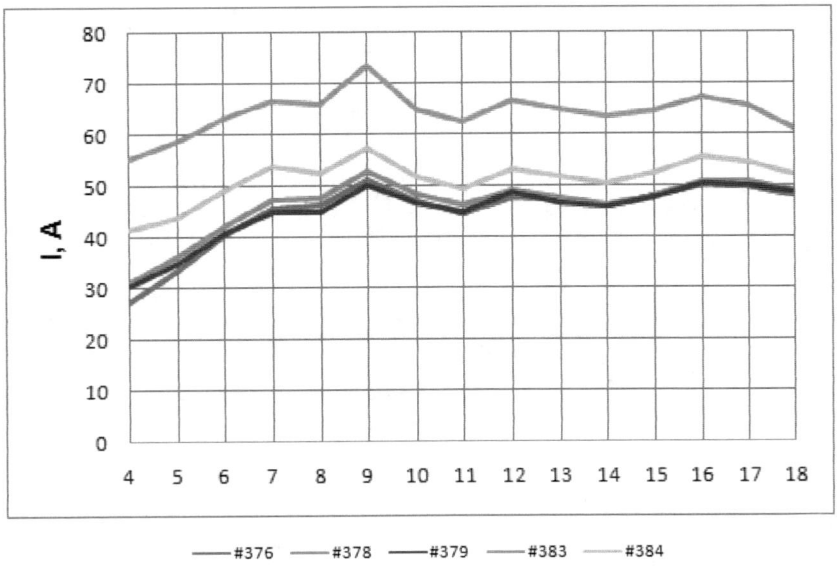

Figure 3 Currents between electrode pairs.

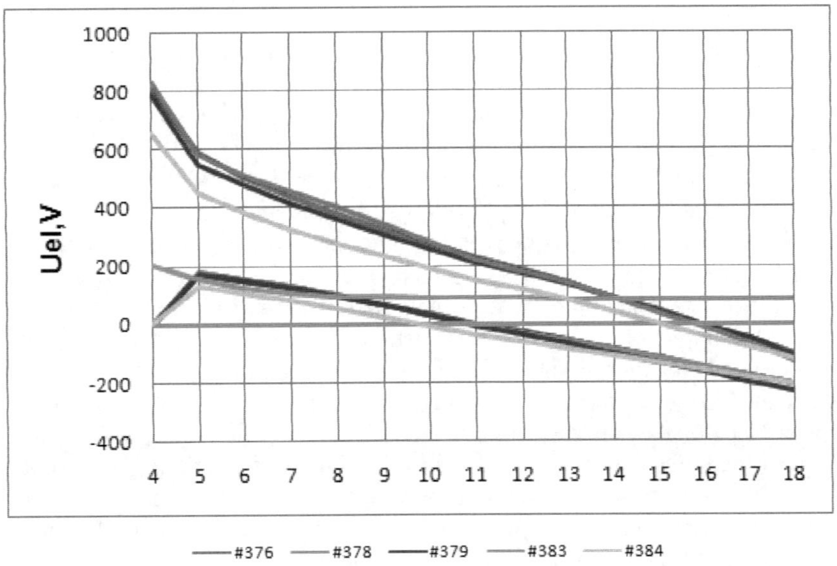

Figure 4 MHD accelerator electrode potentials.

Journal of Physics: Conference Series **1394** (2019) 012018 doi:10.1088/1742-6596/1394/1/012018

The current quantity varied from 30 A for the first of the electrode gaps that loaded to 55 A for subsequent pairs of electrodes while the voltage between them reduced from a maximum level to 100 V. The growth in the electric current between electrode pairs corresponded to increase of current density from 20 A/cm^2 to 35 A/cm^2. The last value from mentioned was observed for most sections of the MHD accelerator.

For each run the lower curve in Figure 4 shows the distribution of the Hall potential measured on the cathode wall of the MHD channel relative to the first of the loaded electrodes (#4). The potentials of unloaded electrodes from the 1st to the 3rd and from the 19th to the 21st were not measured in this series of experiments. On the upper curve, the corresponding anodes potentials are plotted.

The input electric power determined for each electrode pair by multiplying the corresponding currents and voltages is represented in Figure 5.

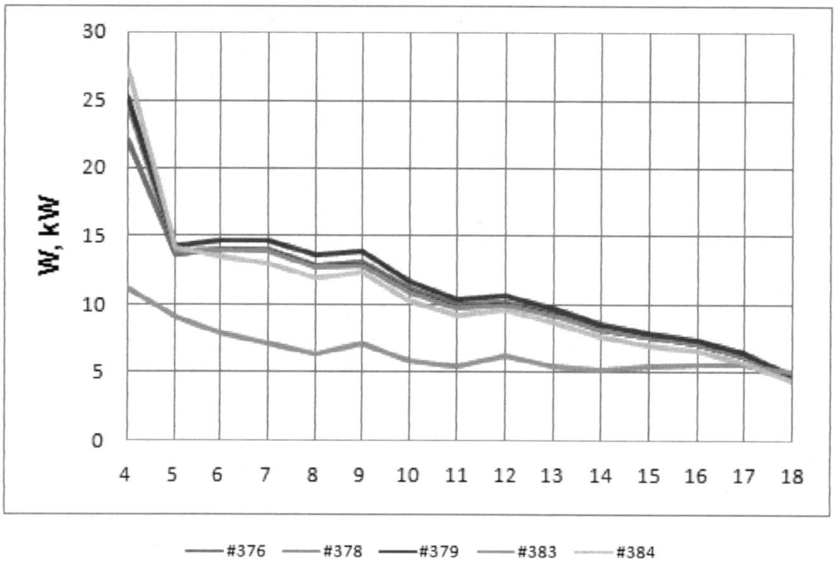

Figure 5 Electric power supplied to each electrode pair from an external source.

Charts marked with the identical color corresponded to the same run. The time of the MHD-acceleration stage was successively increased from 0.6 s (run #376) to 1.5 s (run #378), then to 3.2 s (run #379) and finally to 5.2 s (run #384). It can be seen that a good agreement between the experimental data was observed. The distributions on record in the run #384 slightly fall out of the rest of the curves bundle since this run was carried out another day when the initial conditions for experiments conducting at the SMGDU were slightly different apparently.

A special place in these studies was occupied by the run #383 conducted without the switching-on of a magnetic field. Enlarged interelectrode currents were observed and the Hall potential was close to zero in this run. Since the magnetic field was switched off it can be concluded that all the electrical power about 100 kW supplied from outside was spent on heating the gas, that is, the MHD accelerator worked as an additional air heater in this mode. In the case when the magnetic field was turned on the currents between the electrodes decreased, and the voltages increased so the total power applied to the discharge gaps of 15 pairs of electrodes was 170 kW.

The readings of total pressure sensor installed in the Eiffel chamber also were recorded in addition to electrophysical measurements. Figure 6 illustrates how the total pressure varied during MHD acceleration in a short time run (blue curve) and a long time one (red curve). The pressure sensor showed lower value in the first of these since the signal did not have time to reach its maximum value due to the inertia of the air duct connecting the Pitot tube to the sensor. To illustrate this circumstance a pressure signal recorded in the run of an intermediate duration (green curve) was added to the graphs

Figure 6. It can be seen that the measured value reached the level corresponding to a 5 second run in about 1.5 seconds.

At the end of the experiments channel was opened. The destructions of the side walls were not detected after a thorough examination of the fire surfaces. A weak erosion of the copper electrodes was observed. However, a gradual spalling of ceramic insulators was noticed that eventually led to a decrease in resistance between electrode pairs up to a short circuit.

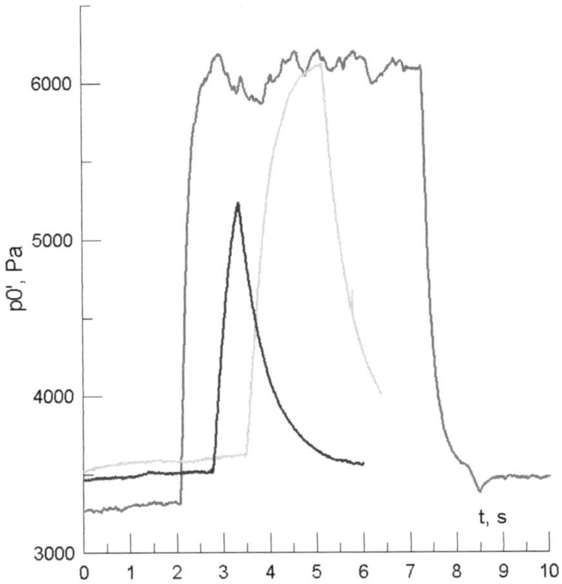

Figure 6 The signals of the total pressure sensor measured in the Eiffel chamber: run #376 – blue curve, run #378 – green curve and run #384 – red curve.

Summing up the research data we can draw the following conclusions. Firstly, it was found that the electrophysical and gas-dynamic characteristics of the MHD accelerator did not depend on the duration of the run. Secondly, a mode with a distribution of the magnetic field modified along the axis of the MHD channel was investigated in which the destruction of the side walls was not detected even in runs lasting several seconds. It is necessary to conduct additional research into the methods of interelectrode insulators protecting in order to further increase the resource of the water-cooled channel which provides for the generation of long in time high-speed flows.

Acknowledgments
The work was carried out with the support of the Russian Foundation for Basic Research (project 18-08-00890 A).

References

[1] V.I. Alferov. Current Status and Potentialities of Wind Tunnels with MHD Acceleration. High Temp. — 2000. Vol.38. No.2. P.300-313.

[2] V.I. Alferov, A.S. Bushmin, and I.V. Egorov. Experimental Investigation of Flow Past Simple Model Bodies in Hypersonic Wind Tunnels at Similar Values of the Mach and Reynolds Numbers but at Different Physical Flow Velocities. Fluid Dynamics. —2015. Vol.50. No.1. P.109–117.

[3] N.I. Batura, E.B. Vasilevsky, N.M. Kolushov, N.G. Zhurkin, S.I. Inshakov, I.S. Inshakov, V.A. Bityurin, D.S. Baranov, S.S. Bychkov, V.A. Grushin, V.I. Zalkind, N.V. Tretyakova. New features in the study of thermal protection of materials on the stand with magnetogasdynamic acceleration // Materials of the XXVIII scientific and technical conference on aerodynamics.

— Zhukovsky: TsAGI, 2017. — P.53-54.

[4] N.I. Batura, N.G. Zhurkin, E.B. Vasilevsky, V.A. Bityurin, A.N. Bocharov, P.P. Ivanov, V.I. Zalkind, V.A. Mozdykov, S.S. Bychkov, V.A. Grushin, D.S. Baranov, N.V Tretyakova. Design and the first test results of the water-cooled MHD accelerator of SMGDU TsAGI // Abstracts of the 16th International Workshop on Magneto-Plasma Aerodynamics. Moscow: JIHT RAS, 2017. — P.15-17.

[5] D.S Baranov, V.A Bityurin, A.N Bocharov, S.S Bychkov, V.A Grushin, N.V Tretyakova, N.I Batura, E.B Vasilevsky, N.G Zhurkin, N.M Kolushov. Flow characteristics in the water-cooled channel of the MHD accelerator // Journal of Physics: Conf. Series — 2018. — Vol.1112. P.012017.

[6] Magnetohydrodynamic energy conversion. Physical and technical aspects. Edited by academician V.A. Kirillina, academician A.E. Sheindlin. — Moscow: Nauka, 1982. 386 p.

Journal of Physics: Conference Series **1394** (2019) 012019 doi:10.1088/1742-6596/1394/1/012019

Numerical simulations on MHD break in hypersonic flow

V.A. Bityurin[1], A.N. Bocharov[1] and N.A. Popov[2]
[1]Joint Institute of High Temperatures, Russian Academy of Sciences
[2]Moscow State University, Moscow, Russia

E-mail: bocharov.ya@yandex.ru

Abstract. One of possible effect of electromagnetic field on the high-speed ionized airflow is considered. Namely, deceleration of the flying body due to interaction of the electric current induced around the body with the magnetic field generated by on-board magnetic system is studied. The principal scheme of magnetohydrodynamic interaction over the flat plate is presented. The goal of the study is to demonstrate the deceleration effect in terms of standard aerodynamic drag. Two test cases are considered: small-scale (lab-scale) interaction zone size and large-scale (space-scale) one.

The primary result of the study is a feasibility to increase the total drag of the body in more than order of magnitude relative to the original one. The effect can be achieved by means of practically reasonable magnetic field amplitudes. Common for both cases is that the total drag grows as the magnetic field increases. The difference between two cases is that significantly smaller magnetic fields are required in space-scale case to increase aerodynamic drag than in the lab-scale case. In any case, the drag control could be used as a potential for high-speed flow control.

Introduction

This paper presents one of possible application of MHD interaction in high-speed air flow over the body called MHD parachute. Original MHD parachute [1] was considered as a surface short-current MHD generator providing a maximal force of interaction. Numerical estimations of MHD parachute effect made under conditions of ground-based experimental facility have shown that aerodynamic drag can be increased in several times. Besides direct flow control by the MHD interaction, the MHD parachute provides a possibility of heat flux mitigation [2] during re-entry due to deceleration of the vehicle in atmospheric layers where the heat loads are still rather low.

In the current paper, MHD parachute effect is considered for typical Earth re-entry flight conditions. In general, the physical and computational model is similar to that reported in papers [3,4]. Real-air thermochemical model is used to numerically simulate a hypersonic flow over the flat plate of finite thickness. The influence of MHD interaction on the air flow is considered in terms of aerodynamic drag. Scaling effects are discussed. The lab-scale wing-like model means that the length of the plate along the flow is 0.32 m, and the thickness of the plate is 0.032 m. The space-scale model implies that both dimensions increase in ten times.

Results and discussion

The principal scheme for studying MHD interaction over the flat plate is shown in Fig.1. Magnetic field is assumed to be generated by two linear conductors with opposite currents. We shall characterize the magnetic field, B^*, by the maximum value at the plate surface (this is the same as those at the surface of one conductor). Also, the value of magnetic field at the center of magnetic system will be used. This mid-point value, B_m, is nearly 5 times less than the characteristic value B^*. Let's consider a flow around the flat plate schematically shown in Fig.1. The following free-stream conditions are specified for both lab-scale and space-scale plate models:

ρ_∞ =2.34·10-4 kg/m3, U_∞ =11137 m/s, T∞=238.5 K, P∞=16.1 Pa, M∞=35.8.

The spatial scale, L^*, will be characterized by the distance between conductor centers. L^* 0.32 m for lab-scale case, and L^*3.2 m for space-scale case.

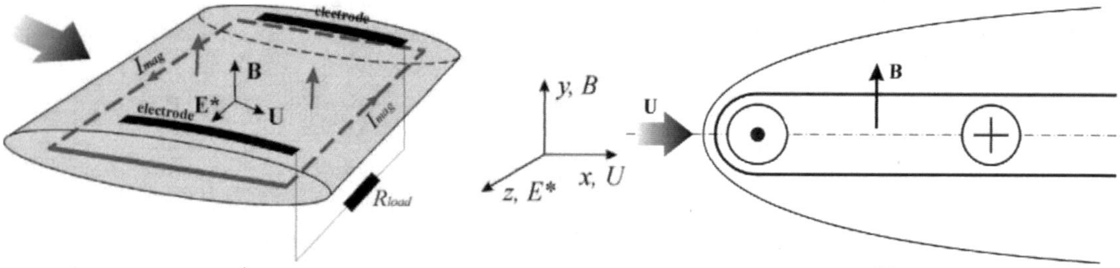

a) b)

Figure 1. Sketch of on-board MHD generator (a) and layout of computational domain in xy plane (b). E*=E+U×B, E is electric field strength.

A pattern of the flow structure can be seen in Fig.2 for lab-scale model and in Fig.3 for space-scale one. In the figures, the electric conductivity (filled contours) and ponderomotive force (vectors) are presented. Original field of conductivity is shown by contour lines for both cases.

Figure 2. Electric conductivity (filled contours) and force (vectors) at B^*=1.8T for lab-scale model. Contour lines show conductivity at B^*=0.

Journal of Physics: Conference Series **1394** (2019) 012019 doi:10.1088/1742-6596/1394/1/012019

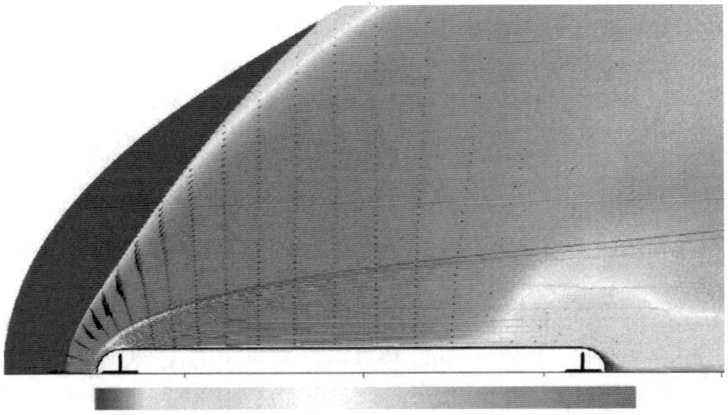

Figure 3. Electric conductivity (filled contours) and force (vectors) at B^*=0.4T for space-scale model. Contour lines show conductivity at B^*=0.

From the viewpoint of flow control, the primary feature of MHD interaction is the increase in the aerodynamic drag. Quantitative characterization of the deceleration is presented in Fig.4 by the relative total drag D/D_0 as function of characteristic magnetic field value. The total drag is treated as the sum of aerodynamic drag (pressure drag and viscous drag) and magnetic drag. The latter is the integral of the component of magnetic force along the flow direction over the domain. D_0 is the original drag at $B^* = 0$.

Possibility to increase the total aerodynamic drag in more than order of magnitude seems to be an important means for the flow control and/or a vehicle's trajectory control. Note, that a deceleration of the body is due to the interaction of electric currents in plasma with the current generating magnetic field. This body force is exactly those induced in plasma bulk.

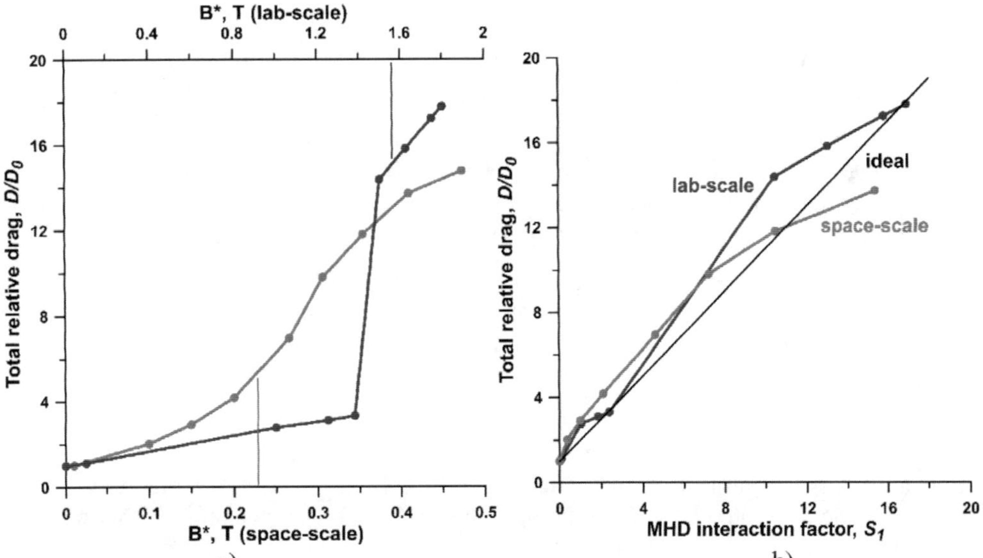

a) b)

Figure 4. Total relative drag: a) drag vs characteristic magnetic field B^*; b) drag vs MHD interaction factor S_1.

The primary result of the study is a feasibility to increase the total drag of the body in more than order of magnitude relative to the original one. The effect can be achieved by means of practically reasonable magnetic field amplitudes. Common for both cases is that the total drag grows as the magnetic field increases. The difference between two cases is that significantly smaller magnetic fields are required in space-scale case to increase aerodynamic drag than in the lab-scale case. In any case, the drag control could be used as a potential for high-speed flow control.

The second difference can be seen from Fig.4a. Namely, there is a jump in relative drag for lab-scale model. This jump was observed for other magnetic systems with close spatial scale. At the same time, the jump in drag was not observed for space-scale models and magnetic systems. It should be noted that a transition from low-drag regime to high-drag one takes typically several milliseconds. So, one can say that for given lab-scale magnetic system there is a critical magnetic field which switches the low-drag flow structure to high-drag one. The jump relates with finite scale of magnetic system and with Hall effect. At small magnetic field, MHD interaction occurs mainly over the plate between conductors. As magnetic field increases, the flow decelerates. The base for MHD interaction diminishes, and zone of electric current flow moves away the magnet. So, real MHD interaction occurs at lower and lower magnetic fields as magnetic field increases (see Fig.2 and 3). For space-scale flow, Hall parameter is small enough because magnetic field magnitude is small enough. Transition from low-drag regime to high-drag one has a smooth behavior (Fig.4a, red line). For lab-scale flow, MHD interaction zone moves from high magnetic field zone (due to decelerated or even stalled flow) with, in general, high Hall effect to the lower magnetic field zone, where Hall effect is small. Transition from Hall effect dominated flow to those with small Hall effect is found to have a critical point. The demonstration lab-scale problem has been solved, in which the Hall effect was neglected. The drag curve is similar to that corresponding to the space-scale case.

Finally, Fig. 4b shows that both drag curves look similar if appropriate characteristic numbers are chosen to specify MHD interaction. As in [3], we take MHD interaction factor S_1 defined as $S_1 = \frac{\sigma_1 B_m^2 L_*}{\rho U}$. Here, characteristic value of conductivity is chosen as $\sigma_1 = \frac{\sigma_0}{(1+\beta_0^2)}$, β_0 is Hall parameter. B_m is mid-point value of magnetic field defined above.

Conclusions

Possible aerospace application of the magnetohydrodynamic method for high-speed flow control known as the MHD parachute has been considered. The dramatic rise of aerodynamic drag has been demonstrated on the real-air flow corresponding to the Earth's re-entry flight for two test models of different spatial size with the built-in magnetic system. It was shown that the drag can be increased by more than order of magnitude by inducing acceptable magnetic fields: ~2 T for the lab-scale model, and ~0.5 T for the space-scale model. Intensity of MHD deceleration can be characterized by the appropriate MHD interaction factor. This quantity seems suitable for both scales considered in the paper despite quite different behavior of the drag for lab-scale and space-scale models. Jump-like drag dependence for lab-scale case is a mutual influence of finite-rate chemistry, decelerated flow structure, and Hall effect in a spatially restricted magnetic field.

References

[1] Bocharov A N, Bityurin V A and Lineberry J 2005 Study of MHD Interaction in Hypersonic Flows The 15th International Conference on MHD Energy Conversion (Moscow, 24–27 May) p 399.

[2] Bityurin V A, Bocharov A N 2005 MHD Flow Control in Hypersonic Flight The 15th International Conference on MHD Energy Conversion (Moscow, 24–27 May 2005) p 429.

[3] V A Bityurin and A N Bocharov 2018 *J. Phys. D: Appl. Phys.* 51 264001 https://doi.org/10.1088/1361-6463/aac566

[4] V A Bityurin *et al* 2019 *J. Phys. D: Appl. Phys.* 52 354001 https://doi.org/10.1088/1361-6463/ab2181

Journal of Physics: Conference Series **1394** (2019) 012020 doi:10.1088/1742-6596/1394/1/012020

Numerical and experimental modelling of MHD interactions at hypersonic flow around blunted body

T.A Korotaeva, V.P. Fomichev, M.A. Yadrenkin, E.K. Shipko

Khristianovich Institute of Theoretical and Applied Mechanics, 4/1 Institutskaya str., Novosibirsk 630090, Russia

korta@itam.nsc.ru

Abstract. The problem of the reducing heat fluxes to the surface of the re-entry capsules when they enter the dense layers of the atmosphere remains an urgent challenge. The purpose of the numerical and experimental study presented in this paper is to check the possibility to decelerate the descent vehicles and reduce the thermal load on their surface in the upper atmosphere using the magnetohydrodynamic (MHD) method. The problem of the flow around a blunt body with a hypersonic air flow ($M = 6$) is considered when a local MHD interaction is realized near a stagnation point. An experimental study has been carried out at the MHD test rig. The test rig is designed to study the fundamental and applied problems of magnetoplasma aerodynamics. At the experimental simulation the free-flow ionization has been implemented using an electric high-voltage discharge, as distinct from the natural conditions when thermal ionization takes place behind shock waves at the flow hypersonic speeds. Numerical simulation of the flow around a blunt body has been carried out in the framework of Navier-Stokes equation with an approximate specification of the sources of force and heat in the region of MHD interaction. The flow parameters have been defined based on the results of the analysis of experimental data. The Stuart number has been obtained in all experiments that made it possible to establish a correlation between the calculated and experimental data. As a result of the studies, the possibility of changing the flow pattern around the model, the motion of the bow shock wave towards the free flow, and the heat fluxes decrease to the body surface have been demonstrated. It has been shown that, as a result of MHD interaction, the pressure and heat flux decrease about 2 times in the vicinity of the stagnation point where recirculated gas flows occur.

Introduction

Development of an effective thermal shield for reentry capsule is one of the most important problems in the advancement of space transportation systems. At high altitudes, capsule speed changes insignificantly. At the same time, shock waves are close to the surface of the vehicle causing the high heat fluxes. Magnetohydrodynamic (MHD) method of heat load reduction and deceleration enhancement is one of the possible methods to control the flow at hypersonic speeds in the upper atmosphere (70–30 km).

MHD methods used to increase the drag of bodies moving with hypersonic speeds and reduce the heat fluxes to their surface has been considered in [1–2]. Theoretical estimation of the altitude-velocity conditions for the realization of strong MHD interaction are presented in the works mentioned. There are many theoretical works devoted to the study of the magnetic field influence on the flow around blunt bodies and heat transfer processes [3, 4]. In the paper [5], the authors presented the results of theoretical studies of flight control of a high-speed vehicle using a magnetic field source in the forebody. It has been shown that the critical part of the typical flight trajectory of the re-entry capsule lies within the range of the MHD control implementation modes. There are works address the problem of using MHD systems on real re-entry capsules. The work [6] is devoted to the numerical simulation of the flow control for the OREX orbital experimental re-entry capsule using the MHD system on the board. The parameters of the free flow and the heating temperature of the model shell were used based on the data obtained during the real flight experiment. It is noteworthy that it is possible to reduce the heat flux at the critical point by about 15%, and reduce the total aerodynamic heating to 40%, depending on the flight altitude of the vehicle at the moderate magnetic fields generated up to 0.5 T. Results of simulation of MHD deceleration of the re-entry capsule Stardust for various modifications of the magnetic system are presented in [7]. It has been shown that is possible to reduce the heat flux to the surface of the vehicle by 2–3 times at the critical point with the total aerodynamic drag body increasing by almost an order of magnitude. Electrodynamic air deceleration of the re-entry capsule Hayabusa released into the atmosphere has been studied in [8]. Total resistance of the body increased by 5%, and the heat flux decreased 15% at the magnitude of magnetic induction B =1 T.

However, there are very few experimental works confirming the effects considered. Results of a study of the influence of a magnetic field on a shock-wave structure at a supersonic flow around a cone with a nitrogen flow and on the change of the heat flux to the model surface (cone-cylinder) are presented in [9]. Ionization of the gas has been implemented using an electric discharge. It has been found experimentally that at the discharge current and the magnetic field induction increasing the thermal load on the model decreases.

That is difficult to implement the flow parameters corresponding to the thermal gas ionization that causes the phase transition at the bow shock wave at the phenomena modeling in aerodynamic test rigs. Under experimental conditions with a lower enthalpy of flow than under natural conditions artificial sources of gas ionization, for example, high-voltage electric discharges are used. Despite the fact that the ionization parameters obtained at the discharge using are not the same as at the natural conditions; and the discharge plasma is significantly nonequilibrium, one can claim the result of the simulated MHD-interaction would give reliable force acting on the flow and the flow pattern.

In the present work we study numerically and at testing the effect of MHD-deceleration of hypersonic flow (M =6) in front of a blunt body in the region located between the detached bow shock wave and the surface of the model blunt part. In this zone, the MHD-force is determined by the current density and magnetic induction, i.e. value $\vec{\boldsymbol{j}} \times \vec{\boldsymbol{B}}$.This value can be used for definition of the parameter of the hydro magnetic interaction or Stuart number. Numerical modeling is carried out at the same Stuart numbers and flow parameters. At the numerical modeling the flow around the body, the region of MHD-interaction can be simulated by a zone with sources of force or force and energy specified. Addition of the force term in the equation of conservation of momentum and thermal terms in the energy equation permits on the one hand to exclude Ohm's law from consideration, on the other hand to evaluate the possible MHD effect on the flow structure and thermal loads acting on the surface models.

Problem statement and techniques

The purpose of this study is to check the concept of the MHD parachute under the test conditions. Compared to the natural conditions where the thermal ionization takes place at the hypersonic flow deceleration, at the testing the free-stream ionization has been implemented using an electric high-voltage discharge. Figure 1 shows the schematic of the experiment (figure 1,*a*) and the problem statement for numerical simulation.

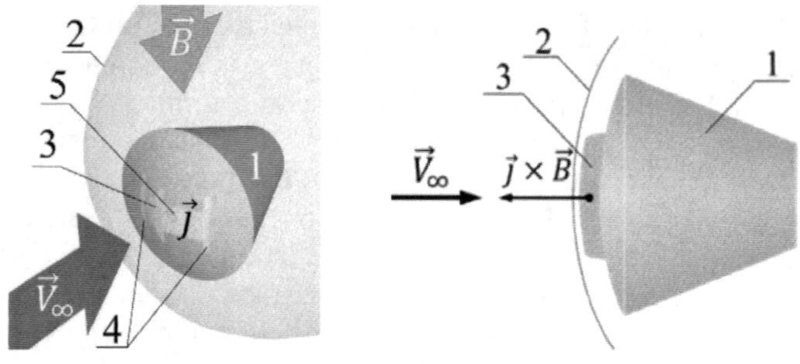

a b

Figure 1. Problem statement: (a) test schematic, (b) numerical simulation scheme.

1 – model, 2 – shock wave, 3 – ionization area (area with sources of force and energy at numerical simulation), 4 – electrodes, 5 – pressure sensor.

1.1. Experiment statement

An air flow with the Mach number M = 6 flows around the body under the conditions simulating a vehicle motion at the altitude of 28-30 km. The arrow shows the flow direction in figure 1. The magnetic field direction is orthogonal to the direction of the electric discharge current ionizing the flow. The region of the MHD interaction is organized in such a way that a decelerating force $\vec{j} \times \vec{B}$ arises in front of the blunt part of the model. The volumetric electromagnetic force is directed towards the free flow.

1.2. Numerical study statement

At the numerical modeling the flow around the body, the region of MHD interaction can be simulated by a zone with sources of force or force and energy specified. In figure 1, b, showing the schematic of the problem for the numerical solution, this zone is indicated by red color. The magnitude of the force $\vec{j} \times \vec{B}$ specified at the calculations can be determined from using value of the Stuart number (1).

$$S = \frac{\vec{j} \times \vec{B} l}{\rho_\infty v_\infty^2} \tag{1}$$

The number has been obtained in all experiments that made it possible to establish a correlation between the calculated and experimental data.

1.3. Test rig and experimental techniques

Experimental modeling has been carried out at the hypersonic air flow M = 6 using the pulsed aerodynamic MHD-test rig [10] based on a shock tube. Test rig is designed to study the fundamental and applied problems of magneto plasma aerodynamics. Electromagnet make it possible to change the magnitude of the magnetic field up to 2.5 T. Ionization of the flow in the vicinity of the stagnation point on the model has carried out using high-voltage electric discharge with the duration of 120 μs and value of about 120 A depending on the experimental conditions. Static pressure is of 1300–1600 Pa, flow density is of about 0.02 kg/m3, temperature is about 300K, flow velocity is about 2000 m/s. After the bow shock wave the static pressure increases by 4 times. At these conditions an electric discharge ionizing flow should be characterized as an arc discharge of low pressure with thermal ionization of air particles.

The stagnation pressure on the model at the critical point has been measured by a high-frequency sensor. Optical diagnostics has included schlieren photographs of the bow shock wave deformation and photographs of the discharge plasma glow. The observation has been carried out at the frequency of 80 kHz with exposure of 1 μs.

The photographs have been used to determine the length of the electric-discharge plasma in the direction along the magnetic field, which made it possible to estimate the magnitude of the discharge current density and the magnitude of the hydromagnetic interaction parameter S. The parameter has been also defined by the magnitude of the pressure change at the critical point as a result of the MHD interaction. These values are in good agreement with ones obtained at numerical simulation, although not in the entire range of parameters.

Experimental results

Measurements of the pressure in stagnation point of the blunted model and current oscillogram have shown the process of flow statement and MHD-influence. A typical pressure oscillogram is given in figure 2. One can see the main stages of the process: the flow settling near the model (I), quasi-stationary mode of the test rig (II) and the moment of the initiation of electric discharge in the magnetic field (III) leading to the gas pressure decreasing at the critical point of the model.

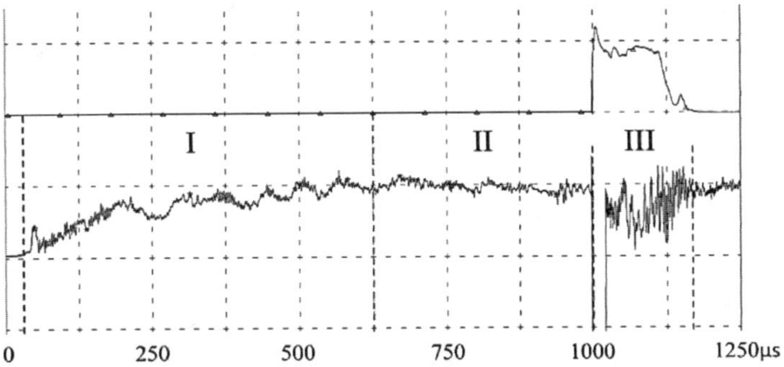

Figure 2. Oscillogram of the discharge current (top) and oscillogram of the pressure sensor (bottom).

Discharge burning occurs behind the bow shock wave where on the axis of the flow the pressure of about 52 kPa under the experimental conditions. Given the high value of current and relatively low energy deposition, one can presumably assume a thermal model of the discharge plasma, in which the separation of the electron temperature from the gas temperature is not significant. Thus, it is possible to estimate the zone of effective gas conductivity and current density using the Schlieren pictures. Assessment of the Stuart number is based on the assumption that the problem under study can be considered in a two-dimensional approximation since the electric current creating the volumetric electromagnetic force $\vec{j} \times \vec{B}$ flows predominantly in a transversal direction normal to the image plane.

Figure 3 shows the typical photographs of the flow and discharge plasma without and with magnetic field. The pictures are taken with frequency 80 kHz.

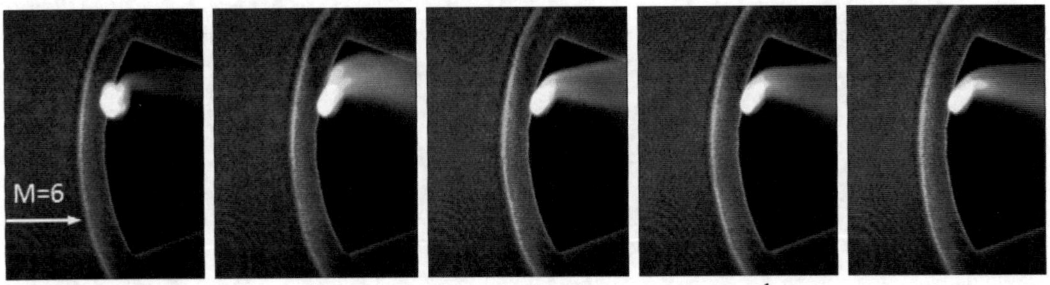

a b c d e

Figure 3 Shock-wave structure with electric discharge and without B-field.

One can see on figure 3 that the hypersonic flow blows of the discharge initiated in the vicinity of stagnation point. So the discharge plasma does not influence on the bow shock structure. Figure 4 shows the electrical gas discharge between the electrodes (figure 4,*b*), the counter flow movement of the discharge region as effect of action of the electromagnetic forces (figure 4,*c-e*) and the transformation of the departed bow shock wave before the MHD interaction zone (figure 4,*c-e*).

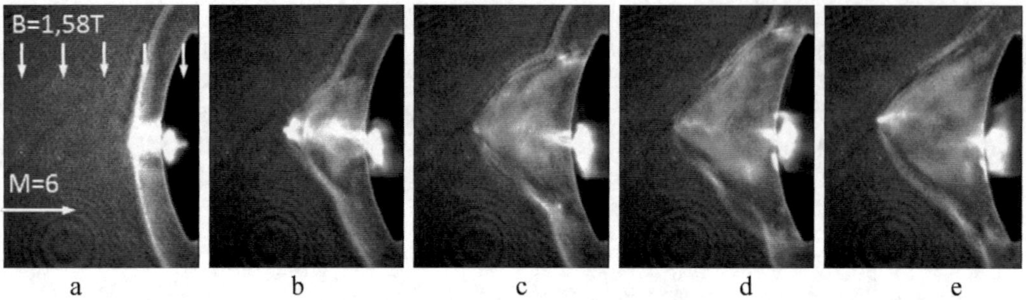

Figure 4. Shock-wave structure evolution at the strong MHD interaction between the flow and discharge plasma in the magnetic field.

Numerical modelling

Numerical simulation of the problem has been performed using the academic version of ANSYS package. The stationary Reynolds-averaged Navier-Stokes equations have been solved for the axisymmetric problem statement. Turbulence model *k-ω SST* has been used at simulation of turbulent regime. In the framework of the problem, a 'density-based' solver has been exploited for an implicit second-order accuracy scheme of Roe.

Figure 4,*a* shows an example of a two-dimensional computational mesh built using ANSYS ICEM CFD (grid generator). The mesh consists of about 200 000 rectangular cells. It is finer in the boundary layer zone (figure 5,*a*). One can see that two cell zones have been defined. The first one is '*fluid*' noted 'I' with air property given, the second one ('II') can contain sources of the force and/or energy.

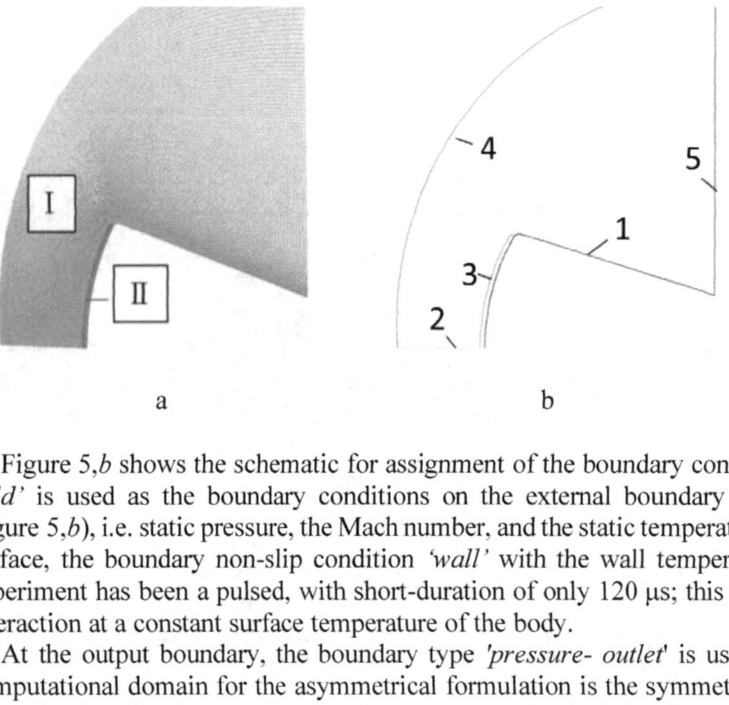

Figure 5 Mesh (a) and boundary condition (b):

I – cell zone-'fluid', II – cell zone-'fluid1', 1 – 'wall', 2 – 'axis', 'interior', 4 – 'pressure-far-field', 5 – 'pressure-outlet'.

Figure 5,*b* shows the schematic for assignment of the boundary conditions. The type '*pressure-far-field'* is used as the boundary conditions on the external boundary of the computational domain, (figure 5,*b*), i.e. static pressure, the Mach number, and the static temperature have been set. On the model surface, the boundary non-slip condition *'wall'* with the wall temperature T_w = 300°K is used. The experiment has been a pulsed, with short-duration of only 120 μs; this allows us to simulate the MHD interaction at a constant surface temperature of the body.

At the output boundary, the boundary type *'pressure- outlet'* is used. The lower boundary of the computational domain for the asymmetrical formulation is the symmetry axis and the condition *'axis'* is specified on it. The boundary between zones I and II is defined as '*interior*'. The air parameters are follows: the '*ideal gas'* option is set as the equation of state, the heat capacity *Cp* is assumed to be constant, the thermal conductivity is determined by the kinetic theory, the viscosity is described by Sutherland's law. To account for electromagnetic forces in the frame of ANSYS FLUENT, a zone has been defined near the model that is either localized as at testing or occupying the entire blunt part (figure 5,a), in which the standard means of FLUENT have been used to set the source of distributed force and energy. Fundamentally, source terms are given for the equations of conservation of motion and energy. We present these equations in the form [11]:

$$\rho\left(\frac{\partial\vec{v}}{\partial t}+(\vec{v}\vec{\nabla})\vec{v}\right)=-\nabla p+\vec{F}+\frac{\partial\vec{\tau}_x}{\partial x}+\frac{\partial\vec{\tau}_y}{\partial y}+\frac{\partial\vec{\tau}_z}{\partial z}, \quad (2)$$

where the source of pondemotor force $\vec{F}$ under the assumption of electric neutrality of the gas can be written as $\vec{F}=\vec{j}\times\vec{B}$. This force can be determined using formula (1). The energy equation must include the power determined by the pondemotor force $\vec{F}\vec{v}$, and the Joule heating $\frac{j'^2}{\sigma}$.

$$\rho\left(\frac{\partial H}{\partial t}+(\vec{v}\vec{\nabla})H\right)=\frac{\partial p}{\partial t}+\vec{F}\vec{v}+div\vec{q}+\frac{\partial\vec{\tau}_x}{\partial x}\vec{v}+\frac{\partial\vec{\tau}_y}{\partial y}\vec{v}+\frac{\partial\vec{\tau}_z}{\partial z}\vec{v}+\frac{j'^2}{\sigma} \quad (3)$$

The velocity $\vec{v}$ is determined as a value averaged over region I at the numerical simulation of the flow around the body taking into account only the source of force. Joule heating is determined using conductivity and current density determined by experimental studies.

In this paper, these source of force and energy for the mentioned flow parameters with the Stuart number of 0.1, for example, is of the order of 0.8 N/cm^3 and 500 W/cm^3, correspondingly.

Numerical simulation has been carried out under experimental conditions but it has not exactly repeated the experiment. The Stuart number, varying from 0.1 to 0.3, allowed finding a correlation between the results of experiments and calculations.

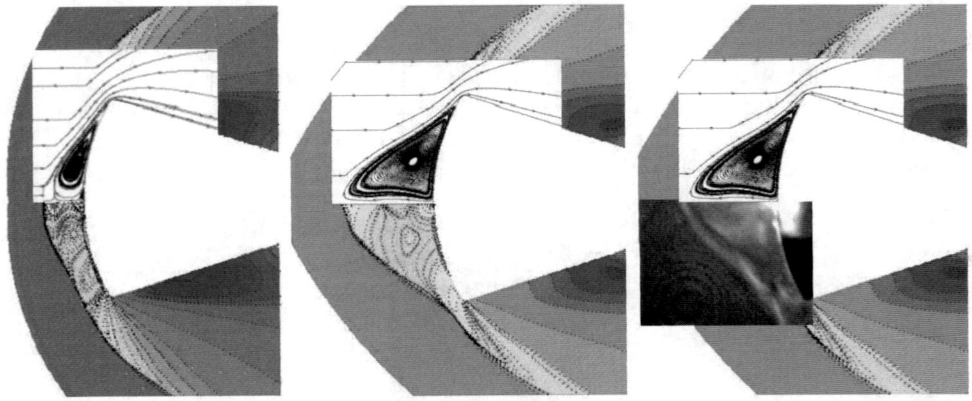

a b c

Figure 6 Numerical and experimental flow visualization(pressure pattern on the bottom, streamlines on the top): (a) the results of the calculation with source of force (S = 0.3); (b) the results of the calculation with source of force and heating; (c) Schlieren visualization compared to numerical data.

In figure 6 one can see an example of the results obtained numerically for a thin, “smeared” on the blunted part of the body imitating the MHD interaction region. Streamlines, pressure pattern and schlieren visualization obtained in one of the experiments described above are shown. A vortex zone in front of the blunt body can be clearly seen in the figure. The zone is significantly smaller in the case when only the force source is taken into account, figure 6,*a*. Figure 6,*b* shows that addition of thermal source leads to the significant stand-off increasing. Numerical flow pattern is close to the experimental one when both force and energy sources are “switched on”. Numerical simulation has shown that the region of MHD interaction leads to heat flux decreasing in to the blunt part of the model.

Figure 7 shows the heat flux distribution along the generatrix of the body of revolution. The solid black line corresponds to the uniform flow around the body. The other lines correspond to the cases of the MHD interaction simulation at the different Stuart numbers.

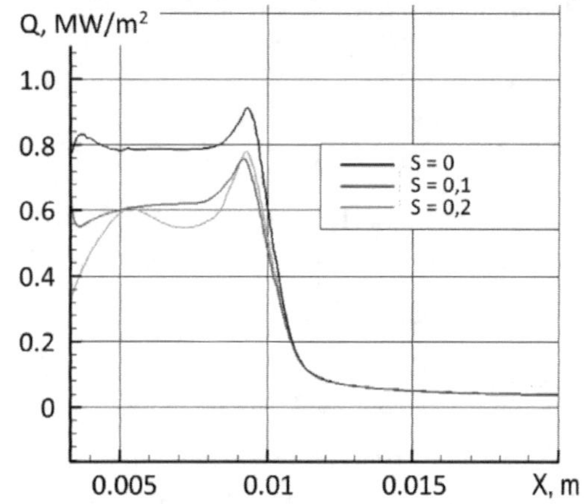

Figure 7. Heat flux distribution along the body generatrix at the uniform flow around the model (black line) and flow with force source corresponding to Stuart number S ≈ 0.1 (red line) and S ≈ 0.2 (green line).

The result obtained agrees with the conclusions of the work [7], where authors showed that that is possible to reduce the heat flux to the surface of the apparatus at its critical point by a factor of 2-3 at certain regimes. There is, practically, a complete deceleration of the flow that leads to the appearance of a vortex zone in front of the body and decrease of the heat flax in it.
The heat flux to the body decreasing is determined by the decrease in the recovery temperature of the flow decelerated in the magnetic field.

In figure 8 one can see an example of the results obtained numerically for a region imitating the region of MHD interaction in the case when it is localized in front of the blunt part of the model, as that is at the testing. Streamlines, pressure isolines and schlieren (flow visualization obtained in one of the experiments) are shown. The position and size of the region can be estimated from the solid black line over the pressure isolines (figures below).

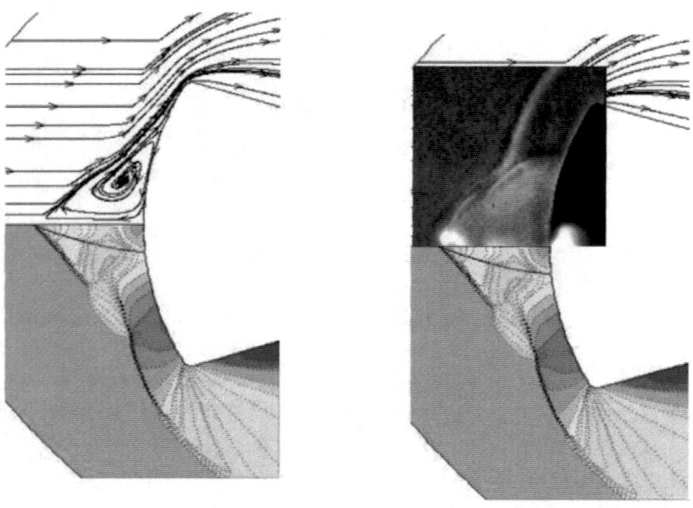

a b

Figure 8. Experimental and numerical flow visualization (S = 0.2): (a) numerical data, (b) – Schlieren visualization compared with numerical results

Discussion and conclusions

The problem of the flow around a blunt body with a hypersonic air flow (M = 6) is considered when a local MHD interaction is realized near the stagnation point. An experimental study has been carried out at the MHD test rig. At the experimental simulation the free-flow ionization has been implemented using an electric high-voltage discharge. Numerical simulation of the flow around a blunt body has been carried out in the framework of the stationary Reynolds-averaged Navier-Stokes equations with an approximate specification of the sources of force and heat in the region of MHD interaction.

The flow parameters have been defined based on the experimental data analysis. The Stuart number has been obtained in all experiments that made it possible to establish a correlation between the calculated and experimental data. As a result of the studies, the possibility of changing the flow pattern around the model, the motion of the bow shock wave towards the free flow, and the heat fluxes decrease to the body surface have been demonstrated. It should be noted that only force effect taking into account (Stuart number) is not enough for adequate simulation of the MHD-influence on bow shock structure.

It has been shown that the MHD interaction leads to the pressure and heat flux decrease about 2 times in the vicinity of the stagnation point where recirculated gas flows occur.

The results of experiments and calculations prove that that is possible to control the shock-wave structure in the high-speed air flow in front of the blunt body with external ionization of the flow in the wide range of external magnetic field induction while reducing thermal loads on the surface of the capsule.

Acknowledgments

The research has been (partly) carried out within the framework of the Program of Fundamental Scientific Research of the state academies of sciences in 2013-2020 (project No. 0323-2018-0006 and 0323-2018-0011) and the grant from the Russian Science Foundation (project No. 17-19-01375).

References

[1] Rosa, R.J., Engineering Magneto-Hydrodynamics *Ph. D. Thesis*, 1956.

[2] Resler, E.L., Sears, W.R., The Prospects for Magneto-Aerodynamics *J. Aeronaut. Sci.*, Vol. 25, April 1958, pp. 235-245, 258.

[3] Bush W.B., The Stagnation-Point Boundary Layer in the Presence of an Applied Magnetic Field, *J. Aerospace Sci.*, Vol. 28, No. 8, Aug. 1961, pp. 610–611.

[4] Kemp, N.H., On hypersonic stagnation-point flow with a magnetic field, *J. Aeronaut. Sci.*, 1958, vol. 25, No. 6, pp. 405-407.

[5] Ericson, W.B., Maciulaitis, A., Investigation of Magnetohydrodynamic Flight Control, *J. Spacecraft*, Vol. 1, No. 3, 1964, pp. 283-289.

[6] Fujino, T., Ishikawa, M., Numerical simulation of control of plasma flow with magnetic field for thermal protection in earth reentry flight, *IEEE Trans. on Plasma Sci.*, 2006. Vol. 34, No. 2. pp. 409-420.

[7] Bityurin, V.A., Bocharov, A.N., Features of electromagnetic thermal protection of space capsule/doi: 10.1134/S106378501105004X

[8] Katsurayama H. et al., Numerical Study of the Electromagnetic Control of a Weakly Ionized Flow around a Blunt Body: Role of a Insulative Boundary in the Flow, *AIAA Paper* 2007-4529.

[9] Bobashev, S.V., et al.,Experimental investigation of magnetohydrodynamic action on a heat flux toward the surface of a model, *Technical Physics,* doi: 10.1134/S106378421012008X

[10] I. P. Ginzburg "Aerogazodinamika (Kratkii Kurs)" Moscow, Izd. Vysshaya Shkola, 1966, p 1–404.

[11] Golovnov, I.A., Pozdnyakov, G.A., Pravdin, S.S., Podzin, V.E., Pushkareva, T.I., Fomin, V.M., Fomichev, V.P., Yakovlev, V.I. A Test rig for experimenatal simulation and study of the MHD-interaction in a hypersonic flow, *Preprint № 7-2003*, ITAM SB RAS, 2003.

Journal of Physics: Conference Series **1394** (2019) 012021 doi:10.1088/1742-6596/1394/1/012021

Study of the ionizing gas flow in the channel of plasma accelerator with different ways of gas inflow at the inlet

A N Kozlov [1], N S Klimov [2], V S Konovalov [1], V L Podkovyrov [2] and R V Urlova [2]

[1] Keldysh Institute of Applied Mathematics, RAS, Moscow, Russia
[2] Institute for Innovation and Fusion Research, Troitsk, Moscow, Russia

E-mail: andrey-n-kozlov@mail.ru

Abstract. Numerical studies of the ionization process are presented for various ways of the gas supplying at the inlet of the coaxial channel of the quasi-stationary plasma accelerator. Simulation of two-dimensional axisymmetric flows of ionizing gas was carried out on the basis of the equations of radiation magnetic gas dynamics taking into account radiation transport. The approximation of local thermodynamic equilibrium in three-component medium consisting of atoms, ions, and electrons was used in the MHD model. It was determined that non-uniform gas inflow at the channel inlet leads to even greater instability of the pulsating unsteady flows of the ionizing gas in the plasma accelerator. The method of long characteristics was used in the 3D model of radiation transport, including the basic mechanisms of emission and absorption for different portions of the spectrum. The radiation field in the ionizing gas flow was determined, including the integral radiation characteristics and the spectral radiation intensities in the narrow ranges of frequencies or photon energies.

1. Introduction

New cycle of theoretical and experimental research in the quasi-stationary plasma accelerators (QSPA) (see e.g. [1-5]) is conducted in order to study the effect of various ways of the gas supplying to the QSPA channel on the gas ionization process. The simplest plasma accelerator, shown schematically in figure 1, consists of two coaxial electrodes connected to the electrical circuit. Breakdown in gas between the electrodes leads to the formation of an ionization front corresponding to the phase transition from the gaseous state of matter to the plasma. In classical installations, the processes occur in the presence of the azimuthal magnetic field H_ϕ, which is generated by the electric current flowing along the inner electrode. Beyond the front, the ionized plasma is accelerated in the longitudinal direction due to the Ampere force $\boldsymbol{F}_A = \frac{1}{c}\boldsymbol{j} \times \boldsymbol{H}$, where $\boldsymbol{j}$ is the current in the plasma, which has predominantly radial direction. These processes take place, in particular, in small accelerators of the first stage of the two-stage QSPA [1]. The QSPA installations are intended for technological applications and thermonuclear research (see e.g. [1, 6, 7]), and are also of interest for the development of the promising powerful electrojet plasma engines, since they are flow-type systems in which gas is continuously injected at the inlet. The latter circumstance becomes especially relevant in view of the developed small-sized atomic reactors.

Journal of Physics: Conference Series **1394** (2019) 012021 doi:10.1088/1742-6596/1394/1/012021

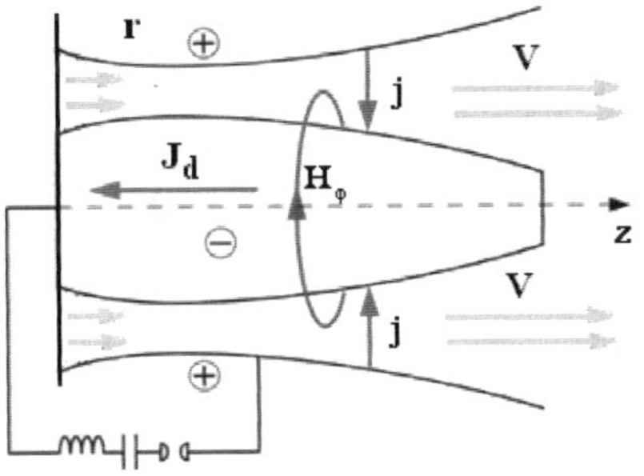

Figure 1. Scheme of the simplest coaxial plasma accelerator

Theoretical and numerical studies of the ionization process in the QSPA are carried out using models of various levels of complexity. The ionizing gas flow in the framework of the quasi-one-dimensional approximation was considered in [8, 9] for narrow cylindrical channel. The foundations of the theory of processes at the ionization front for stationary flows have been developed in [10]. Three-dimensional numerical model of radiation transport in streams of ionizing gas is presented in [11] taking into account a number of factors related to the accuracy of description of the geometry of the radiating volume, the exclusion of shadow regions from the calculations, the details of description of the radiation spectrum and the main mechanisms of emission and absorption.

This paper presents the results of numerical studies of two-dimensional axisymmetric flows of ionizing gas in the QSPA-T installation modernized at the Troitsk Institute for Innovation and Fusion Research to study the possibility of different ways of the gas supplying at the inlet to the accelerator channel.

2. Equations of radiation magnetic gas dynamics

The MHD model of two-dimensional axisymmetric flows of an ionizing gas is based on the transfer equations for three-component medium [12], consisting of atoms, ions and electrons, as well as the diffusion equation of the magnetic field, which follows from the Maxwell's equations and the Ohm's law, if the inertia of electrons and the displacement current are neglected. The ionization process is being investigated for the hydrogen used in the experiments at this stage. The masses of atoms and ions are equal $m_a = m_i = m$. The temperature at the ionization front increases up to a level $1 \div 3\ eV$. Concentration of gas entering into the channel has fairly high values $n = 10^{16} \div 10^{18}\ cm^{-3}$. We can assume that the medium is quasi-neutral $n_i = n_e$, and the velocities of components of medium are equal to each other $\boldsymbol{V}_i = \boldsymbol{V}_e = \boldsymbol{V}_a = \boldsymbol{V}$. Estimates and experiments also show that we can confine ourselves to the case of single temperature mixture.

Transformation of the initial equations with regard to the above assumptions leads to the following set of the modified MHD equations:

$$\frac{\partial \rho}{\partial t} + \nabla \cdot (\rho \boldsymbol{V}) = 0, \qquad \rho \frac{d\boldsymbol{V}}{dt} + \nabla P = \frac{1}{c} \boldsymbol{j} \times \boldsymbol{H}, \qquad \frac{d}{dt} = \frac{\partial}{\partial t} + (\boldsymbol{V}, \nabla),$$
$$\frac{\partial}{\partial t}(\rho\, \varepsilon) + \nabla \cdot (\rho\, \varepsilon\, \boldsymbol{V}) + P\, \nabla \cdot \boldsymbol{V} = \frac{j^2}{\sigma} - \nabla \cdot \boldsymbol{q} - \nabla \cdot \boldsymbol{W}, \qquad (1)$$
$$\frac{\partial \boldsymbol{H}}{\partial t} = \nabla \times (\boldsymbol{V} \times \boldsymbol{H}) - c\, rot \frac{\boldsymbol{j}}{\sigma}, \qquad \boldsymbol{j} = \frac{c}{4\pi} \nabla \times \boldsymbol{H},$$

$P = P_a + P_i + P_e = (1 + \alpha)(c_P - c_V)\rho\, T$, $\varepsilon = (1 + \alpha)\, c_V\, T + \varepsilon_I$, $k_B / m = R = c_p - c_v = c_v\, (\gamma - 1)$, $\alpha = n_e / (n_a + n_i)$, $\boldsymbol{q} = -\, \kappa_{e \to a}\, \nabla T$. The equations contain the density of heavy particles $\rho = m(n_a + n_i)$, the degree of ionization α, the heat flux $\boldsymbol{q} = -\, \kappa_{e \to a}\, \nabla T$ caused by electron-atomic heat conduction $\kappa_{e \to a}$, and the density of the radiation energy flux $\boldsymbol{W}$. The equation for internal energy includes the Joule heating $Q_{ei} = \boldsymbol{j}^2/\sigma$, which considerably exceeds the heat generated by friction with other components. The concept of internal energy per unit mass $\varepsilon = (1 + \alpha)c_V T + \varepsilon_I$ includes an additional term $\varepsilon_I = \varsigma \alpha I / m_i$ that corresponds to the energy losses due to ionization, where value I is the ionization energy of atom. The electrical conductivity of three-component medium in equations (1) is equal to $\sigma = e^2 n_e / m_e \nu_e$, where the average frequency of electron collisions with other particles is composed of the frequencies of collisions with atoms and ions $\nu_e = \nu_{ea} + \nu_{ei}$, where $\nu_{ea} = n_a \langle V_e \rangle S_{ea}$, $\nu_{ei} = n_i \langle V_e \rangle S_{ei}$, S_{ea} and S_{ei} are the effective collision cross sections. Heat transfer at high degrees of ionization is determined by electron thermal conductivity across the magnetic field including dependence on parameter $\omega_e\, \tau_e$ [12], and at low degrees of ionization, atomic thermal conductivity plays a certain role.

In the case of local thermodynamic equilibrium (LTE), the concentrations of all three components of the medium and the electron temperature are connected by the Saha relation

$$\frac{n_i\, n_e}{n_a} = K_1(T) = \frac{2\, \Sigma_i}{\Sigma_a} \left(\frac{m_e\, k_B\, T}{2\, \pi\, \hbar^2}\right)^{3/2} exp\left(-\frac{I}{k_B\, T}\right), \quad (2)$$

where the value K_1 is the constant of the ionization balance, Σ_a and Σ_i are the statistical sums of atom and ion. The Saha relation (2) corresponds to the direct process of ionization of atoms from the ground state as a result of the electron impact and the reverse process of recombination in the collision of ions with electrons: $A + e \leftrightarrow A^+ + e + e$. The equation (2) under the condition of quasi-neutrality determines the degree of ionization

$$\alpha = -K_1(T) \,/\, 2\, n + \sqrt{(K_1(T) \,/\, 2\, n)^2 + K_1(T) \,/\, n}\,. \quad (3)$$

The heat transfer due to the radiation energy flux, which can play a certain role in the redistribution of energy, is taken into account in the framework of radiation magnetic gas dynamics. The density of the radiation energy flux $\boldsymbol{W}$ and the radiation energy density U, which is substantially less than the thermal or internal energy of the medium $U << \rho\, \varepsilon$, are determined by radiation intensity $I_\nu(\boldsymbol{r},\, \boldsymbol{\Omega})$ along the chosen direction and for a certain part of spectrum using the following relations

$$U(\boldsymbol{r}) = \frac{1}{c} \int_0^\infty \int_0^{4\pi} I_\nu(\boldsymbol{r},\, \boldsymbol{\Omega})\, d\, \Omega\, d\nu, \quad \boldsymbol{W}(\boldsymbol{r}) = \int_0^\infty \int_0^{4\pi} I_\nu(\boldsymbol{r},\, \boldsymbol{\Omega})\, \boldsymbol{\Omega}\, d\, \Omega\, d\nu. \quad (4)$$

The radiation intensity is calculated by means of the stationary equation of the radiation transport

$$\boldsymbol{\Omega} \cdot \nabla\, I_\nu(\boldsymbol{r},\, \boldsymbol{\Omega}) = \eta_\nu(\boldsymbol{r}) - \kappa_\nu(\boldsymbol{r}) \cdot I_\nu(\boldsymbol{r},\, \boldsymbol{\Omega})\,, \quad (5)$$

in which the absorption coefficient $\kappa_\nu(\boldsymbol{r})$ and emissivity $\eta_\nu(\boldsymbol{r})$ depend on the state of the medium, its density and temperature, as well as the spectral parameter ν associated with the photon energy $h\nu$. The whole spectrum is divided into more than 300 spectral groups. The absorption coefficient and emissivity are determined using relations in [11] that contain three parts corresponding to absorption and emission in lines, photo-ionization and photo-recombination, and also scattering (see e.g. [13–17]). The coefficients in equation (5) also depend on the relative concentration of k -th state of atom $x_k = n_k/n$, the oscillator strength f_{kj} for $k \rightarrow j$ transition in atom, the line profile $\varphi_{kj}(\nu)$ corresponding to the bound-bound transition, the cross sections of photo-ionization and bremsstrahlung. The line profile, which takes into account the different mechanisms of broadening of spectral lines, is determined using the Voigt formula. The populations of states under the LTE condition are related by the Boltzmann formula $x_k = \frac{n_a}{n}\frac{g_k}{\Sigma_a} exp(-(E_k - E_1)/k_B T)$, where values g_k are the statistical weights of k -th state of atom, $E_k = Z^2 e^2/2\, a_o\, k^2$ is the energy levels of hydrogen atom and a_o is the Bohr radius.

3. Formulation of problem and numerical solution methods

To find the numerical solution of the problem with equations (1) and (3), dimensionless variables are used. The units of measurement are the characteristic gas concentration n_o and temperature T_o at the inlet to the accelerator channel, the length of the channel or its part L, and the characteristic value of azimuthal magnetic field H_o at the inlet, which is determined by the discharge current J_d in the system so that $H_o = 2J_d/cR_o$, where R_o is the characteristic channel radius. The units of measurement of other variables are formed using the specified parameters, including units of pressure $P_o = {H_o}^2/4\,\pi$, velocity $V_o = H_o/\sqrt{4\,\pi\,\rho_o}$, time $t_o = L/V_o$, electric field $E_o = H_o V_o/c$, current in the plasma

$j_o = cH_o/4\pi L$, and density of the radiation energy flux $W_o = V_o\ H_o{}^2/4\ \pi$. As a result, the set of the MHD equations in dimensionless variables contains the dimensionless parameters presented in [8–11].

The gas at the inlet to the channel for $z = 0$ is supplied with the known values of density $\rho(r) = f_1(r)$ and temperature $T(r) = f_2(r)$. The discharge duration is significantly greater than the particle transit time. Therefore, we believe in the numerical model that the current in the system is kept constant and flows only through the electrodes. Hence we have $j_z = 0$ or $r\ H_\phi = r_o = const$ $(r_o = R_o/L)$ at $z = 0$. Coaxial electrodes $r = r_a(z)$ and $r = r_\kappa(z)$, forming the channel walls, are equipotential ($E_\tau = 0$) and impermeable ($V_n = 0$) surfaces.

The algorithm for the numerical solution of the problem with equations (1) in the profiled channel, shown in figure 2 in the plane of variables $(z,\ r)$, includes the mapping of the computational domain onto a rectangle in the plane $(y,\ z)$ using the relation

$$r = (1 - y)\ r_\kappa(z) + y\ r_a(z)\ . \tag{6}$$

The numerical model of two-dimensional axisymmetric flows of ionizing gas and plasma suggests the splitting in the coordinate directions and physical factors. The calculation of the hyperbolic part of the MHD equations is based on the difference scheme with correction of flows (see e.g. [18]). The effect of electrical conductivity and heat conduction in the parabolic part of the set of the MHD equations is taken into account by means of the flux variant of the back-substitution method [19].

The three-dimensional formulation of the problem of radiation transport for the axisymmetric flow of ionizing gas and plasma in the accelerator channel is necessary within the framework of the integral relations (4), on the basis of which the density and flux of radiation energy are calculated at each node or cell of the coordinate grid. Three-dimensional grid is built by rotating the initial grid in variable plane $(z,\ r)$ by 360 degrees around the axis of the channel with a certain step. An additional angular grid along the azimuthal and polar angles is built at the nodes of the coordinate grid, providing the uniform distribution of the rays along the directions in the complete spatial angle $\Omega = 4\ \pi$. Ray tracing is carried out in accordance with the method of long characteristics [16, 17] in order to determine the points of intersection of the rays with the faces of the cells of the 3D grid. We assume that the absorption coefficient κ_ν and emissivity η_ν in equation (5) are constant within a single cell. The method of characteristics allows to take into account the geometry of the accelerator channel in details. The invisible shadow regions are excluded from the calculation of the radiation energy flux. The matching of the known solutions of equation (5), corresponding to constant coefficients κ_ν and η_ν within each cell, is performed at their boundary. As a result, we obtain a solution on a characteristic passing through an arbitrary number of homogeneous regions. Details of the algorithm for solving the problem of radiation transport in the accelerator channel are presented in [11]. The developed algorithm makes it possible to determine the emission spectrum in any spectral range for any ray emerging from the plasma volume.

The heated walls of the accelerator channel can serve as an additional source of radiation. To evaluate the effect of radiation coming from the channel walls, we used the data of [20] on the degree of blackness of various materials, including copper. Calculations taking into account this factor showed that even in the case of an increase in wall temperature, the radiation of the electrodes does not significantly affect the radiation intensity inside the channel.

4. Results of numerical experiments

Calculations of two-dimensional axisymmetric flows of ionizing gas in the channel of the modernized plasma accelerator QSPA-T were carried out on the basis of the presented modified MHD model for various ways of the gas supplying at the inlet. One of the calculations shown in figure 2 for the following characteristic parameters of the problem: $n_o = 4 \cdot 10^{17} cm^{-3}$, $T_o = 750\ {}^oK$, $J_d = 50\ kA$, and $L = 10\ cm$ corresponds to the non-uniform supply of gas at the inlet in accordance with curve 1 in figure 3. Possible ways of specifying the density at the channel inlet are shown in figure 3 at $z = 0$. Here, line 2 corresponds to uniform inflow of gas at the inlet, curves 1 and 3 correspond to non-uniform inflow. Concentration of gas for curve 1 in figure 3 is less in the vicinity of the inner electrode.

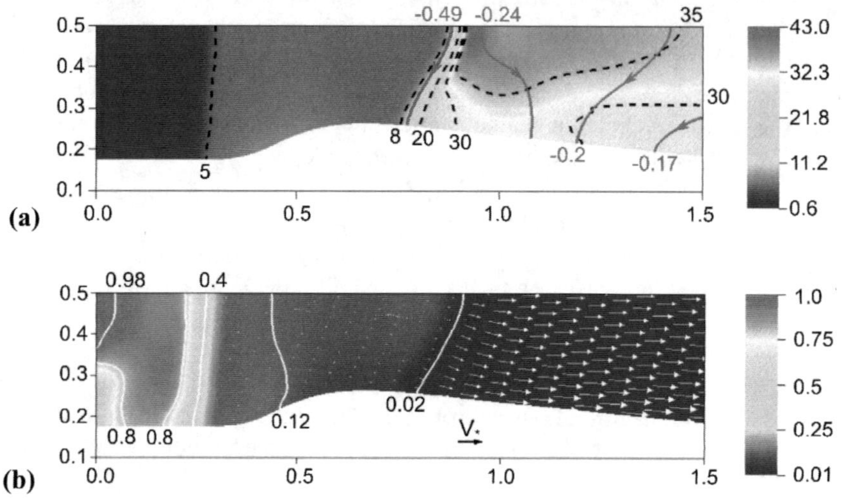

Figure 2. Distributions of (a) temperature and plasma current, (b) density and velocity vector field in the ionizing gas flow under condition of non-uniform inflow at the inlet

The characteristic temperature distribution $T(z,\ r)$ in axisymmetric plasma flow is given in figure 2(a) in the plane $(z,\ r)$. Level dashed lines complement the temperature distribution. The solid curves or level lines for function $r\ H_{\phi} = const$ in figure 2(a) correspond to the current in the plasma. We have $r\ H_{\phi} = r_o$ for $z = 0$. Direction of the arrows for plasma current in figure 2(a) corresponds to the selected polarity of electrodes. We assume that external electrode is anode.

It can be seen that the ionization front is formed in the narrow part of the channel at $z \approx 0.8$. The sharp increase in temperature occurs in the vicinity of the front. Plasma current mainly flows in the volume of the ionized gas behind the ionization front. The sharp acceleration of the plasma occurs behind the front due to the Ampere force. Figure 2(b) shows the characteristic density distribution and the velocity vector field in the ionizing gas flow under condition of non-uniform inflow of gas at the inlet. The scale of the velocity vectors is determined by the vector $V_* = 15 \cdot\ V_o$ indicated in figure 2(b). Here, the unit of velocity is equal to $V_o = 0.7 \cdot 10^6\ cm\ /\ s$ for the initial parameters indicated above.

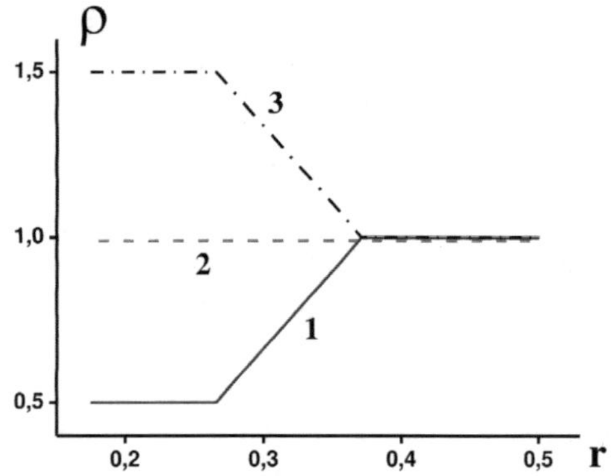

Figure 3. Possible distributions of gas density at the inlet to the channel

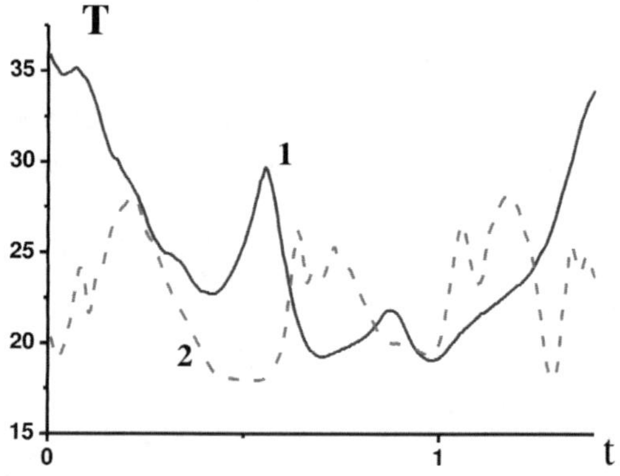

Figure 4. Temperature change over time behind the front

It is known that the ionizing gas flow can be pulsating or non-stationary at certain values of discharge current J_d and mass flux $\dot{m}$. The empirical condition for stationarity of the gas ionization process was formulated and presented in [9] within the framework of the quasi-one-dimensional flow model by means of the inequality $J_d^2 \;/\; J_{\dot{m}} > K$, where $J_{\dot{m}} = e\dot{m}/m_i \;\; (kA)$ is the mass flux expressed in current units, and the constant K generally depends on the geometry of the accelerator channel.

The calculation variant presented in figure 2 corresponds to the unsteady flow of ionizing gas. In this case, the most noticeable changes over time are observed for temperature and degree of ionization. Curve 1 in figure 4 characterizes the temperature change with time at $z = 0.9$ on the middle coordinate line for $y = 0.5$ in accordance with (6) and for the above parameters corresponding to figure 2.

The characteristic one-dimensional distributions of the MHD variables corresponding to figure 2 are presented on figure 5 along the middle coordinate line at y = 0.5 taking into account relation (6). Figure 5(a) shows that a sharp increase in temperature at the ionization front at $z \approx 0.8$ is accompanied by a sharp increase in the degree of ionization (solid curve). Figure 5(b) illustrates the change in the flow velocity (solid curve), the gas-dynamic sound velocity $C_g = \sqrt{\gamma\; P \;/\; \rho}$ (dashed curve) and the speed of the fast magnetosonic wave C_s (dash-dotted curve) along the middle coordinate line. The speed of the fast magnetosonic wave in the absence of a longitudinal magnetic field or in the presence of a single azimuthal field component corresponds to the signal speed [1], which is equal to $C_s = \sqrt{C_g^2 + C_A^2}$, where $C_A^2 = H^2/\;\rho$. This figure shows that the acceleration process is accompanied by a sequential transition of the flow velocity values through the values of the gas-dynamic sound velocity and the signal velocity.

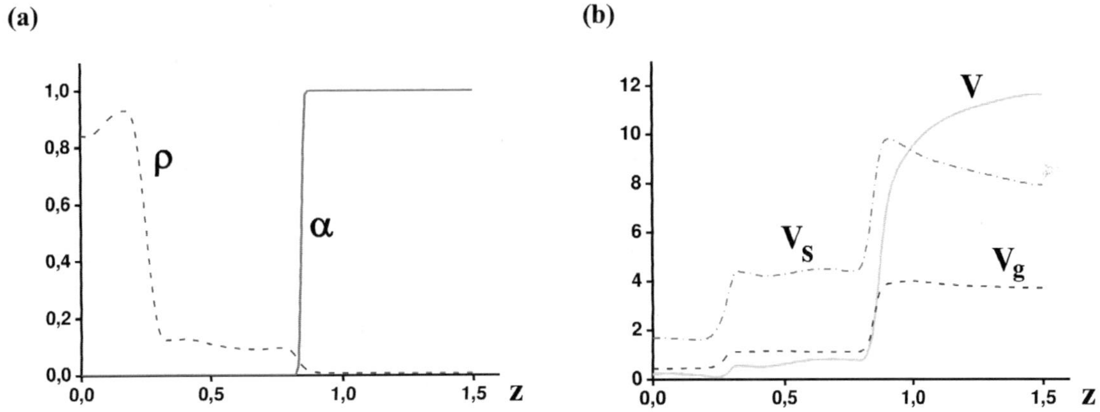

Figure 5. Changes in (a) density and degree of ionization, (b) flow velocity (solid curve), gas dynamic sound velocity (dashed curve), and fast magnetosonic wave velocity (dash-dotted curve) along the middle coordinate line

Numerical studies have been conducted for various ways of gas supply at the inlet. In some cases, there is a breakdown of the ionization process. This corresponds to curve 3 in figure 3, for example. Uniform gas supply at the inlet, corresponding to curve 2 in figure 3, also led to the unsteady flow of ionizing gas (see curve 2 in figure 4) for the above parameters of the problem. Accordingly, the transition to the non-uniform gas supply at the inlet does not contribute to the stabilization of the ionizing gas flow for the specified parameters and the discharge current $J_d = 50 \; kA$. On the contrary, the comparison of curves 1 and 2 in figure 4 shows that the non-uniform supply of gas at the inlet leads to an even greater destabilization of the pulsating flows of ionizing gas, obtained under the condition of the uniform supply of gas.

The integral flow characteristics, which are calculated in the outlet section of the channel, are also of some interest. In this case, mass flux is equal to$\dot{m} \approx 2\ g/s$, and thrust of installation is $P \approx 173\ н$.

5. Integral and spectral characteristics of radiation

The radiation field in the ionizing gas flow is determined on the basis of the solution of the radiation transport equation in the 3D formulation of the problem. The two-dimensional distribution of the radiation energy density U and the vector field of the radiation energy flux density $\boldsymbol{W}$ in the axisymmetric ionizing gas flow are presented in figure 6. This radiation field corresponds to the density and temperature distributions shown in figure 1 under the condition of the non-uniform gas supply at the inlet to the channel.

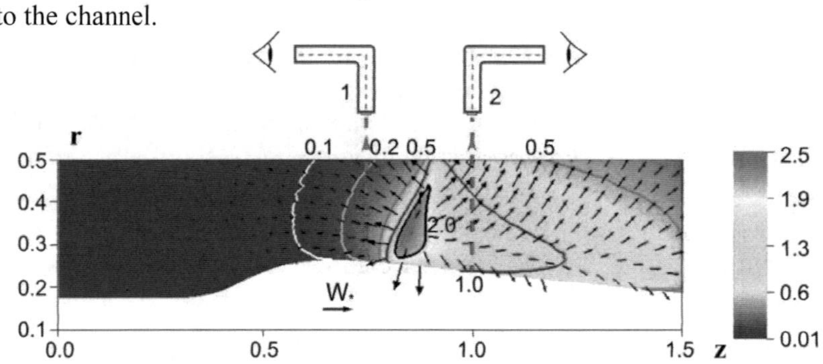

Figure 6. Distribution of the radiation energy density and vector field of the radiation energy flux density in the ionizing gas flow

The figure 6 shows the distribution of the magnitude $\widetilde{U}$, which is related to the radiation energy density by means of relation $\widetilde{U} = 10^{-8} \cdot c \cdot\ U\ erg\ /\ cm^2 s$. The scale of the vectors $\boldsymbol{W}$ is determined by the modulus of the vector $W_* = 7 \cdot 10^7\ erg\ /\ cm^2 s$, which is indicated in the figure.

There is a region with relatively high values of the radiation energy density U in the vicinity of the ionization front. The radiation energy flux $\boldsymbol{W}$ is directed in all directions from this region, including the direction of the incoming flux of weakly ionized gas. This leads to its preliminary ionization directly ahead of the front, as well as in the depth of volume of the incoming gas.

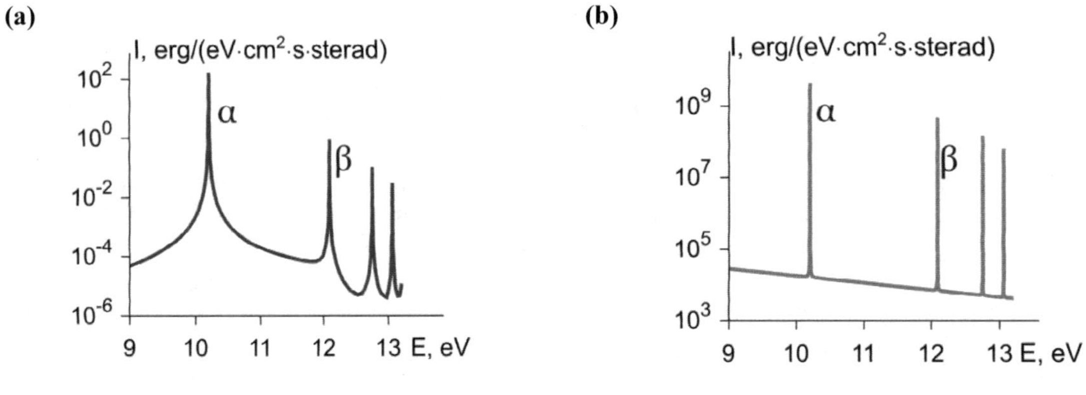

Figure 7. Spectral intensities of radiation (a) in direction of ray 1 and (b) for ray 2, presented in figure 6

The 3D model of radiation transport allows determining the spectral radiation intensity in any frequency range or photon energy for any ray emerging from the plasma volume in the direction of the detector. This circumstance opens up new opportunities for carrying out the complex theoretical and experimental studies based on flux spectroscopy. For example, figure 7 shows the spectral intensities of radiation in the frequency range, which contains the α and β Lyman lines for radial rays 1 and 2, presented in figure 6 and exiting the plasma volume in the direction of the detectors.

6. Conclusion

New cycle of theoretical and experimental research for the modernized quasi-stationary plasma accelerator QSPA-T, developed in the Troitsk Institute for Innovation and Fusion Research, is being conducted to study the influence of various factors on the gas ionization process. Computational research of the phase transition in the flow of the ionizing gas and plasma in the coaxial channel of the plasma accelerator have been carried out on the basis of the MHD equations in the approximation of local thermodynamic equilibrium. The results of numerical studies of two-dimensional axisymmetric flows of the ionizing gas under various conditions of gas supply at the inlet to the accelerator channel were presented. It is shown that non-uniform gas inflow at the inlet leads to the still greater destabilization of the pulsating unsteady flows of ionizing gas, obtained under the condition of uniform supply of gas for the considered parameters of the problem.

Acknowledgement
Research was supported by the Russian Foundation for Basic Research (project No. 18-29-21007).

References

[1] Morozov A I 2008 Introduction to plasmadynamics (Moscow: Fizmatlit)
[2] Klimov N *et al* 2009 *J. Nuclear Materials* **390-91** 721-26
[3] Garkusha I E *et al* 2017 *Nuclear Fusion* **57** 116011
[4] Astashynski V M et al 2004 J. Surface and Coating Technology **180-81** 392-95
[5] Kozlov A N, Drukarenko S P, Klimov N S, Moskacheva A A and Podkovyrov V L 2009 *Problems of Atomic Science and Technology. Series: Plasma Physics* **1** 92-4
[6] Kozlov A N 2017 Plasma Physics and Controlled Fusion **59** 115004
[7] Kozlov A N 2019 Plasma Physics and Controlled Fusion **61** 035008
[8] Kozlov A N 2000 *Fluid Dynamics* **35** 784-90
[9] Brushlinskii K V, Kozlov A N and Konovalov V S 2015 *Computational Mathematics and Mathematical Physics* **55** 1370–80
[10] Barmin A A and Kozlov A N 2013 *Fluid Dynamics* **48** 556-66
[11] Kozlov A N and Konovalov V S 2017 Communications in Nonlinear Science and Numerical Simulation **51** 169-79
[12] Braginsky S I 1965 Transport processes in a plasma *Reviews of Plasma Physics* vol 1 (New York: Consultants Bureau) p 205
[13] Zel'dovich Y B and Raizer Y P 2002 *Physics of shock waves and high-temperature hydrodynamic phenomena* (New York: Dover Publication Inc. Mineola)
[14] Mihalas D 1978 *Stellar atmospheres. Second edition* (San Francisco: W. H. Freeman)
[15] Veinstein L A, Sobel'man I I and Yukov E A 1979 *Excitation of atoms and broadening of spectral lines* (Moscow: Nauka)
[16] Chetverushkin B N 1985 Mathematical modeling of radiating gas dynamic problems (Moscow: Nauka)
[17] Castor J I 2000 *Lectures on radiation hydrodynamics* (Livermore: Lawrence Livermore National Laboratory)
[18] Oran E S and Boris J P 1987 *Numerical simulation of reactive flow* (New York: Elsevier)
[19] Degtyarev L M and Favorskii A P 1969 Sov. J. Computational Mathematics and Mathematical Physics **9** 285-94
[20] Siegel R and Howell J 1972 *Thermal Radiation Heat Transfer* (New York: McGraw-Hill)

Kinematics, dynamics and energy balance of electric arc in a magnetic field

V A Bityurin, A N Bocharov, I A Moralev, T N Kuznetsova

Joint Institute of High Temperatures, Russian Academy of Sciences
Izhorskaya 13. Bd 2, Moscow, Russia, 125412

valentin.bityurin@gmail.com

Abstract. Electric arcs (EA) are widely used in various applications as a high-density energy source, providing heating of the environment to temperatures of several thousand degrees. The presence of an external magnetic field perpendicular to the arc current vector causes a ponderomotive force normal to the plane of the electric current and magnetic field vectors. Thus, in addition to the thermal effect, there is also a dynamic one. It is these properties that motivate the use of EA as an MHD actuator to control the flow characteristics of gaseous working bodies in various technological processes. MHD actuator is most often considered as a control element (1) in combustion processes to intensify the mixing of fuel-air mixtures and maintain combustion due to the volume and constant high-temperature energy supply and (2) to control the characteristics of the boundary layer in the flow of the working surfaces of the devices. In both cases, the key issue is the efficiency of mixing (relative displacement of neighbouring elements of the medium). Within the framework of the standard "Eulerian" description of the gas-dynamic fields evolution in the laboratory reference frame, the movements of individual elements of the medium are not explicitly available. In this regard, this work is focused on the trajectories analysis of the individual mass elements exposed to the MHD of the actuator.

1. Introduction

Electric arcs, in addition to the thermal effect, could provide a dynamic effect, which predetermines the possibility of developing various technologies for converting energy and momentum: MHD generators and accelerators, current breakers, railguns, mixers, heaters, etc. These features make it possible to consider the use of EA as an MHD actuator to control the characteristics of the flow of gaseous working bodies in various technological processes.

At the present time the MHD actuator is considered most often as a means of control either in combustion processes to intensify the mixing of fuel-air mixtures and maintain combustion due to the volume and constant high-temperature energy supply [1-3], or to control the characteristics of the boundary layer when flowing around the working surfaces of devices [3-5]. In both cases, the key issue is the efficiency of mixing (relative displacement of neighboring elements of the medium). Within the framework of the standard "Eulerian" description of the gas-dynamic field evolution in the laboratory reference frame, the movements of individual elements of the medium are not explicitly available. In this regard, this work, initiated by large-scale experimental and computational-theoretical studies of MHD actuators for various applications in modern aerodynamics [3-8], is focused on the trajectories analysis of individual mass elements affected by the MHD actuator. To this end, a standard numerical experiment using a complete two-dimensional unsteady system of Navier-Stokes equations for laminar flow of real air with simplified equations of magnetic gas dynamics (an extended characteristic of the model used with the implementation in the PlasmAero package is presented, for example, in [8]), supplemented by a "Lagrangian" stage that allows tracking the trajectories of individual elements of the medium. Note that Lagrangian processing does not affect the solution of the main problem in any way. Formally, the volumetric ponderomotive force $\mathbf{F} = \mathbf{J} \times \mathbf{B}$ appears together with the pressure gradient and the viscous stress tensor. In the considered configuration with a limited area of current existence in a transverse magnetic field, the ponderomotive force is fundamentally vortex, while the pressure gradient is a potential vector. Consequently, these two factors cannot be balanced, and the vortex viscous terms are small, at least at the initial point in time after the interaction is switched on. The result is the appearance of an inertial force $\rho dV/dt$. It begins the formation of the velocity field up to the date of termination of the ponderomotive force (the external current or the magnetic field is switched off). It is obvious that the stabilization of the process can be provided only by viscous forces, which seems unlikely because of the fundamentally different structure of these two dynamic factors. In addition, for environments with temperature dependence of conductivity, the energy balance, determined primarily, Joule heat dissipation is an essential factor. Below it will be shown that the heat exchange of the arc channel with the external environment is one of the determining factors of the evolution of the arc channel in the transverse magnetic field. It is precisely these properties that motivate the use of EA as an MHD actuator to control the characteristics of the flow of gaseous working bodies in various technological processes .An important factor is also that the initiation of motion of a medium outside the scope of the ponderomotive force is caused by inhomogeneous pressure field, as a reaction to applied ponderomotive force just inside the arc channel. It is clear that the target task in the problems of mixing intensification is initially the relative displacement of neighboring elements of the medium, which is achieved at a certain time interval. The measure of this criterion can be extracted from trajectories analysis of the certain family of mass elements motion, for this it is necessary to integrate the velocity field in time with the registration of the individual mass elements position of medium. It is the construction of trajectories on the velocity field obtained in a standard way that forms the basis of the "Lagrangian" stage declared above. As our experience has shown, the direct use of mass coordinates to construct a solution seems inefficient because of strong deformations of the mass grid.

2. Problem statement

To reduce multi-factor effects (such as effects of wall close, the effect of initial velocity field of the medium, etc.) with the desire to match a particular experiment, we consider here the movement of the electric arc in a homogeneous initially stationary air at normal conditions. It is assumed that a cylindrical arc channel sufficiently extended along the current is placed in a uniform transverse magnetic field. Below, it is believed that the arc current is controlled by an external circuit. As a rule, a constant current pulse at a finite time interval is considered (the values of these parameters are chosen from the conditions of compliance with the experiment [1]). It can be shown that under these conditions there is constant everywhere the external electric field generated by the power supply and the arc current is determined by the integral conductivity of the arc cross-section:

$$I = E \iint \sigma\,(P(x,y), T(\,x,y))\,dxdy$$

at a given current, the electric field value is defined from the previous as

$$E = I / \iint \sigma\,(P(x,y), T(\,x,y))\,dxdy$$

The initial parameters of the arc channel should ensure the maintenance of the required current (usually several tens of amperes) at reasonable values of the voltage on the power source-several hundred volts. The used procedure requires setting the initial perturbation in some small region to initiate an electric discharge. Usually one of two methods is used: isobaric-temperature at the level T = 6000-9000 K and density $\rho = P/RT, P = 10^5$ Pa, or T isochoric, $p = \rho\ RT$ with density under normal conditions.

This paper does not aim to study systematically the characteristics of the arc channel evolution in a transverse magnetic field in a wide range of changes in the determining parameters of the process as a whole. Therefore, referring to empowerment in the interpretation of the results of experiment presented in [1], will stay in this stage for the example of [1]: the arc current is given as a rectangular pulse I_{arc} = 30 - 60 with duration 130 - 260 μs, the magnetic field is constantly and uniformly $\mathbf{B} = (0, By, 0)T$ = 0.33 – 0.66 T, environment – air at normal conditions $p = 10^5$ Pa, T = 300 K

3. Numerical simulation

The following assumptions are made in the numerical analysis. The modeling is performed in a 2D approach. It means that all flow variables are assumed to be homogeneous in z-direction and all ∂/∂z derivatives are zero. Electric field strength is assumed to be uniform in xy-plane. The arc plasma is assumed to be in local thermodynamic equilibrium (LTE), with the conductivity being a function of temperature only. Finally, the Hall current in z direction, that can lead to the 3D effects onset, was omitted in current approach. With these assumptions, the mathematical formulation for the MHD actuator reads as follows.

3.1. Governing equations

The flow field can be found from the solution of two-dimensional Navier-Stokes equations enclosed with the appropriate equations of state (1) – (4).

$$\frac{\partial \rho}{\partial t} + \nabla(\rho U) = 0 \tag{1}$$

$$\frac{\partial \rho U}{\partial t} + \nabla(\rho U \otimes U) + \nabla \tau = \frac{-\partial P}{\partial r} + J \times B \tag{2}$$

$$\frac{\partial \rho e^0}{\partial t} + \nabla(\rho U h^0) + \nabla(U\tau) + \nabla q = JE + Q_{rad} \tag{3}$$

$$h^0 = e^0 + \frac{P}{\rho}, e^0 = e + \frac{U^2}{2}, (\gamma - 1)e = \frac{P}{\rho}, P = \frac{\rho}{M} R^0 T \tag{4}$$

$$\frac{\partial}{\partial r} = \eta_x \frac{\partial}{\partial x} + \eta_y \frac{\partial}{\partial y}, \nabla = \frac{\partial}{\partial x} + \frac{\partial}{\partial y} \tag{5}$$

In equations (1) – (4) t is time, ρ is the mass density, $\boldsymbol{U}(\boldsymbol{U_x}, \boldsymbol{U_y})$ is the velocity, $\boldsymbol{P}$ is the termodynamic pressure, $\boldsymbol{T}$ is the temperature, $\boldsymbol{e_0}$ is the specific total energy, e is the specific internal energy, $\boldsymbol{h_0}$ is the specific total enthalpy, $\boldsymbol{h}$ is the specific internal enthalpy, $\boldsymbol{M}$ is the molecular weight. Operators in (5) designate gradient and divergence, respectively. ηx, ηy are unit base vectors.

To specify LTE air plasma characteristics the approximations [9] are used:

$$h = h_{eq}(P,T), M = M_{eq}(P,T), \gamma = \gamma_{eq}(P,T) \tag{6}$$

The heat flux q and viscous stress tensor τ read as:

$$q = -\lambda \frac{\partial T}{\partial r} \tag{7}$$

$$\tau_{ij} = \frac{2}{3}\eta \delta_{ij} \nabla U - \eta \left(\frac{\partial U_i}{\partial x_j} + \frac{\partial U_j}{\partial x_i}\right) \tag{8}$$

In (7), (8) η is the molecular viscosity coefficient, λ is molecular heat conductivity, δ_{ij} is Kronecker-Capelli symbol. Transport coefficients λ and η are calculated from approximations [9].

3.2 Electrodynamic part

Electrodynamic part of the problem reduces to the relations for total arc current and electric field strength assuming Ohm's law. Namely,

$$J_z = \sigma E_z \tag{9}$$

$$I_z(t) = \int J_z dS = \int \sigma E_z dS = E_z \int \sigma dS, \tag{10}$$

3.3. Real gas properties
Total arc current $\boldsymbol{I_z(t)}$ is set as an external parameter, so (10) is used for determining average electric field strength $\boldsymbol{E_z}$. After this, source terms in equations (2) and (3) can be found. Electric conductivity $\boldsymbol{\sigma(x,y)}$ is calculated in accordance with [9]. Radiative losses in (3) are estimated as follows: $\boldsymbol{Q_{rad}} = \boldsymbol{\chi\sigma_{SB}T^4}$, where χ is the absorption coefficient, σ_{SB} is the Stefan-Boltzman constant. Absorption coefficient is calculated from tables given in [10].

3.4 Computational details
Equations (1) – (3) are discretized on a structured grid in conservative variables (ρ, $\rho\mathbf{U}$, ρe^0). AUSM ([11],[12]). The technique [13] is used to obtain spatial reconstruction of higher order flow variables. Viscous flows (7), (8) are calculated, as, for example, in [13,14]. The explicit Euler method is used for time integration of equations (1) - (3).
The solution was constructed on a rectangular grid. To reduce the influence of far field conditions, an exponentially increasing grid belt was introduced along the perimeter of the base grid. Initialization of the counteragent current channel was carried out either by Isobaric, isochoric or perturbation, respectively, of temperature or pressure in a small (usually ~ 100·100 mm^2) region. The kinematics of the process is studied in the framework of the proposed here the option "Euler-Lagragian" approach. For this purpose, along with the main stationary grid, which is the integration of two-dimensional nonstationary Navier-Stokes equations, is introduced moving "Lagrangian" mesh, which is associated with the mass elements of the medium at some initial time, then the position of nodes of the moving grid is determined by integrating in time the velocity field, taught from the solution of the original problem on the Euler grid. Note that the Lagrangian stage has no influence on the solution of the main problem and is only an additional information element.

Dynamic and energy characteristics: the fields of pressure, velocity, density and different types of energy are the main result of the solution on a stationary grid.

4. Results
The qualitative interaction process characteristic of the current channel with the external magnetic field and with the medium is given in Fig.1, which shows the temperature, pressure and velocity fields evolution in the vicinity of the initially initiated isochoric perturbation of temperature and density in the region of 120·120 mm^2, $T_0 = 10000$ K, $P_0 = 3.33$ MPa over the entire range of the current pulse. The main features of the process are determined, firstly, by the initial conditions-highly inhomogeneous distributions of pressure and temperature and, secondly, determined by the current dynamic and thermal effects on the current channel. Changes in gas-dynamic fields outside the current channel are determined by the transfer of momentum and energy by "fast" shock-wave and relatively slow diffusion processes.

Under the position of the arc here and further we will mean the position of the maximum temperature point, which qualitatively corresponds to intuitive experimental practice. Differentiating the time of this ham allows you to determine the" speed " of the arc. In addition to a certain value of the arc velocity, the numerical solution gives the true velocity of the medium at the point of maximum temperature. The evolution of all these parameters is shown in Figure 2. The most important fact is almost two-fold excess of the medium true velocity above the "phase" speed the movement of "hot" points, and specifies the displacement of the hot spot on the medium, i.e. the speed of the hot spot are the sum of the velocity environment and movement of high temperature in the environment. It is this circumstance that motivated the use of "Lagrangian" procedures to increase the information content of the computational and the theoretical analysis.

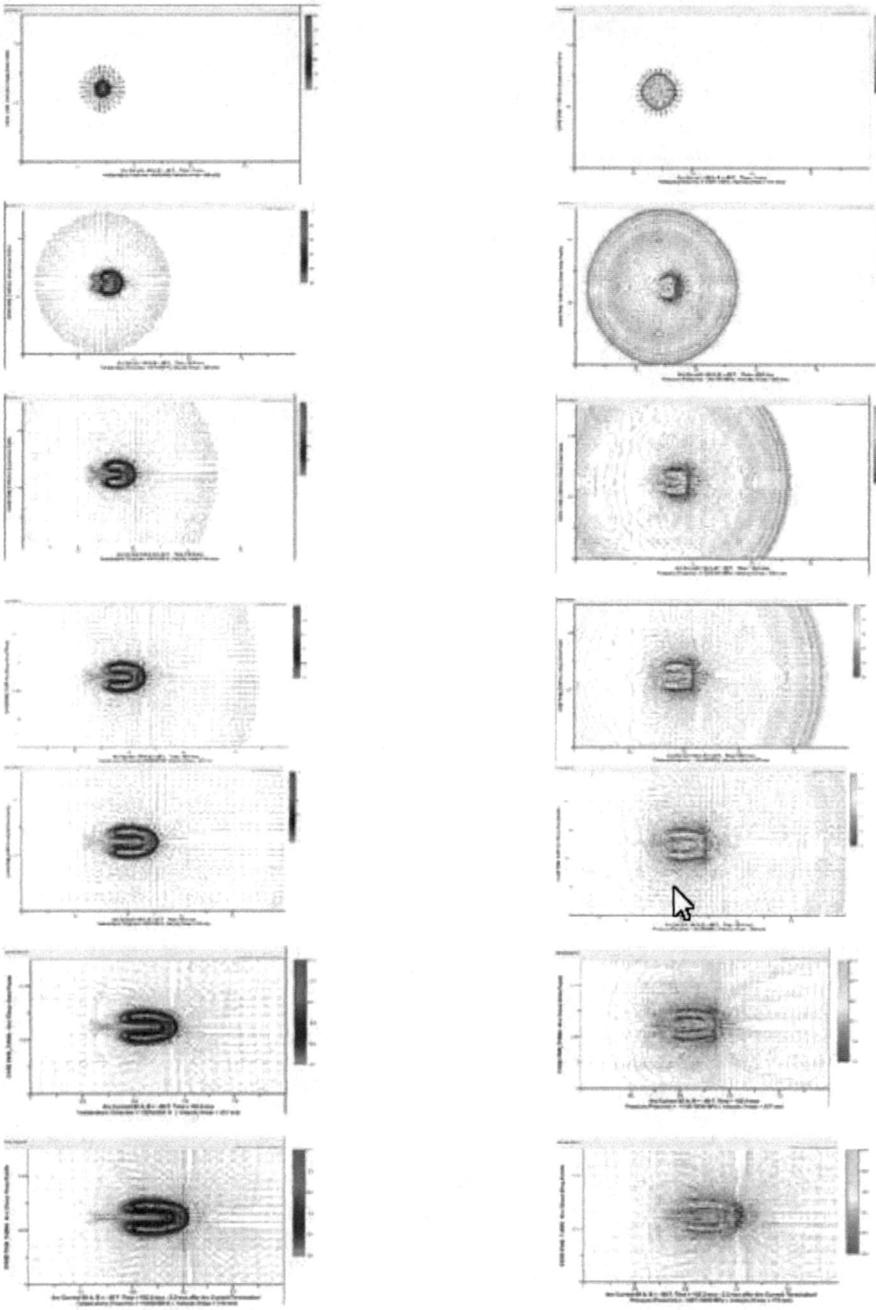

Figure 1. Time evolution of temperature (contours) and velocity (arrows) on left, and pressure on right.

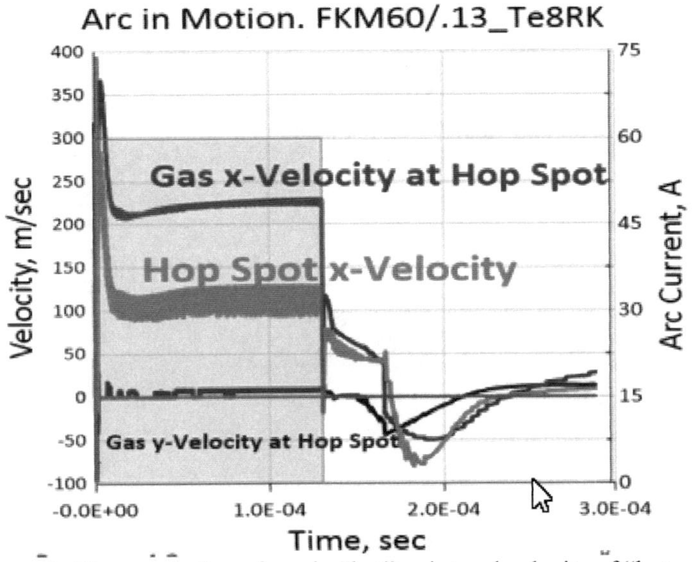

Figure 2. Time evolution of gas in "hot" point and velocity of "hot point".

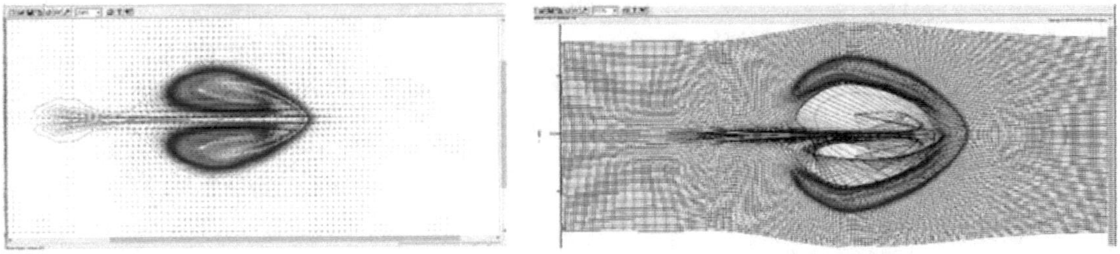

Figure.3. Temperature field on Euler grid (left) and on Lagrangian one (right) 150 μs after current switch off (t_{break}=130μs).

Figure 3 shows an example of the temperature distribution and velocity vector on the fixed (left) and mobile (Lagrangian – right) grids at some point in time after switching off the arc current. Unfortunately, the strong deformation of the moving mesh in the evolution process makes it almost impossible to visually analyze the process. In this regard, we will continue to use only a small fragment of the movable grid, in which the arc was initialized. Figure 4 is an example of arc evolution in Lagrangian coordinates is presented. At the initial moment of time, the place of the isochoric perturbation initiating the arc is fixed. In this example, this region is a rectangular element of the Euler (fixed) grid size 5·5 cells, which is given a 30-fold increase in temperature and pressure. The gas mass with disturbing parameters is marked for further tracking simply by the temperature field, at this initial point in time. This temperature field is represented here and in the future by a blue-pink color palette. The actual field of true temperature (changing during the evolution of the arc) is represented in a green-red color palette. Note that the upper figure represents the situation not at the initial, but at a somewhat later (+ 10 microseconds) point in time. The next two frames will represent the moment of magnetic field shutdown-approximately 65 microseconds after initialization, and at the moment of arc dock shutdown – approximately 130 microseconds. Figure 4 clearly shows that the active current-carrying mass is outside the zone of initial shock-wave expansion, and the initially superheated high-pressure mass has cooled to temperatures below 6000 K, at which the air ceases to be any significant conductor.

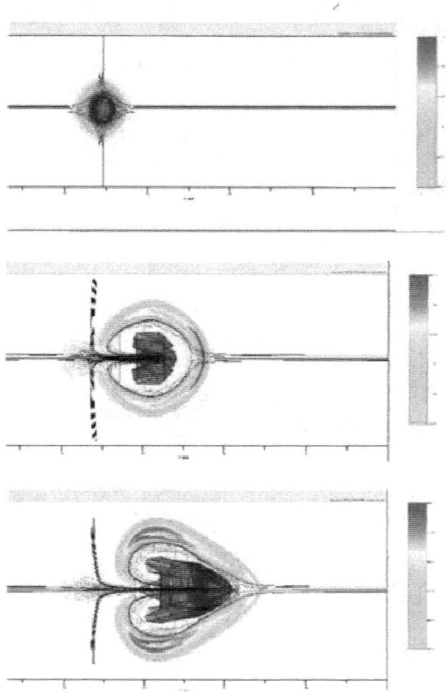

Figure 4. Position of arc-related mass (pink-blue) and active mass (red-green) at three time instances (from top to bottom): 10μs, 65 μs, 130 μs.

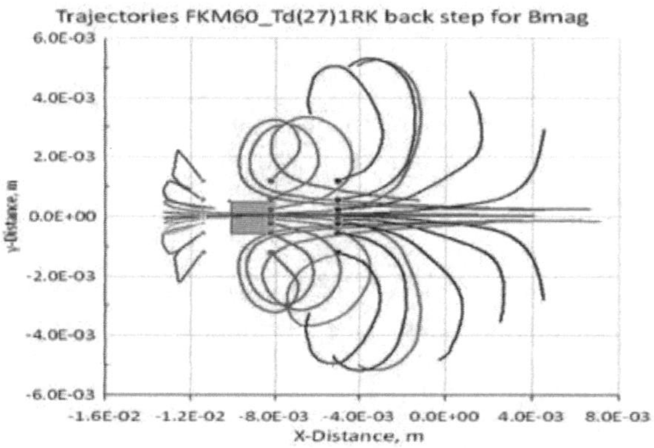

Figure 5. Trajectories of mass-elements of current channel at 410 μs after initialization of the arc.

Another kinematic characteristic is the mass element trajectory of the medium. Figure 5 presents three beams of trajectories, the starting points of which are selected as follows. Three groups of initial coordinates are located on vertical segments of straight lines, which are symmetric with respect to the horizontal axis of the current spot, the central group lies on the vertical diameter of the current channel, and the left and right - symmetrically with respect to the Central one. The initialization area is conventionally shown in Figure 5 painted rectangle. The central (red) trajectories are divided into two groups. The two central trajectories are almost rectilinear and limit the "through channel", in which there are several, also almost rectilinear trajectories of mass elements of the external undisturbed medium. The external trajectories of the initially hot volume undergo a certain circulating phase, similar to the trajectories of the mass elements of an ideal liquid when flowing around a circular cylinder. The group of right elements moves similarly to the last, just to passage free for the axial jet of hot and cold elements. Apparently, the main contribution to the mixing function is made by this axial tube, the contribution of external remote elements to the mixing seems negligible.

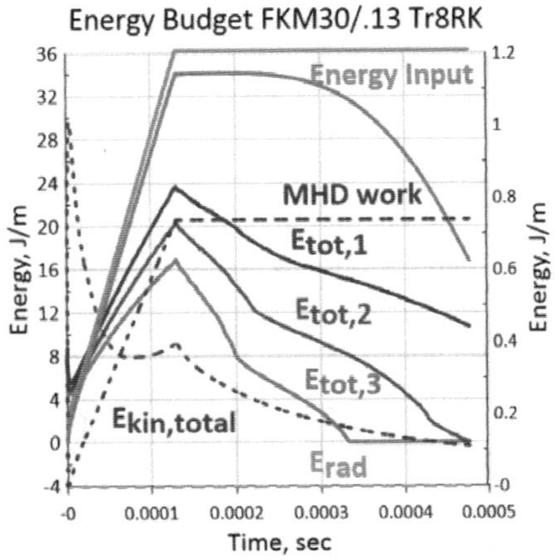

Figure 6. Integral power characteristics of the discharge.

The energy efficiency of the MHD actuator can be estimated from the data shown in Figure 6. In this picture, the solid curves represent the values scaled on the left axis and the dashed ones on the right. The modification of the velocity field (in the case of the initial state of rest) can be estimated simply as the total kinetic energy of the entire domain of consideration. It is useful to compare this value with the full operation of the ponderomotive force, which is several times greater by the time the current is turned off, as can be seen. Moreover, the contribution of the total energy (the upper envelope curve) is about 50 times greater than the work of the ponderomotive force. The feasibility of this technology should be justified by additional criteria. In addition, a detailed analysis of the gas-dynamic field formation and peculiarities of the trajectories of mass elements of the environment and possible ways of process of optimization to control the MHD actuator functioning.

5. Concluding remarks

This preliminary study of the arc evolution in transverse uniform magnetic field reveals that the experimentally visualized the arc motion velocity is rather a kind of "phase" propagation velocity of the maximal temperature (or maximal intensity of radiation). Thus, such a value couldn't be used directly in arc dynamics characterization. The momentum and energy balances require the accurate accounting for the convection fluxes through some arc border.

The analysis of the Lagrangian grid deformation and the corresponding trajectory of mass elements shows that the strongest relative displacements of the media are located inside the arc and its close environmen. In far field the Langrangian grid deformation is rather low that means that mixing effectiveness there is also low.

Energy balance of the arc is only 2-3 % related to the creation of motion generation, but mostly connected to the heat deposition into the media.

Euler-Lagrangian approach used in this study has shown rather new promising features in the fast running process with intensive momentum and energy conversion studies.

Acknowledgement
This work is supported by the PROJECT ГР AAAA-116051810079-7

References

[1] F P. N. Kazansky, I. A. Moralev, A. A. Firsov, V. A. Bityurin, A. N. Bocharov, S. B. Leonov, experimental study of vortex generation with a pair of counter-moving pulse Arcs, AIAA Scitech forum 2019, January 7-11, 2019, San Diego, California; DOI 10.2514/6.2019-0051. Yang et al., "Low-voltage circuit breaker arcs - Simulation and measurements," J. Phys. D. Appl. Phys., vol. 46, no. 27, 2013.

[2] V. A. Bityurin, A. . Likhachev, S. A. Medin, and G. A. Lyubimov, "On the Dynamics of a nonUniform Conducting Flow in an MHD Generator Channel."

[3] V. Bityurin and A. Bocharov, "Advanced MHD assisted mixing of reacting streams," in 39th Aerospace Sciences Meeting and Exhibit, American Institute of Aeronautics and Astronautics, 2001.

[4] A. N. Bocharov, V. A. Bityurin, I. B. Klement, and S. B. Leonov, "A STUDY OF MHD ASSISTED MIXING AND COMBUSTION," in 41th AIAA Aerospace Sciences Meeting and Exhibit, 2003, p. AIAA-2003-0378.

[5] A. N. Bocharov, V. A. Bityurin, I. B. Klement, and S. B. Leonov, "EXPERIMENTAL AND THEORETICAL STUDY OF MHD ASSISTED MIXING AND IGNITION IN CO- FLOW STREAMS," 2002, pp. 1–12.

[6] J. Sirohi, B. Pafford, and L. L. Raja, "Experimental Characterization of the RailPAc Plasma Flow Actuator," 44th AIAA Plasmadynamics Lasers Conf., pp. 1–16, 2013.

[7] T. Fujino, S. Hirayama, and M. Ishikawa, "Numerical Simulation of Unsteady Behavior of Rotary Arc Plasma under Magnetic Field," in 41st Plasmadynamics and Lasers Conference, 2010, no. July, p. AIAA 2010-4641.

[8] A. N. Bocharov, V. A. Bityurin, Studies of Hypersonic MHD Flows / / Lambert Academic Publishing, 2017, pp. 219(In Russian)

[9] A. D'Angola, G. Colonna, C. Gorse, and M. Capitelli. Thermodynamic and trans temperature range. Eur. Phys. J. D 46, 129–150 (2008), DOI: 10.1140/epjd/ e2007 -00305-4.

[10] Yasuhiro Wada and Meng-Sing Liou. An Accurate and Robust Flux Splitting Scheme for Shock and Contact Discontinuities // SIAM J. Sci. Comput., May 1997. V.18, No.3, pp.633-657

[11] Meng-Sing Liou. A Sequel to AUSM: AUSM+ // J. Comp. Phys, 1996. 129, pp.364-382.[12]. Barth T.J., and Jesperson D.C. The Design and Application of Upwind Schemes on Unstructured Meshes // AIAA Paper 89-0366, Jan 1989.

[12] Mathur S.R., Murthy J.Y. All Speed Flows on Unstructured Meshes Using a Pressure Correction Approach // AIAA Paper 99-3365, 1999.

[13] Ferziger J.H., Peric M. Computational Methods for Fluid Dynamics // Springer-Verlag Berlin, 1996.

Journal of Physics: Conference Series **1394** (2019) 012023 doi:10.1088/1742-6596/1394/1/012023

The pre-breakdown characteristics of weakly ionized media in the high non-uniform electric field

M S Apfelbaum, R A Syrovatka and V I Vladimirov

Joint Institute for High Temperatures of the Russian Academy of Sciences (JIHT RAS) 125412, Izhorskaya st. 13 Bd. 2, Moscow, Russia

msa@ihed.ras.ru

Abstract. The theoretical model of pre-breakdown ions formation in a liquid dielectric and their flows, caused by high non-uniform electric field is represented. The 3D system of macroscopic pre-breakdown electrohydrodynamic equations is written. The electric field influence on the molecule dissociation rate is taken into account. The system are included the Poison equation for electric field potential, equation of ions formation and Navier-Stokes equations with electric force The our up-steady analytical electrodynamic solution of these equations for potential of electric field distribution at spherical high voltage capacitor with liquid transformer oil type dielectric is described deviations from Ohms law, observed in many experiments. The analytical non-stationary solution for velocities distributions calculations of considerable liquid dielectric flows are obtained. The theoretical and experimental micro-breakdown radius zone dependences on applied voltage at air corona discharge are obtained for edge high voltage electrode.

1. Introduction

Deviations from Ohms law for weakly ionized solid media in pre-breakdown uniform electric fields experimentally discovered Poole about 100 years ago [1]. In a weakly conductive liquid media the same experimental effect was obtained by Wien about 100 years ago too [2]. For gaseous deviations from Ohms law were also obtained in high electric field many years ago.

Theoretically exponential effect for considerable media was explained by Frenkel for solid dielectrics and Onsager [2] for liquid weak electrolytes and for weakly conductive liquid dielectrics. The space charge and electrohydrodynamic flows observed in these dielectrics at a pre-breakdown condition [3].

The space charge formation, according [4], may be in pre-breakdown fields by up steady all electrohydrodynamic characteristics. And steady conduction may be as unipolar (corona discharge type) or as quasi-neutral {plasmas or electrolyte type).The last case considered us early in [4] .The pre-breakdown current-voltage theoretical and experimental characteristics of considerable media in non-uniform electric fields are described us in [5]. The purpose of present work is researches of the electrohydrodynamic flows, caused by these high non-uniform electric field .These intense flows are observed in transformer oil type liquids [6, 7] with complex molecular structure [8].The hydrodynamic transfer of high voltage space charge, appeared in considerable liquids, is described in [9]. In review [7] the surface high voltage electrode effects influence on considerable pre-breakdown electrohydrodynamic flows is researched. This influence must be taken into account at plan high voltage

Journal of Physics: Conference Series **1394** (2019) 012023 doi:10.1088/1742-6596/1394/1/012023

electrodes. And when applied high voltage field is non-uniform the volume effects influence on considerable flows is dominated

2. Theoretical model

The following inequalities are valid for bulk charge concentrations $n_{\pm}$, impurities concentrations n_p and neutrals concentrations $n_{a,}$ in a low-conducting medium

$$\begin{cases} n_{\pm} \ll n_{a,} \\ n_p \ll n_a. \end{cases} \tag{1}$$

The upper relation can be considered as the condition of weak ionization (dissociation) or low conductivity in considered medium. The rates of volume ionization (or dissociation) and recombination are supposed to be known thermodynamics functions of the above mentioned concentrations, temperature and field intensity $\boldsymbol{E}$ (up to breakdown values). They can be represented in following form:

$$W_I = W_I\big(n_a, n_p, T, |\boldsymbol{E}|\big) = W_I = W_I\big(n_a, n_p, T, 0\big) f(|\boldsymbol{E}|) \tag{2}$$

$$\begin{cases} W_r = K_r n_+ n_- \\ K_r = \frac{(b_+ + b_-)|e|Z}{\varepsilon\varepsilon_0} \end{cases} \tag{3}$$

Here W_I denotes the volume rate of ionization (dissociation), W_r denotes the volume rate of ion recombination, b is the mobility of the ions, e is the elementary charge and T is the absolute temperature. The expression for K_r (recombination constant) was obtained by Langevin in 1903 and by Onsager in 1934 for particular case of weak electrolytes. It is two-particle ion-ion recombination, when the energy excess is absorbed by a medium. Onsager had shown as well the weak influence of sub-breakdown electrical fields on K_r when direct and reverse processes are ionization and recombination correspondingly. But in case when these processes are the dissociation and reverse dissociation the influence of $\boldsymbol{E}$ on K_r can not be disregarded. Besides weak electrolytes the Langevin formula is valid for dense gases with chemical reactions, where ionic conductivity is much greater than electronic one. Corresponding conductivity can be due to various processes. Among them are the neutral molecules (with ionic chemical bonds) dissociation and the neutral molecules (with covalent chemical bonds) ionization. The latter goes through the ionization of the electron from some neutral particle with relatively low ionization potential and attaching it to another with relatively high electron affinity.

For the ion diffusion coefficient we have used more known Einstein-Nernst relation:

$$ZD_{\pm} = \frac{k_B T b_{\pm}}{|e|}, \tag{4}$$

where Z is the ion valence, D is the diffusion coefficient, k_B is the Boltzmann constant.

The function $f(|\boldsymbol{E}|)$ describes the dependence of the ionization (dissociation) rate on the electric field intensity. The expression for it was obtained by Frenkel for the solid dielectric and generalized by Ostroumov in [3] for the liquid weakly conductive dielectric case ($Z > 1$) with ion chemical bonds in molecules. It was used in present research and has form:

$$f(|\boldsymbol{E}|) = \exp(\beta|\boldsymbol{E}|^{0.5}); \quad \beta = \frac{|Ze|^{1.5}}{\sqrt{\varepsilon\varepsilon_0} k_B T}. \tag{5}$$

Corresponding equations, describing creation and annihilation of the space charge, high voltage conductivity and electric field distributions, can be written as in [4]:

$$\frac{\partial q}{\partial t} + (\boldsymbol{V}, \nabla q) - \frac{k_B T b}{|eZ|}\Delta q + (\boldsymbol{E}, \nabla\sigma) = \frac{q\sigma}{\varepsilon\varepsilon_0}, \quad \Delta\phi = -\frac{q}{\varepsilon\varepsilon_0} \tag{6a}$$

$$\frac{\partial\sigma}{\partial t} = (\boldsymbol{V}, \nabla\sigma) + b^2(\boldsymbol{E}, \nabla q) - \frac{\frac{k_B T b}{|eZ|}\Delta\sigma}{Z} - \frac{\sigma_0^2}{\varepsilon\varepsilon_0}\exp(\beta|\boldsymbol{E}|^{0.5}) + \frac{q\sigma}{\varepsilon\varepsilon_0} = 0, \tag{6b}$$

where $\boldsymbol{V}$ is the velocity, q is the volume charge density, σ is the conductivity, ϕ is the electric potential.

The well-known hydrodynamics equations should be added to Eqs. (6) to construct the closed system of equations. The first of them is the continuity equation for incompressible media:

$$\mathrm{div}\boldsymbol{V} = 0. \tag{7}$$

We used equation (7) for obtaining electrodynamic equations (6). Due to incompressibility it is enough to consider the law of momentum conservation (without the law of energy conservation). The momentum conservation equation can be written as:

$$\boldsymbol{\chi} = \mathrm{rot}\boldsymbol{V}. \tag{8}$$

$$\rho\frac{d\boldsymbol{\chi}}{dt} - \rho(\boldsymbol{\chi},\nabla)\boldsymbol{V} - \rho\nu\Delta\boldsymbol{\chi} = 2[\nabla q, \boldsymbol{E}], \tag{9}$$

where $\boldsymbol{\chi}$ is the vortex, ρ is the density, ν is the kinematic viscosity.
The initial conditions for system (6) – (9) are

$$q(t=0) = \boldsymbol{V}(t=0) = 0;\ \sigma(t=0) = \sigma_0 \tag{10}$$

The quasi-stationary analytical electrodynamic solution of Eqs. (6) for the spherical symmetry electrical potential distribution ϕ in quasi-neutral medium can be obtained from charge conservation law (this well-known law may be obtained from equations (6) too [10]). This solution looks like

$$\phi(r) = \left(\frac{I}{4\pi\sigma_0}\right)^{0.5}\left[\frac{8}{\beta} - \left(\frac{\varepsilon\varepsilon_0}{\tau\sigma_0}\right)^{0.5}\left(|\boldsymbol{E}|^{0.5} + \frac{8}{\beta}\right)\right]\mathrm{sign}(\phi(r_0)); \tag{11a}$$

$$|\boldsymbol{E}|\exp\left(\frac{\beta}{2}|\boldsymbol{E}|^{0.5}\right) = \frac{I}{4\pi\sigma_0 r^2};\quad |\phi(r_0)| = U,$$

where I is the electric current, σ_0 is the low voltage conductivity and U is the applied voltage.
For the flat geometry one can obtain nonstationary solutions of Eqs. (6) [10]:

$$I = \frac{\sigma SU}{d},\quad q = 0,\quad |\boldsymbol{E}| = \frac{U}{d},$$
$$\sigma = \frac{\varepsilon\varepsilon_0\left[\sigma_0\tau + \varepsilon\varepsilon_0 - (\varepsilon\varepsilon_0 - \sigma_0\tau)e^{-2t/\tau}\right]}{\tau\left[\sigma_0\tau + \varepsilon\varepsilon_0 + (\varepsilon\varepsilon_0 - \sigma_0\tau)e^{-2t/\tau}\right]}, \tag{11b}$$
$$\tau = \frac{\varepsilon\varepsilon_0}{\sigma_0\exp(0.5\beta|\boldsymbol{E}|^{0.5})}.$$

One can see from (11b) that relaxation time $\tau = \varepsilon_0/\sigma_0$.
And quasi-neutral solution (11) is the zero approaching of hydrodynamic space charge transfer differential operator serious [9]:

$$q = \sum_{i=0}^{\infty}(\tau v\nabla)^i\sigma\boldsymbol{E}\nabla\tau, \tag{12}$$

where τ is the charge relaxation time.

The mathematical space of differential operators, obtained in [9], is not Banach space. The quasi-exponential dependence as dependence [5] for volt-ampere pre-breakdown characteristics calculations of high voltage spherical capacitors may be obtained from (11). These deviations were explained early in [2] for plan capacitors. The Laplace condition of pre-breakdown electric field, obtained in [6], may be obtained from (11) too.

For cylindrical symmetry the analytical solution of equations (6) as (11) cannot be obtained. For 2D and 3D high voltage electrodes configurations geometry the solutions of equations (6) may be obtained only numerically. Graphic of volt-ampere characteristics (11) for different distances between spherical electrodes is shown in figure 1. As the distance between electrodes decreased the pre-breakdown current with same high voltage increased

And in contrary for the unipolar conduction solutions of our equations (6) the analytical formulae for pre-breakdown volt-ampere characteristics calculations may be obtained analytically only for cylindrical symmetry case. This solution was obtained early for weakly ionized gaseous in [11]. This solution more often used for weakly ionized gaseous media. Due to conductivity of the air is much smaller that conductivity of the transformer oil one should use empirical constant of corona ignition voltage in the solutions of equations (6). And the simplest formulae for such characteristics was obtained by Townsend about 100 years ago:

$$I = \frac{2bU(U-U_c)L}{(r_0+d)^2\ln(1+d/r_0)}, \tag{13}$$

where U_c is the corona ignition voltage, L is the electrodes length, d is the distance between electrodes, r_0 is the radius of the corona electrode. According [12] formula (13) can be used for point-plate electrode.

Based on the law of conservation of energy one can obtain expression for the size r_d of the corona region where elementary processes occur:

$$r_d = \left(\frac{2bU^2(U-U_c)t_iL}{4\pi(r_0+d)^2\ln(1+d/r_0)\delta}\right)^{0.5}, \tag{14}$$

where t_i is the corona ignition time, δ is the surface tension coefficient.

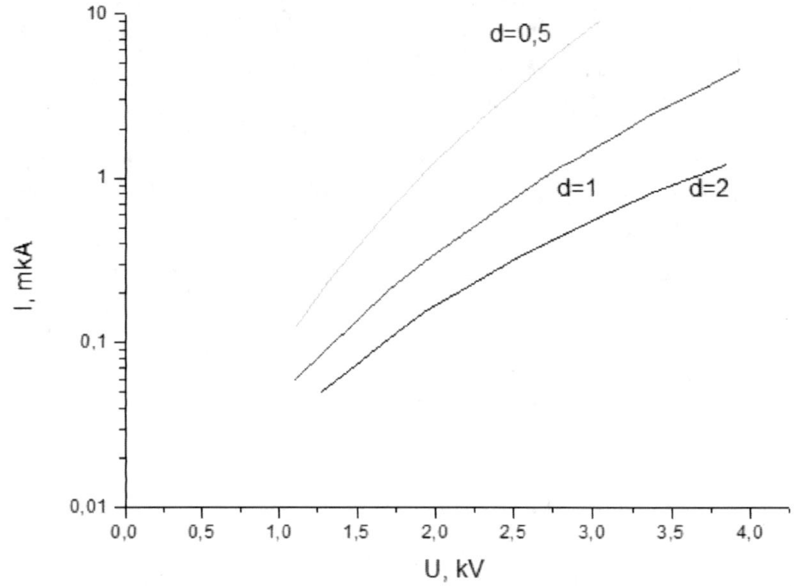

Figure 1. Quasi-exponential volt-ampere characteristics of the spherical capacitors with transformer oil according to (11). (Z = 2, according to [8]). Distance between electrodes is given in cm. The radius of high voltage inner electrode is 1.5 mm.

Our non-stationary hydrodynamic solution of equations (6) – (9) with using (10), (11) and [13] for development of weakly conductive liquid jet flows from high voltage point electrode has the following form:

$$\Psi = \frac{\varepsilon\varepsilon_0 U^2(t-\tau)\sin^2\theta}{32\pi\rho r}, \tag{15}$$

where Ψ is the stream function.

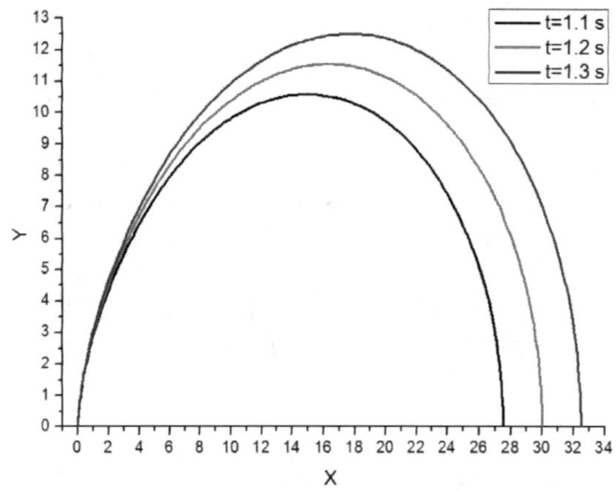

Figure 2. Graphic of the constant transformer oil jets stream lines (13) for U = 4 kV and debit 0.1 L/s for different times moment. The sizes of Cartesian coordinates are given in mm.

3. Experimental results of prebreakdown characteristic in the case of the corona discharge in air

Stationary corona discharge without additional ionization source was investigated in this work. The experiments were carried out in air at atmospheric pressure. The needle with an edge radius equal to 70 μm was placed above the center of a flat round electrode with a diameter of 100 mm at a distance of 13 mm. In this case, the finite size of the flat electrode can be neglected. High dc voltage was applied to the needle. The flat electrode was grounded. The dc voltage was measured using voltage divider with the voltage ratio of 1:1000 and multimeter APPA 505. The error of measuring the dc voltage at the limit of 10 V was equal to ±0.17 V. The current was measured using ammeter with the accuracy class of 0.5.

We believe that the corona discharge region where the ionization processes occur corresponds to the region emitting in the visible range. Photography was used to determine the size of this area. Nikon 1 V2 (4608×3072) camera with Nikon 60mm f/2.8D AF Micro-Nikkor lens was used for taking pictures. The photography was carried out with a fully open aperture, the exposure was equal to 5 s, and the photosensitivity was equal to 400 ISO units. A pixel of the matrix corresponded to a 0.0044×0.0044 mm^2. The position of the camera during the experiment did not changes.

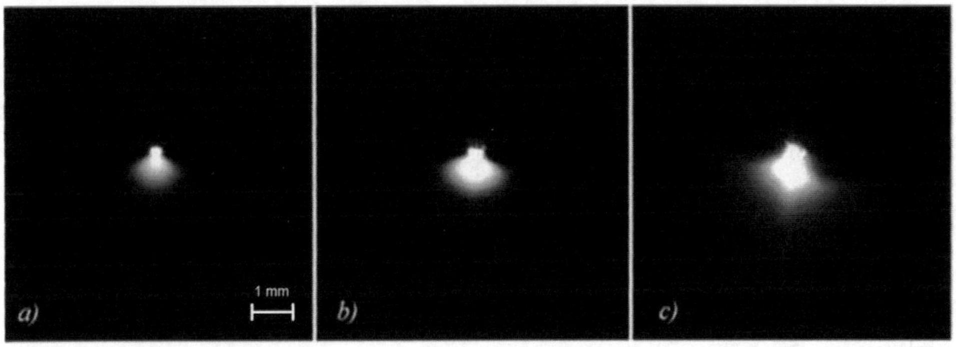

Figure 3. Photos of the discharge with the following parameters: *a)* $I = 6$ μA, $U = 5.01$ kV; *b)* $I = 27$ μA, $U = 8.01$ kV; *c)* $I = 68$ μA, $U = 11.04$ kV.

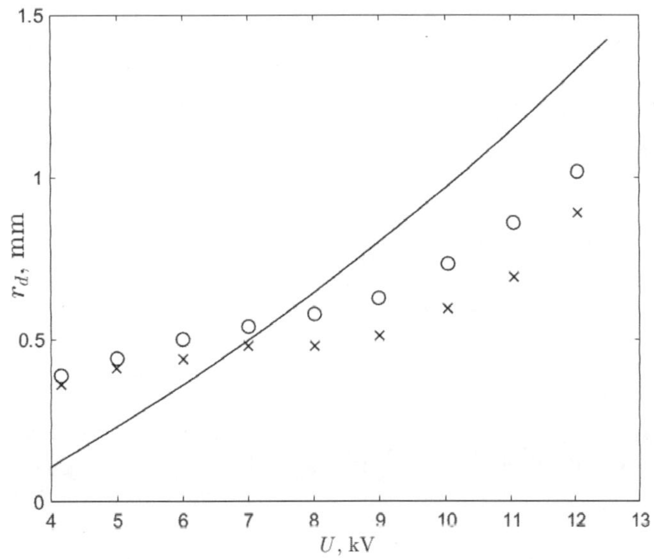

Figure 4. The distance from the end of the needle to the boundary of the ionization region along the vertical axis (circles) in comparison with the square root of the cross-sectional area (crosses) of the ionization region and theoretical calculations (14) in the case of the corona discharge in air.

Figure 2 shows photographs of the discharge of negative polarity for various parameters. At the discharge voltage less than 7 kV the glow area is stationary and vertically symmetrical and expands with increasing discharge current. At higher discharge currents, the glow area loses symmetry and begins to fluctuate. The breakdown occurred at the discharge voltage more than 12 kV.

It can be seen that the size of the ionization region increases monotonically with increasing discharge voltage.

Figure 4 shows size of the ionization region in the case of the corona discharge in air.

4. Conclusion

This study presents in the form of graphs and formulas the monotonously growing dependences of the prebreakdown stationary volt–ampere characteristics of weakly conductive liquid dielectrics as well as weakly ionized gases. Obtained characteristics differ from the traditional volt-ampere characteristics for point electrodes by a sharper current increase with increasing applied voltage. In the case of the interelectrode gap of the high-voltage spherical capacitor being filled with viscous heat-conducting liquid dielectrics such as transformer oil independent of the change in polarity of the electrodes, the calculation formula testifies to the growth of the prebreakdown current with a decrease in the interelectrode gap.

It was found that in the case of the corona discharge in weakly ionized gases like air volt- ampere characteristics for positive and negative discharge are significantly different in contrast to weakly conductive liquid dielectrics. The size of the ionization region in the case of corona discharge in air increases monotonically with increasing discharge voltage.

An analytical solution for the evolution of flows of viscous incompressible fluid (transformer oil) for the point-plane geometry was obtained.

Acknowledgments

This work was supported by the program of the Presidium of the Russian Academy of Sciences "Thermophysics of high energy densities".

References

[1] Poole H H 1916 *Phil. Mag.* **2** 112–20
[2] Onsager L 1934 *J. Chem .Phys.* **2** 599–615
[3] Ostroumov G A 1978 *Interaction of electrical and hydrodynamic field. Physical funddemantal of electrohydrodynamics* (Moscow: Nauka)
[4] Apfelbaum M S 1986 *Russ .J. Electrochem.* **22** 1473–81
[5] Apfelbaum M S, Vladimirov V I and Pecherkin V Ya 2015 *Surf. Eng. App. Electrochem.* **51** 260–8
[6] Apfelbaum M S and Polyansky V A 1982 *Magn. Hydrodyn.* **1** 51–6
[7] Zhakin A I 2012 *Phys. Usp.* **182** 495–520
[8] Arakelyan V G 2012 *Physico-Chemical Basis of Oil-Filled Equipment, Reference Data, Analysis of the Study, Diagnosttics, Monitoring* (Moscow: Tetraprint)
[9] Apfelbaum M S 1987 *Electron. Obrab. Mater.* **23** 59–62
[10] Apfel'baum M S 2011 *Surf. Eng. Appl. Electrochem.* **47** 35–43
[11] Tikhodeev N N 1955 *J. Tech. Phys.* **25** 1449–1457
[12] Raizer Yu P 1991 *Gas Discharge Physics* (Springer-Verlag Berlin Heidelberg)
[13] Sozou C and Pickering W M 1977 *J. Fluid Mech.* **10** 673–83

Journal of Physics: Conference Series **1394** (2019) 012024 doi:10.1088/1742-6596/1394/1/012024

Investigation of Q-DC discharge influence on fuel jet mixing with supersonic airflow

A A Firsov[1,*], E V Dolgov[1], A V Efimov[1], I A Moralev[1], D A Yarantsev[1] and S B Leonov[1,2]

1 Joint Institute for High Temperatures (JIHT RAS), 125412, Moscow, Russia
2 University of Notre Dame, Notre Dame, Indiana 46556, United States

* af@jiht.org

Abstract. This paper presents the results of experimental study of plasma-based actuator for mixing of the fuel with air in a supersonic flow. The long unsteady plasma filament of Q-DC electrical discharge is located along the fuel jet from a wall-arranged supersonic injector with a wall-normal direction of injection. The location of plasma filament over the fuel jet was controlled by using the nozzle of injector as one of discharge electrodes. Three different gases were tested for simulation of fuel injection: air, CO_2 and C_2H_4. It was found that the discharge position in respect of the fuel jet significantly depends on the injected gas. The activation of discharge within all injected gases results in development of fluid instability at fuel-air interface which leads to the mixing enhancement.

Introduction

Ramjet/scramjet efficient operation in a wide range of flow parameters (pressure and temperature) is one of the most challenging technical problems in design of a hypersonic vehicle. A combustion chamber operates, as a rule, in a supersonic flow regime with a Mach number M ~ 2. Key issues in the arrangement of supersonic combustion process are stabilization of the flame front (flameholding) and maintaining of high combustion efficiency. In most schemes, the fuel feeding is organized using wall-mounted injectors or pylons [1–5] that makes the acceleration of the injected fuel mixing with incoming airflow to be a critical task to achieve a high value of combustion completeness. A typical solution employed for mixing enhancement and stabilization of combustion in scramjet is the use of a backward wallstep, wall cavity [6], or a more complex configuration with mechanical elements in the flowpath. For example, in [7], a thin wedge, witch width is substantially less than the height, is used for optimization of injection into a supersonic flow. The injection of a fuel occurs from the combustor's wall behind a wedge, which leads to an increase of fuel injection depth into the flow, to stretching of interface between the gases, and, finally, results in increase of the mixing rate. A similar flow pattern is realized when the fuel comes from pylons. Along with high efficiency, such systems have significant weaknesses, which are the loss of total pressure or the need for an active cooling system, especially in the case of a high-enthalpy flow. Due to the a slow diffusion process in comparison with the presence time of mixture components in the zone of chemical reactions, the intensification of mixing in a supersonic flow can only be produced by a kinematic transfer of components in vortex structures formed at the fuel-oxidant boundary. A promising way for the mixing acceleration is an extended electric spark that introduces significant gas-dynamic perturbations into the flow [8, 9]. With sufficiently high amplitude of energy deposition to the electric discharge, it is possible not only to introduce initial perturbations into the flow, but also change the increments of their development as a result of flow structure modification in the core airflow [10]. During formation of a localized zone of energy release at shear layer of the fuel jet, the mixing intensification happens as a result of the Richtmyer–Meshkov hydrodynamic instability, which occurs within area of non-collinear gradients of gas density and pressure. This leads to formation of deterministic turbulence and, then, to the appearance of small-scale stochastic perturbations that accelerate the stretching of the boundary and produce the kinematic mixing. Initialization of initial perturbations in the interaction zone may be performed by a pulse-periodic energy deposition in the jet or at some distance from it. Such configurations were considered theoretically in [11], where a significant intensification of mixing in the flow with a Mach number M = 2.5 and a relatively weak energy source (0.1 J per pulse) was demonstrated. Quasi-direct current (q-dc) discharge can be used as an alternative to the spark discharge. It was considered as an effective instrument for the fuel ignition and the flame front stabilization in the supersonic flow [4]. In one of the tested configurations of the electric discharge and fuel injection, the injector, flush-mounted on the combustion chamber wall, was combined with one of discharge electrodes [12]. It was found that the plasma filament follows the boundary between the jet and the oxidizer, coming off the wall. In this case, the behavior of the discharge becomes even more unsteady than during its operation near the wall of the combustion chamber. It was suggested that such unsteady behavior should also positively affect the fuel and oxidizer mixing. The current work proceeds those studies and is devoted to explore in details the discharge effect on the fuel jet and supersonic flow mixing in the geometric configuration described in [12]

Scheme of experiment

The experiments were performed in supersonic wind tunnel PWT-50 of JIHT RAS. Flow parameters were as follows: Mach number $\boldsymbol{M} = 2$, static pressure $\boldsymbol{P_{st}} = 170$ torr, and stagnation temperature $\boldsymbol{T_0} = 300$ K. Detailed description of experimental facility can be found in Ref. [12]. As it is shown in figure 1, fuel injector (1) was flush-mounted in the ceramic insertion (2) installed on the metal wall of the test section and supplied with a gaseous fuel. In these experiments, a supersonic injector Ø4 mm was used with Mach number $\boldsymbol{M_j} = 1.8$, at typical mass flow rate of a secondary gas $\boldsymbol{G_{fuel}} = 2$–4 g/s injected perpendicular to a core flow. Three different gases were used for imitation of fuel injection: air, CO_2 and C_2H_4. The conditions of injection and discharge parameters were selected to prevent the ignition of ethylene. The fuel injector was combined with one of discharge electrodes (anode) for generation of q-dc filamentary discharge (3) nearby the air-secondary gas interface. The discharge was powered by 5 kV source connected through fast high-voltage solid-state switch and current limiting resistor.

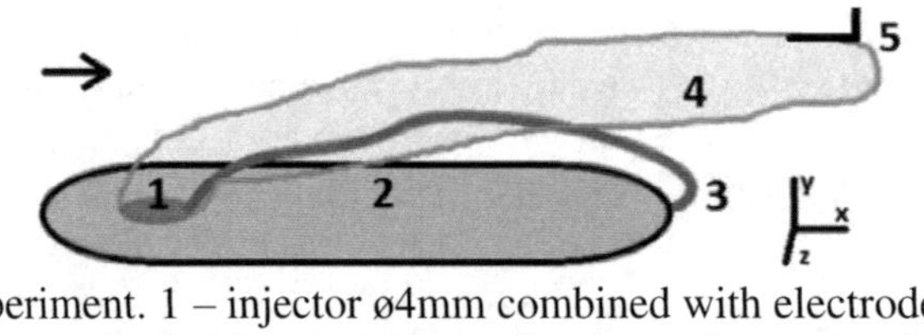

Figure 1. Scheme of experiment. 1 – injector ø4mm combined with electrode; 2 - ceramic insertion; 3 - Q-DC electrical discharge; 4 - fuel jet; 5 - Pitot probe, arrow shows the direction of M = 2 air flow

Experimental results

Typical electrical characteristics of discharge are presented in figure 2. The averaged discharge current was 3 A and averaged voltage of discharge was 1.5 kV. The discharge behavior is highly transient: instant values of current and voltage significantly varies in time with frequency 3–15 kHz. Such a behavior is resulted from the discharge interaction with secondary gas jet which carries the discharge away from wall and leads to significant increase in discharge length. Typical realizations of discharge shape at ethylene injection taken with high-speed camera are shown in figure 3. In these experiments, the discharge length was 60–70 mm that is controlled mostly by dimensions of the ceramic insert. It is shown that the plasma filament can have a significant variation in length and degree of curvature. Note, the plasma filament has a complex three-dimensional curved shape in both xy and xz planes of view (in figure 3, c-d), that may significantly affect the air-fuel mixing.

High-speed schlieren visualization of flow structure at jet injection was performed using high power pulsed laser diode as a light source and Basler camera. Pulse duration of laser was 100 ns that limits the exposure of schlieren image. The discharge image obtained by schlieren system has an exposure of 10 μs based on the camera shutter. Schlieren visualization of the fuel jet injection in supersonic flow allows demonstrating the influence of discharge on the character of flow. As it is shown in figure 4, activation of discharge makes the flow downstream the injector significantly unstable with increased gradient of the gas density on the boundary of jet, which is corresponded to the discharge curvature. To confirm the discharge effect on the flow of injected gas the Pitot probe was installed 100 mm downstream of injection port on the boundary between the jet and the core flow. Activation of the jet injection to the supersonic flow results in pressure decrease measured by the probe. It was found that after activation of discharge the character of pressure measured by the probe changes: oscillations of pressure are observed, which correlate with change in discharge shape and electrical characteristics, as it is shown in figure 2 with arrows.

To visualize the central plane of the jet simultaneously with the discharge operation, the particle visualization method (PIV) was applied. The width of the laser sheet was about 0.5 mm, the duration of the laser pulse was 15 ns, and the frame exposure was 1 μs. The jet was seeded with oil droplets with a diameter of about 2 μm, and the oncoming supersonic flow was seeded with a natural condensate of water vapor. The measurement system used made it possible to obtain instant images of the discharge and the simultaneous distribution of oil droplets visualizing the shape of the jet. Such visualization in the case of CO_2 injection is shown in figure 5. It can be seen that the discharge is localized along the jet below its lower boundary. Averaging the PIV measurements over 50 starts with identical parameters showed that the discharge activation causes deeper penetration of the jet into the main flow and the increase of angle of oblique shock wave caused by jet. The interface between the fuel and the oxidizer increases due to the use of a discharge.

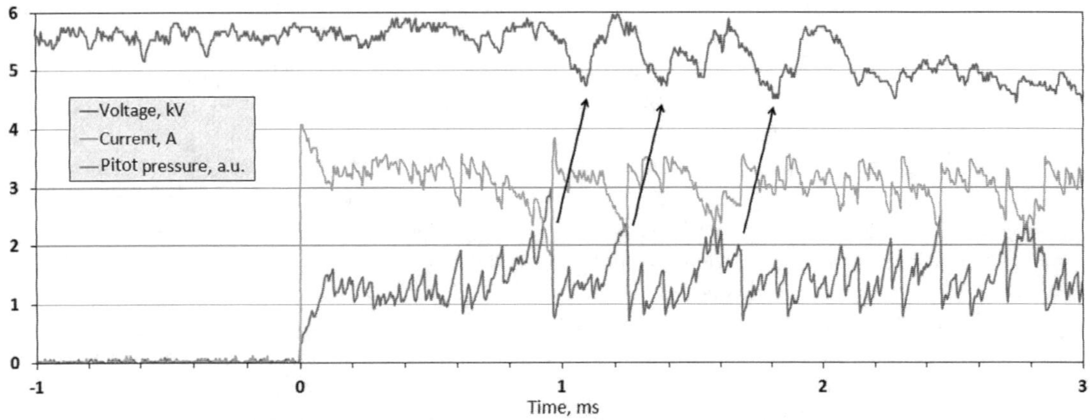

Figure 2. Electrical characteristics of discharge and its impact on Pitot pressure.

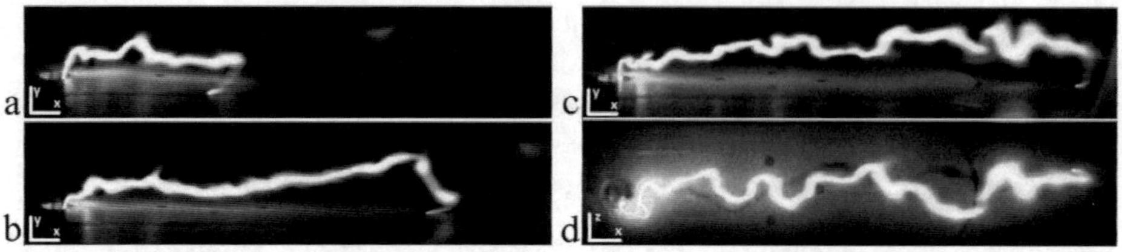

Figure 3. Typical images of discharge at ethylene injection. a - c – different length of discharge, images c and d are presented for the same moment of time at different planes of view.

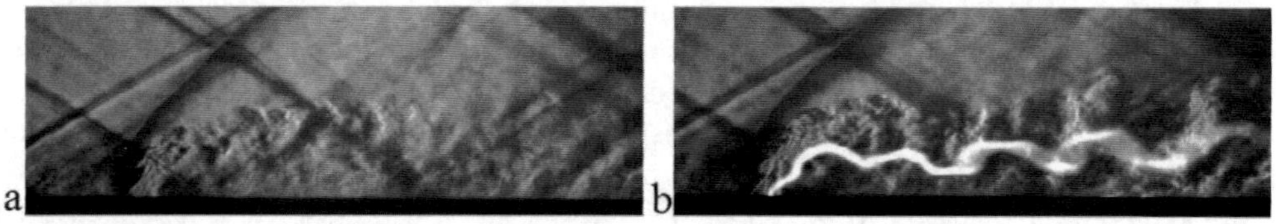

Figure 4. Schlieren visualization of the fuel jet injection in supersonic airflow. a - discharge off, b - discharge on.

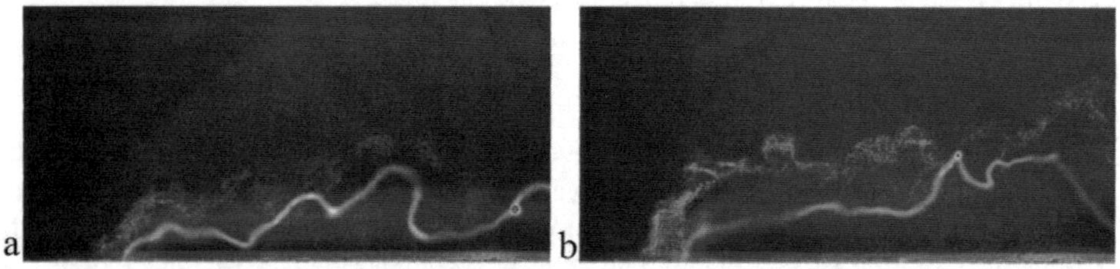

Figure 5. Photos of the discharge at the time of PIV visualization of the CO_2 jet at different injection pressures: a - 3 atm; b - 5 atm.

Summary

From previous studies [12], it was known that the electric discharge prefers localization in the region with high gradients of gas density or species concentration, especially if the initiation of the discharge is geometrically associated with the injection of a secondary gas. In the latter case, the discharge is localized in the mixing layer of the fuel jet in a supersonic incoming flow. In the current work, the analysis of experimental data, including high-speed schlieren visualization, high-speed video recording, electrical measurements, and pressure measurements using a Pitot probe, demonstrates that the $q - dc$ discharge operation significantly affects the flowfield in the region of jet injection into a supersonic flow, creating high gradient of gas density and pressure oscillations in the wake of the jet. For all considered injected gases, an increase in the unsteadiness of jet was observed. In the case of ethylene injection, the plasma filament dynamics was observed to be especially unstable comparing to the air or CO_2 jets. This fact is explained by the discharge specific localization during the injection of ethylene. It was shown by PIV visualization that the discharge activation causes deeper penetration of the jet into the main flow. The interface between the fuel and the oxidizer increases due to the use of a discharge. Further efforts are planned to obtain a quantitative data on the mixing dynamics and enhancement by means of the electric discharge generated within the shear layer.

Acknowledgments

This work is supported by RFBR grant No. 18-08-01452. The work of Efimov AV on discharge visualization are also supported by President grant №075-15-2019-633 (MK -2610.2019.8).

References

[1] Sabelnikov V A, Voloschenko O V, Ostras V N, Sermanov V N and Walter R 1993 AIAA 2145

[2] Vinogradov V A, Goldfeld M A and Starov A V 2013 Combustion, Explosion, and Shock Waves 49 383–91
[3] Zhang L, Yang V and Lin K C 2013 AIAA 011
[4] Firsov A A, Savelkin K V Yarantsev D A and Leonov S B 2015 Phil. Trans. 373
[5] Gaitonde D V, Riley L P, Hagenmaier M A and Donbar J M 2016 AIAA 1901
[6] Barnes F W, Qiuya T and Corin S 2016 Journal of Propulsion and Power 32 305–10
[7] Camilo A and Kenneth H Yu 2015 Journal of Propulsion and Power 31 1532–43
[8] Leonov S B, Firsov A A, Isaenkov Y I, Shurupov M A, Yarantsev D A and Shneider M N 2011 AIAA 2327
[9] Shurupov M A, Leonov S B, Firsov A A, Yarantsev D A and Isaenkov Y I 2014 High Temperature 52 169–78
[10] Zudov V N, Tretyakov P K and Tupikin A V 2009 AIAA 3586
[11] Zheltovodov A A and Pimonov E A 2013 Tech.Phys.s Let. 39 1016–8
[12] Savelkin K V, Yarantsev D A, Adamovich I V and Leonov S B 2015 Combust. Flame 162 825–835

Journal of Physics: Conference Series **1394** (2019) 012025 doi:10.1088/1742-6596/1394/1/012025

Experimental study of gas dynamics caused by spark discharge near wall

E V Dolgov, N S Kolosov, I A Moralev and A A Firsov*

Joint Institute for High Temperatures (JIHT RAS), 125412, Moscow, Russia

* af@jiht.org

Abstract. Experimental research of the high-speed flow caused by the decay of the thermal cavern after spark discharge located on the wall in ambient air is performed using schlieren technique. Parametric investigations include varying of discharge length in a range of 3-7mm and energy in a range of 60-200mJ. It was found that the rise of discharge energy results in insignificant increase of the size of disturbance height. For tested spark gaps at fixed total energy impact the maximal height of disturbance corresponds to minimal length of discharge (3mm). The results obtained are in a good qualitative and quantitative agreement with the simulation data previously published by the authors.

1. Introduction

The spark electrical discharge has been thoroughly investigated [1] and is actively used in various fields of technical applications, including ignition in internal combustion engines and generation of NOx for biomedical applications, etc.

Recently, an interest was arisen for a long (3-40mm) pulsed (100ns-5μs) spark discharge due to its ability in generation of gasdynamic perturbations potentially beneficial for a flow control and mixing intensification [2]. In several publications, the results of studies on the features of the spark discharge thermal cavern decay were presented: it was shown that large-scale jet flows are formed in the direction perpendicular to the channel due to the spark channel bends during the decay of the long spark discharge [3]. The effect of discharge shape (bending) on pattern of the gasdynamic perturbations was shown by using additional electrodes to distort the shape of the discharge, and also by numerical simulation [2].

One of promising directions of the use of gasdynamics caused by a spark discharge is development of plasma actuators for suppression of a transonic buffet - the normal shock wave oscillations that takes place in transonic flow around a wing at high Mach number or high angle of attack [4-6]. The spark discharge acted as a vortex generator in the boundary layer on a flat surface [4-5], and in the work published in paper [6] the spark discharge was combined with protrusions on the wing surface to create a directed transverse action. In paper [4], it was shown that the spark discharge near the surface leads to the formation of a jet of a rather complicated shape: after the shock wave has run away, the thermal cavern is significantly modified in geometry - a gas flows from the ends of the discharge channel to the center. Then the region of the low-density hot gas moves into plane perpendicular to the initial position of the spark, into a wide sector of about 180 degrees in the form of a half of toroid, the center of which takes a gas from the surrounding area. In paper [5], it was shown experimentally that such a discharge on the surface of the wing can lead to decrease the buffet amplitude. By the numerical simulation it was shown an entire suppression of the oscillations of the shock wave upon action by a pulse-periodic spark discharge.

Journal of Physics: Conference Series **1394** (2019) 012025 doi:10.1088/1742-6596/1394/1/012025

There is of a practical interest to study a very short nanosecond (~ 100ns) discharges due to potential applications for the airflow control, including applied aerodynamics problems (flow around various bodies) and problem of combustion/ignition of lean mixtures. In paper [7], it was shown that the decay of a thermal cavern of a nanosecond spark discharge with a spark gap of 1 mm results in the transfer of active particles into the surrounding volume by a distance of ~ 10 mm from the discharge axis. To use the discharges for aerodynamic applications, it is important to optimize the discharge parameters in terms of the flow control authority, the effect on the airflow pattern depending on the discharge parameters, such as interelectrode distance and energy deposition [8]. In this paper, the major focus is made on the experimental parametric study of structure and dynamics of gas perturbations caused by the long spark discharge.

2. Description of experimental setup

Electrical scheme and voltage-current measurements

Electrical scheme of experiment was modified in comparison to another one used in our previous experiment [2] for increasing the operating voltage and improving of discharge jitter. Principal electrical scheme is presented in Fig. 1,a. Main capacitor C1 (1.1-4.4nF) was charged using power supply U0 (8-16kV) through ballast resistance R1=10 MΩ which limits the charging current. Two spark gaps were connected inline, first one (1) was an externally triggered spark gap with operating voltage 4-10kV, second discharge gap (2) consists of two copper electrodes flush-mounted into ceramic BN (boron nitride) plate with spark gap 3, 5 or 7 mm. Resistance R2=1.7Ω was used to increase of resistive component of scheme during discharge pulse and reduce current oscillations caused by stray inductance. Resistive voltage divider based on R3=200 MΩ and R4=100 MΩ was used to increase the operating voltage range for spark gap (2). All lines of discharge circuit were as short as possible. Voltage U on spark gap was measured close enoughtto discharge using high voltage probe Tektronix P6015A, and discharge current was measured using Pearson current monitor mod.411. Maximum current and voltage amplitudes were the following - U up to 18 kV, and I up to 700 A. Typical voltage and current on time dependencies are presented in Fig.1,b for 5mm spark gap and energy E =100mJ.

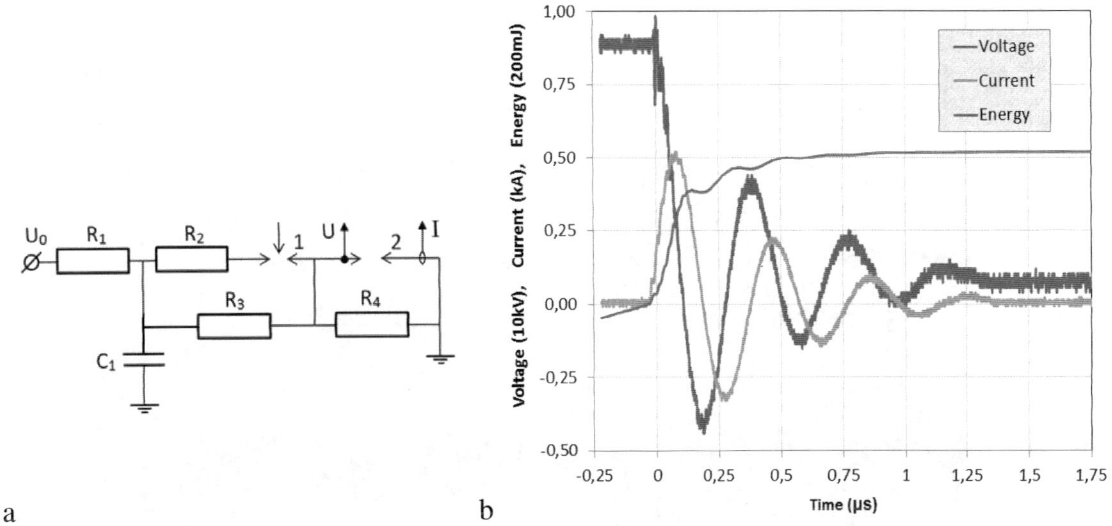

a b

Figure 1. a - principal electrical scheme of experiment, b – typical voltage and current on time dependencies (5mm spark gap and energy E=100mJ)

Journal of Physics: Conference Series **1394** (2019) 012025 doi:10.1088/1742-6596/1394/1/012025

Schlieren system details

Schlieren visualization of gas-dynamic disturbances caused by the discharge was performed using a highly sensitive shadow device tuned as a schlieren system. The shadow device consists of two lenses with a diameter of 150 mm with a focal length of 1.5 m and a swivel mirror that facilitates the adjustment of the system. The infrared laser source is installed in the focus of first lens, and the Foucault knife-edge is located in focus of second lens, and behind the edge the Basler acA2000-340km camera is installed. The experimental system was synchronized using an optical pulse generator, which gives high accuracy of synchronization of various equipment involved in the experiment. To prevent the camera from overexposing by discharge an interference filter is installed in front of the camera. This filter passes only a narrow range of wavelengths, into which the laser light of the shadow system with a wavelength of 812 nm. Laser pulse duration of 92 ns actually determines the exposure of our shadow device. After the camera exposure starts, a spark discharge breakdown is initialized, and then with an adjustable delay relative to the discharge, the laser pulse is activated. Because the electric circuit for triggering the discharge has a certain jitter, the real delay between the discharge and the laser pulse was measured directly by recording the pulse of the discharge current and the pulse from the IR (infrared) radiation detector.

For one breakdown of the discharge in the experiment, one frame of the shadow shooting was obtained. To obtain the dynamics of a thermal cavity, the experiment was repeated many times. At the first stage, it was necessary to choose the optimal position of the Foucault knife (see Fig.2) to observe a clear picture of the thermal cavity expansion and the development of jet instability caused by the operation of spark discharge. We tested the vertical and horizontal position of the knife, at an angle of 45° and shooting in the absence of a knife (direct shadow method). Because optical deviations in the vertical plane are considered in this experiment, the vertical position of the knife provides limited information about the plane passing through the middle of the spark channel. In this case, the disturbance on one side of the secant plane were darkened and lightened on the other, while the best picture was observed in the illuminated region which corresponds to the side part of thermal cavern. An attempt was also made to visualize the experiment in the absence of a knife, i.e. direct shadow method. Due to the strong gradient of the thermophysical parameters of air in the spark region in the time interval 0 - 20 μs, the disturbance that arises can be visualized by this method. At later times, the gradient decreases and the boundary of instability becomes less distinguishable. Then the knife was set in position at an angle of 45 degrees. It is clearly seen that in this case the cavity boundary was sharp only in a thin region close to the electrode. Also two horizontal positions of the edge knife were tested. Best images were obtained using bottom horizontal position of the knife. Using such settings of schlieren system the experiments were visualized over the all range of delay interval.

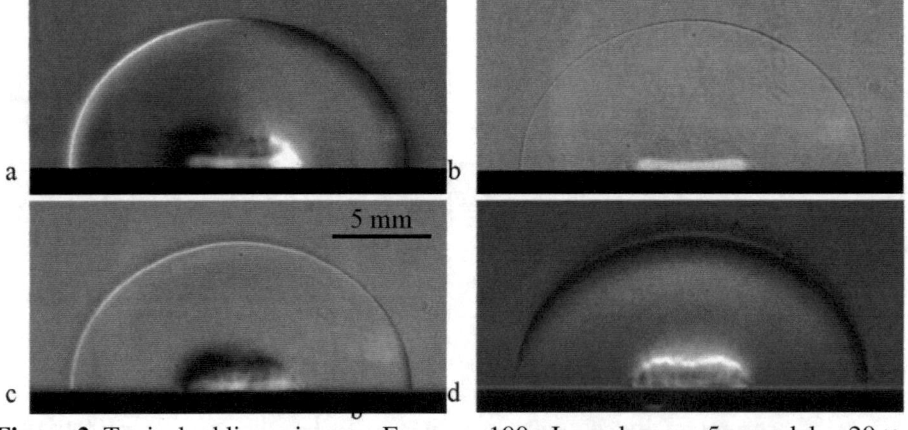

Figure 2. Typical schlieren images. Energy = 100mJ, spark gap = 5 mm, delay 20 μs; a - vertical knife; b - no knife; c - 45° knife; d - horizontal knife.

To accelerate the processing of a large number of shadow images obtained in the experiment, and to reduce the systematic error of data processing, a computer program was prepared that simplifies image analysis. The algorithm of the program was next: an array of images was loaded into the program, for the first of them a line perpendicular to the wall was selected manually, and dissecting the image of the thermal cavity, approximately in half (see Fig.3,a). Along this line, the intensity profile of the shadow picture was maid (see Fig.3,b). Local maxima corresponding to the contact of the thermal cavity with the wall (1), the upper boundary of the thermal cavity (2) and the position of the shock wave (3) are clearly visible on the obtained intensity profile. The first two local maxima were detected by the program and the size of the thermal cavity was determined as the distance between these maxima. The obtained distance was recorded in the memory, and the process was repeated for all obtained images. In addition to the images, a file with a list of shooting delays relative to the discharge breakdown was loaded into the program. For all images with the same delay (3-5 pieces), the averaging procedure was performed, and the obtained dependence of the average size of the thermal cavity on time was displayed on the screen and loaded to the text file.

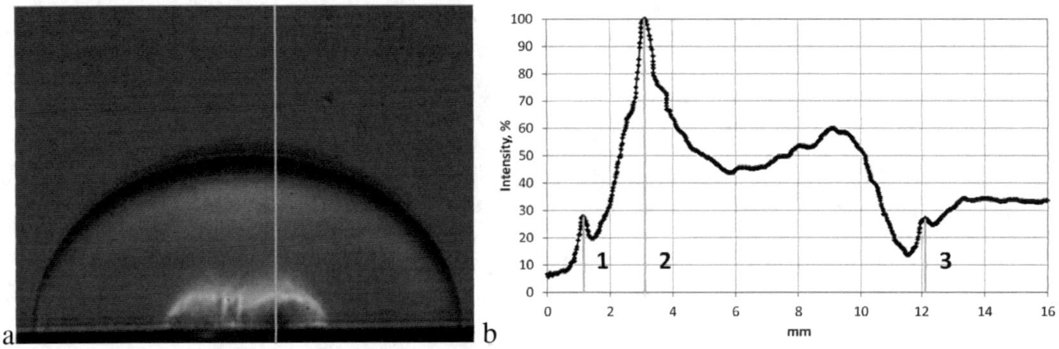

Figure 3. Typical schlieren image during post processing - a; and b - intensity profile along the vertical line in image (a): 1 – wall surface, 2 – cavity boundary, 3 – shock wave

3. Results of experiments and discussion

The study was performed for three spark lengths: 3, 5 and 7 mm. For each length, the spark input energy was varied: E= 60, 100, 150 and 200 mJ. The upper value was limited due to the spontaneous breakdown of the spark gaps. But from the preliminary numerical modelling [9], higher energy input does not lead to the significant increase of the induced flow structures intensity. For all considered cases, the evolution of the cavern was analyzed at the time delay in a range from 0 to 10 μs with a step of 2 μs and up to 100 μs with a time step of 10 μs. The quantitative data on the flow structure were extracted from shadow images. The main analyzed parameter was the distance from the top of the hot cavern to the wall. The dynamics of the cavern top position for three values of energy release is shown in Fig. 4, a. It can be seen that the increase of the pulse amplitude leads to the increase of the total cavern boundary displacement: at a fixed delay t=100 μs the maximum height of the disturbance is obtained for E=200 mJ. Still, the effect of the energy input on the disturbance magnitude decreases at high energy values: increase of energy from 60 to 200 mJ leads to the increase of final cavern height from 3.6 to 5.5 mm.

Figure 4, b shows the cavern boundary dynamics for E =60 mJ and three various interelectrode gaps. One can see that the decrease of the discharge length leads to the monotonous increase of the boundary velocity, with the maximal value obtained at d=3mm. Based on this parametric study, one can conclude that the optimal actuator parameters in the studied parametric space are the minimal interelectrode distance (d=3 mm) and relatively low energy input on the level of 50-60 mJ. The observed data were compared to the previously obtained modelling results [9]. It can be seen that the good quantitative correlation of the numerical and experimental results can be seen. This means that the significant part of the electrical power, induced by the discharge, is indeed transferred into the heating at the submicrosecond time scale. Some difference between presented experimental and computational data could be explained by the fact that real discharge size is larger than spark gap.

The further analysis of the velocity field, induced by the surface spark discharge, will be performed using PIV (Particle Image Velocimetry) method. These measurements will verify the characteristic gas velocities inside the vortices, produced during the decay of the spark cavity.

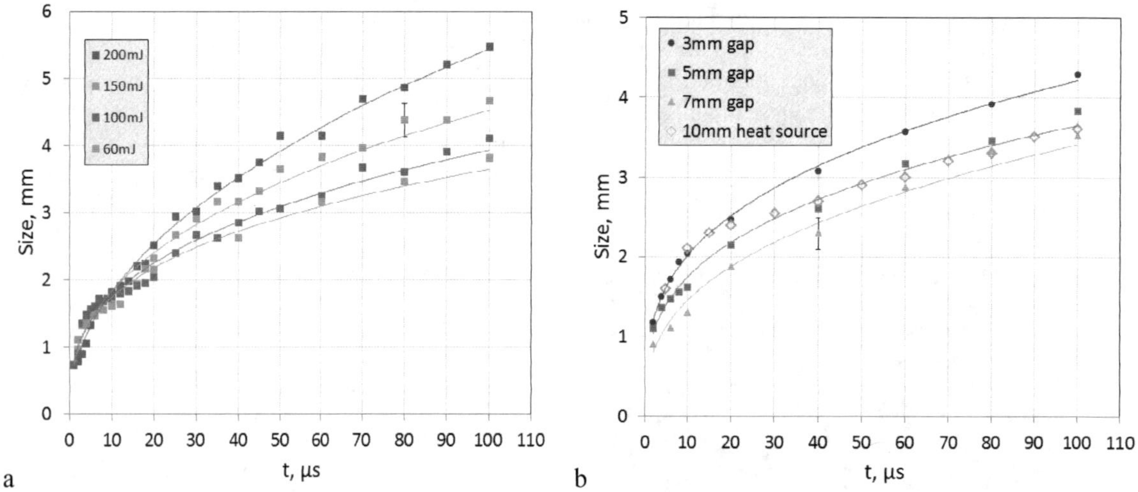

Figure 4. Experimental data about dynamics of afterspark cavity boundary: a - variation of discharge energy for 5 mm spark gap. b - variation of spark gap at fixed discharge energy 60mJ. Heat source means data of CFD (computational fluid dynamics) simulation [9]

4. Conclusions

The experimental study of the gas-dynamic perturbations caused by the decay of a thermal cavern after submicrosecond long spark discharge arranged near the surface in the ambient air at standard conditions was performed using schlieren technique. Parametric investigations included varying of discharge length in a range of 3-7mm and energy in a range of 60-200mJ. Spark discharge induces gas motion which could be analyzed by schlieren visualization of the boundary of thermal cavern. It was found that the rise of discharge energy at fixed spark gap results in low increase of the size of thermal cavern height. For tested spark gaps at fixed total energy impact the maximal height of the disturbance corresponds to minimal length of discharge (3mm). The results obtained are in a good qualitative and quantitative agreement with the simulation data previously published by the authors.

Acknowledgments

This work is supported by grant of Russian Scientific Foundation № 18-19-00740.

References

[1] Bazelyan E M and Raizer Y P, Spark Discharge, CRC Press, **1998**

[2] Shurupov M A, Leonov S B, Firsov A A, Yarantsev D A and Isaenkov Y I, Gasdynamic instabilities during decay of the submicrosecond spark discharge channel, *High Temperature*, **2014**, V.52, I.2, pp.169-178

[3] Leonov S B, Isaenkov Y I, Firsov A A, Nothnagel S L, Gimelshein S F and Shneider M N, Jet regime of the afterspark channel decay, *Phys. Plasmas*, **2010**, V.17, I.5, paper 053505

[4] Firsov A A, Isaenkov Y I, Shurupov M A and Leonov S B, Development of plasma actuator based on surface sparks for a buffet control, 54th AIAA Aerospace Sciences Meeting, AIAA SciTech Forum, *AIAA 2016-169,* **2016**

[5] Firsov A A, Moralev I A, Isaenkov Y I, Leonov S B and Soudakov V G, Suppression of transonic buffet phenomenon by spark plasma actuator, AIAA Aerospace Sciences Meeting, AIAA SciTech Forum, *AIAA 2018-2050*, **2018**

[6] Sidorenko A A, Budovsky A D, Vishnyakov O I, Polivanov P A and Maslov A A, Transonic buffet control by plasma actuator with spark discharge, 7th European Conference For Aeronautics And Space Sciences (EUCASS), 3-6 july 2017, paper *EUCASS2017-345*, **2017**

[7] Stepanyan S, Hayashi J, Lovascio S, Stancu G D and Laux C O, Hydrodynamic effects induced by nanosecond sparks in air and air/fuel mixtures // 55th AIAA Aerospace Sciences Meeting, AIAA SciTech Forum, *AIAA 2017-1581*, **2017**

[8] Singh B, Rajendran L K, Bane S P and Vlachos P, Characterization of Fluid Motion Induced by Nanosecond Spark Plasmas: Using Particle Image Velocimetry and Background Oriented Schlieren // 2018 AIAA Aerospace Sciences Meeting, AIAA SciTech Forum, *AIAA 2018-0680*, **2018**

[9] Dolgov E V and Firsov A A, Numerical study of hydrodynamic perturbations caused by filiform spark discharge near wall // *IOP: Conference Series*, **2018**, 1112(1), 012026

Journal of Physics: Conference Series **1394** (2019) 012026 doi:10.1088/1742-6596/1394/1/012026

Magnetosphere response to stationary solar wind forcing at various plasma resistivities in 2D MHD simulation

E V Gubanov[1], A P Likhachev[1,3] and S A Medin[1,2]

[1] Joint Institute for High Temperatures of the Russian Academy of Sciences, Izhorskaya 13, Bldg 2, Moscow 125412, Russian Federation
[2] Moscow Institute of Physics and Technology, 9 Institutskiy per., Dolgoprudny, Moscow Region, 141701, Russian Federation

[3] E-mail: apl2211@mail.ru

Abstract. The magnetosphere response to the stationary solar wind forcing at various values of the plasma resistivity is studied. The simulation is performed within two-dimensional (2D) MHD resistive model; the meridional plane is considered. The solar wind and southward interplanetary magnetic field (IMF) are supposed to be stationary. The calculations were carried out for two plasma resistivity values $\eta_p = \{10^6;\ 10^4\}\ \Omega{\cdot}\mathrm{m}$, assumed constant in the whole flow field. At higher plasma resistivity the steady magnetosphere convection (SMC) takes place; the solution is symmetric relative to the x-axis. At lower resistivity the sawtooth event is realized; the solution is quasi-periodic and asymmetric. The asymmetry is connected with the formation of the vortex street of Karman type. The sequence of phases of the sawtooth cycle, as well as the evolution of magnetic lines of force and the flow structure in each phase is analyzed. In both solutions the day side magnetic reconnection occurs through the two x-points disposed symmetrically at the top and bottom peaks of the magnetic island.

1. Introduction.

The magnetic reconnection is an essential element of the magnetosphere dynamics causing magnetic substorms – the main factors of space "weather" (see, i.e., books [1], [2]). In MHD approach the solar wind acts as the reconnection driver, different parameters of which (i.e. velocity, density, resistivity, and interplanetary magnetic field (IMF)) can onset reconnection in various ways. According to modern concepts, in a number of physical mechanisms influencing the reconnection process, the leading role belongs to the tearing-instability [3]. This instability develops at the condition $Lu > 1$ fulfilled for the current sheet (the Lundquist number $Lu = \mu v_a l/\eta$, where $\eta,\ v_a = B(\mu\rho)^{-1/2}$, l are characteristic values of the resistivity, Alfven speed, and sheet thickness, respectively) [1]. Thus, to know local value of the Lundquist number needs to perform global modeling of the magnetosphere dynamics at given solar wind parameters. Although the MHD approach is based on the macroscopic fluid dynamics and cannot describe all the microscopic kinetic details of the magnetospheric response to solar wind forcing, it remains an important tool for studying the dynamics of the magnetosphere as a system as a whole.

There is a number of works devoted to the problem of magnetosphere MHD activity under the solar wind forcing. In [4] the role of the two solar wind parameters, the velocity and the southward magnetic field, was studied with the use of the global MHD simulation. It was concluded that the solar wind velocity is the leading factor controlling the state of the geomagnetic tail under the same other driving conditions (see also [5], [6]). In [7] in the framework of the 3D MHD simulation the onset of reconnection in the magnetotail at the southward IMF was investigated. The resistivity was varied in calculations only to suppress numerical oscillations. Its value remained close to the order of the Spitzer magnitude (i.e., practically collisionless plasma) and did not affect the physical results.

As argued in [8], many different attempts have been made to incorporate microscopic physics into MHD approach (hybrid and kinetic models - see, for example, [9], [10]), but all of them are inferior in versatility and efficiency to global MHD models. In addition, the use of kinetic approaches does not always give the best result when model is validated on satellite observations ([11], [12]). Therefore, global MHD models continue to be used in magnetospheric studies. Numerous references to relevant works can be found in reviews and books (see cited above [1]. [2])

In this paper, the driving impact of the solar wind resistivity on the magnetosphere behavior is considered. To select basic value of resistivity, it was estimated according to the formula for the Sagdeev's resistivity [13] and to the measured width of the MHD bow shock [14]. These two estimations give equal order of magnitude, $\eta \sim 10^5$ Ω·m. In section 1, we describe the setup used for global MHD simulation. Results of global MHD simulations at high resistivity relating to steady magnetospheric convection are briefly presented in section 2. Simulation data on low resistivity describing sawtooth events are analyzed in section 3. These data were given special attention.

2. Problem formulation.

The model problem of the solar wind flow past the Earth's magnetosphere is studied. An infinitely long cylindrical body carrying magnetic dipole is flowed over by a hypersonic flow of fully ionized hydrogen plasma. The body has a square cross section; the velocity vector of the incoming flow is directed normal to the side face of the body. The computational area occupies the rectangle extending from −600 R to +100 R in the x direction and from −300 R to 300 R in the z direction. The scale parameter R equals to $6\cdot10^6$ m. The side of the body cross section with the central point at $x = 0$ and $z = 0$ is 2.8 R. The magnetic field is created by oppositely directed currents flowing in two infinitely thin parallel wires. Wires are laid along the normal to the body cross section and pass through it symmetrically relative to the center of coordinates at the distances of 1R. In each wire the electrical current of 170 MA is supplied. At the sunward boundary the solar wind velocity, pressure, density and the interplanetary magnetic field are given. At other boundaries of the calculation area the "soft" conditions are applied. At the body boundaries the zero velocity normal component is preset.

We accept the following values of solar wind parameters: velocity $u_0 = -4\cdot10^5$ m/s, $w_0 = 0$; interplanetary magnetic field $B_{z0} = -1$ nT, $B_{x0} = 0$; pressure $p_0 = 10^{-10}$ Pa; density $\rho_0 = 2\cdot10^{-20}$ kg/m^3. The plasma resistivity is assumed to be constant and is evaluated on the base of the satellite measurements of the bow shock width. The data processing performed by [14] indicates that the magnitude order of the shock width amounts to $\Delta \sim 10^5$ m. Estimating the plasma resistivity as $\eta_p = \mu u_0 \Delta$ for the the solar wind velocity u_0 the result turns out $\eta_p \sim 10^5$ Ω·m. The same value gives Sagdeev's formula [13] at the conditions of high overheating of ions relative to electrons. For the analysis of the resistivity influence on the reconnection mode the two resistivity values of 10^6 and 10^4 Ω·m are accepted. The magnetic Reynolds numbers corresponding to these solutions are 6.4 and 640, respectively. The resistivity of the rod matter equals to 10^{11} Ω·m.

The gasdynamic part of the problem is described by the Euler equations with the MHD source terms:

$$\frac{\partial \rho}{\partial t} + \frac{\partial}{\partial x}(\rho u) + \frac{\partial}{\partial z}(\rho w) = 0 \tag{1}$$

$$\frac{\partial}{\partial t}(\rho u) + \frac{\partial}{\partial x}(\rho uu) + \frac{\partial}{\partial z}(\rho uw) = -\frac{\partial p}{\partial x} + j_y B_z \tag{2}$$

$$\frac{\partial}{\partial t}(\rho w) + \frac{\partial}{\partial x}(\rho uw) + \frac{\partial}{\partial y}(\rho ww) = -\frac{\partial p}{\partial y} - j_y B_x \tag{3}$$

$$\frac{\partial}{\partial t}\left(\rho\left(\varepsilon + \frac{u^2 + w^2}{2}\right)\right) + \frac{\partial}{\partial x}\left(\rho u\left(\varepsilon + \frac{u^2 + w^2}{2}\right)\right) + \frac{\partial}{\partial z}\left(\rho w\left(\varepsilon + \frac{u^2 + w^2}{2}\right)\right) = -\frac{\partial}{\partial x}(pu) - \frac{\partial}{\partial y}(pw) + j_y E_y \tag{4}$$

$$\varepsilon = \frac{1}{\gamma - 1}\frac{p}{\rho} \tag{5}$$

The medium is considered as an inviscid non-heat-conducting perfect gas consisting of protons and electrons with the specific heat ratio γ =5/3.

The electrodynamics is described by the Maxwell equations and the Ohm's law (the Hall effect is neglected):

$$j_y = \frac{1}{\mu}\left(-\frac{\partial B_z}{\partial x} + \frac{\partial B_x}{\partial z}\right) \tag{6}$$

$$\frac{\partial B_x}{\partial t} = \frac{\partial E_y}{\partial z} \tag{7}$$

$$\frac{\partial B_z}{\partial t} = -\frac{\partial E_y}{\partial x} \tag{8}$$

$$E_y = \frac{j_y}{\sigma} - wB_x + uB_z \tag{9}$$

All quantities in equations (1)-(9) are given in usual physical notation.

The computational domain was divided by an orthogonal uniform grid with a step of 0.2R, i.e. the grid dimension is 3500×1500. The hydrodynamic equations have been integrated with the usage of the explicit second order accurate TVD scheme [15]. The electrodynamic equations have been solved using the described in [16] explicit scheme with second order in space and first order in time resolution.

3. Steady Magnetospheric Convection event. Symmetric solution

At the higher resistivity, η_p=10^6 Ω·m, the simulation represents magnetosphere steady-state response to the stationary solar wind forcing (i.e. steady magnetospheric convection event is realized). The solution is symmetric relative to the x-axis. Figure 1 presents magnetic force lines with colored electric current density (left panel) and flow streamlines with colored plasma pressure (right panel). Due to the symmetry of the solution, the figure shows its upper half.

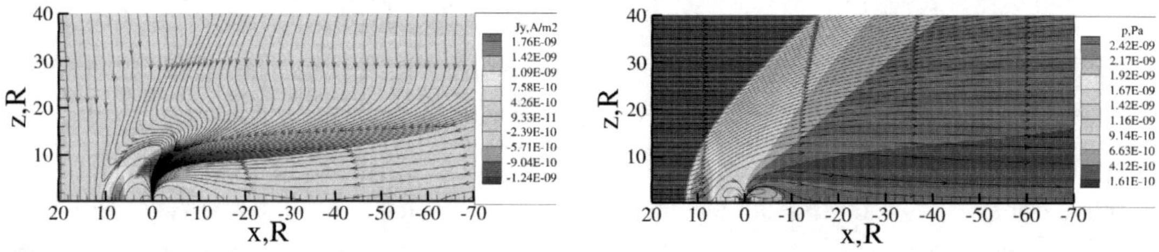

Figure 1. Magnetic force lines with background distribution of the electric current density (left panel) and flow streamlines with background distribution of the pressure (right panel). The vector lines and the background contours are symmetric relative to the x-axis.

The classic structure of magnetosphere in meridian plane is realized with the all standard elements. The bow shock is detached from the body at the distance x =12 R. The velocity ratio at the central part of the bow shock is near to the limiting value $(\gamma + 1)/(\gamma - 1) \approx 4$. At the day side and the night side of the body the attached fluid double vortices are formed.

The balanced reconnection occurs with the rate E_y = 0,4 mV/m both at the day and night sides. The dayside reconnection occurs via magnetic island (not seen in Figure 1, left panel, because of tiny dimensions). The magnetic island O-point is located at x =11R, the two X-points are shifted vertically relative the O-point. The nightside X-point is located at x =-35 R and implements contact between the dipole and the IMF.

4. Sawtooth event. Asymmetric solution

In this section the lower resistivity, $\eta_p = 10^4\ \Omega\cdot\text{m}$, is accepted. This leads to arising of the quasi-periodic substorm or the sawtooth event. The sawtooth event cycle contains the all substorm phases. In Figure 2 the temporal variation of the open magnetic flux per cylinder unit length Φ_0 is shown. The open flux Φ_0 is determined as the difference between values of magnetic vector potential component A_y at the dayside and the nightside x-points.

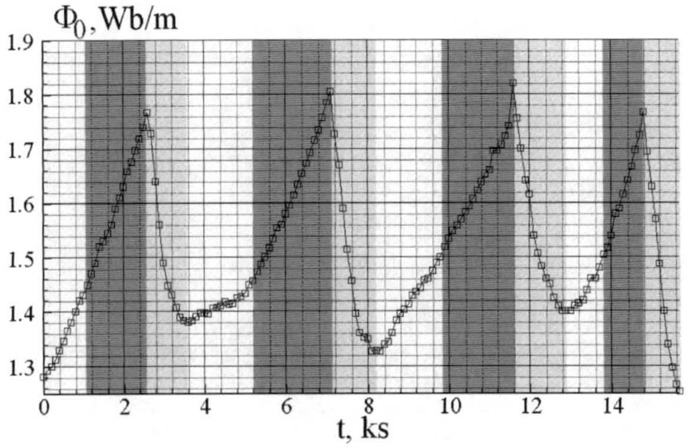

Figure 2. The sawtooth cycle series. The temporal variation of the open magnetic flux per cylinder unit length Φ_0 is shown. The growth, expansion and recovery phases are indicated by shaded vertical bands in yellow, light violet and azure, respectively.

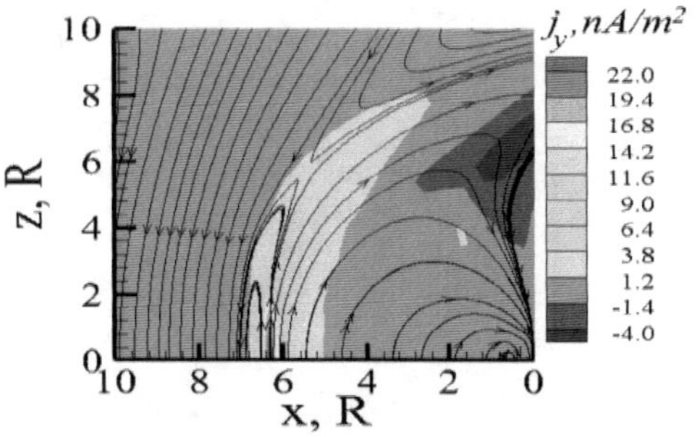

Figure 3. The dayside magnetic field force lines and the colored current density (upper part of the symmetrical picture is shown).

The temporal evolution of magnetic field and flow velocity topology is being considered for the third sawtooth cycle of 8.2-12.8 ms interval (the period is approximately 4.6 ks). The cycle starts at the minimum value of the open magnetic flux. The flux Φ_0 is monotone increasing with the reconnection moment of 9.9 ms switching to the expansion phase. Reaching the sharp peak it drops to the next minimum value what terminates the cycle.

The topology of the magnetic field in the dayside of magnetosphere is presented in Figure 3. The structure of magnetic field is practically steady and symmetric and the only its northern part is shown. Just in the front of the bow shock the magnetic island is formed. It is elongated vertically with the X-points located at the coordinates $x = 6.5\ R, z = \pm 5.6\ R$. The formation of the open magnetic force lines and the magnetopause is seen. The positive current generated in the shock layer is balanced by the negative current in the cusp.

In Figure 4 the tooth variation of the magnetic field topology with the background current density j_y is shown. The sharp change in the tail dynamics is observed comparing to Figure 1. At first the asymmetric longitudinal oscillaions of the magnetic force lines and the current layer appear. Then the formation of plasmoids and their intrainment with plasma flow takes place. The moving dowstream large plasmoid, magnetic island, born in the previous tooth is seen in Figure 4a. The plasmoid has been separated from the growing dipole night sector by the IMF. The IMF lines are strongly streched sunward resulting in the intense current sheet. In the sheet the secondary magnetic island arises (see Figure 4b).

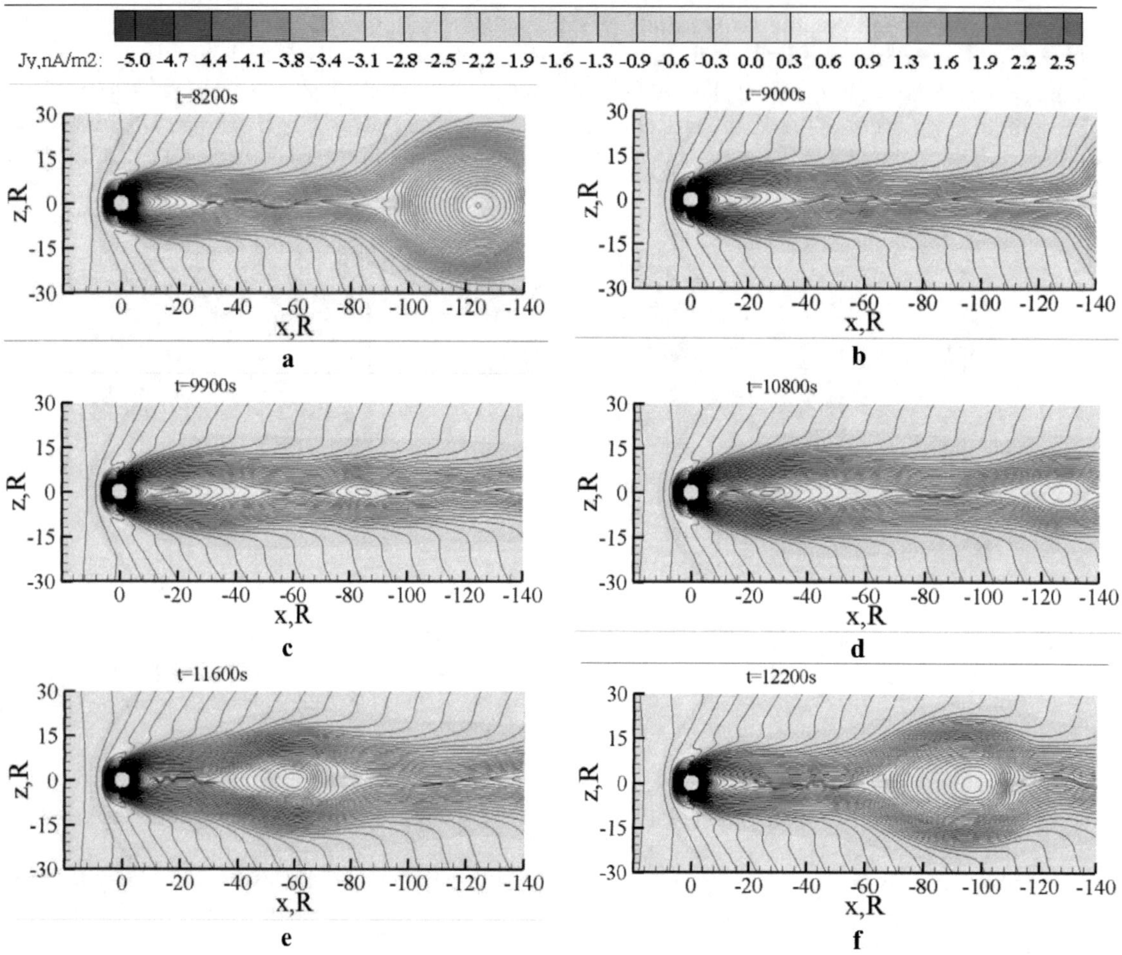

Figure 4. The magnetic field vector lines and the colored current dencity for the tooth cycle of 8.2-12.9 ks.

In the IMF interval the transverse oscillations of the current sheet appear. On top and below the oscillations extremums the moderate spots of the current dencity are exposed. They produce the lateral force $\mathbf{j}\times\mathbf{B}$ which generates the transverce oscillations of plasma flow (see below Figure 5).

At the time moment of 9.9 ks the magnetic field reconection sets in indicating the transition from the growth phase to the expansion phase. Inside of the dipole night sector the magnetic island is growing what increases the sector dimentions. The magnetic island separates from the dipole night sector at the time 11.6 ks starting the recovey phase. The island is bulging and speeding off by the plasma mantle.

The streamlines and the colored vorticity $\boldsymbol{\omega} = \mathrm{rot}\,\mathbf{v}$ are shown in the Figure 5. In the growth phase (see Figure 5a,b) the beginnig of the tail vortex street is seen. In the first frame the splitting of the upper attached vortex occurs. The splitting originates at the point where the plasma velocity reduces to zero and the length-wise net force $(-grad\ p + \mathbf{j}\times\mathbf{B})$ alternates its sign what stretches the vortex out and tears it in two parts. This mechanism is argued in [17]. Also we see the streamlines oscillations in front of the wortex wake which are consistent with the transverce oscillations of the magnetic force lines in Figure 4a,b. The vortex shedding is continuing in the expansion phase, Figure 5c,d. The vortex street is similar to that of the classic Karman vortex street. The vortex wake is confined by the the dipole magnetic field with the implanted magnetic island.

Journal of Physics: Conference Series **1394** (2019) 012026 doi:10.1088/1742-6596/1394/1/012026

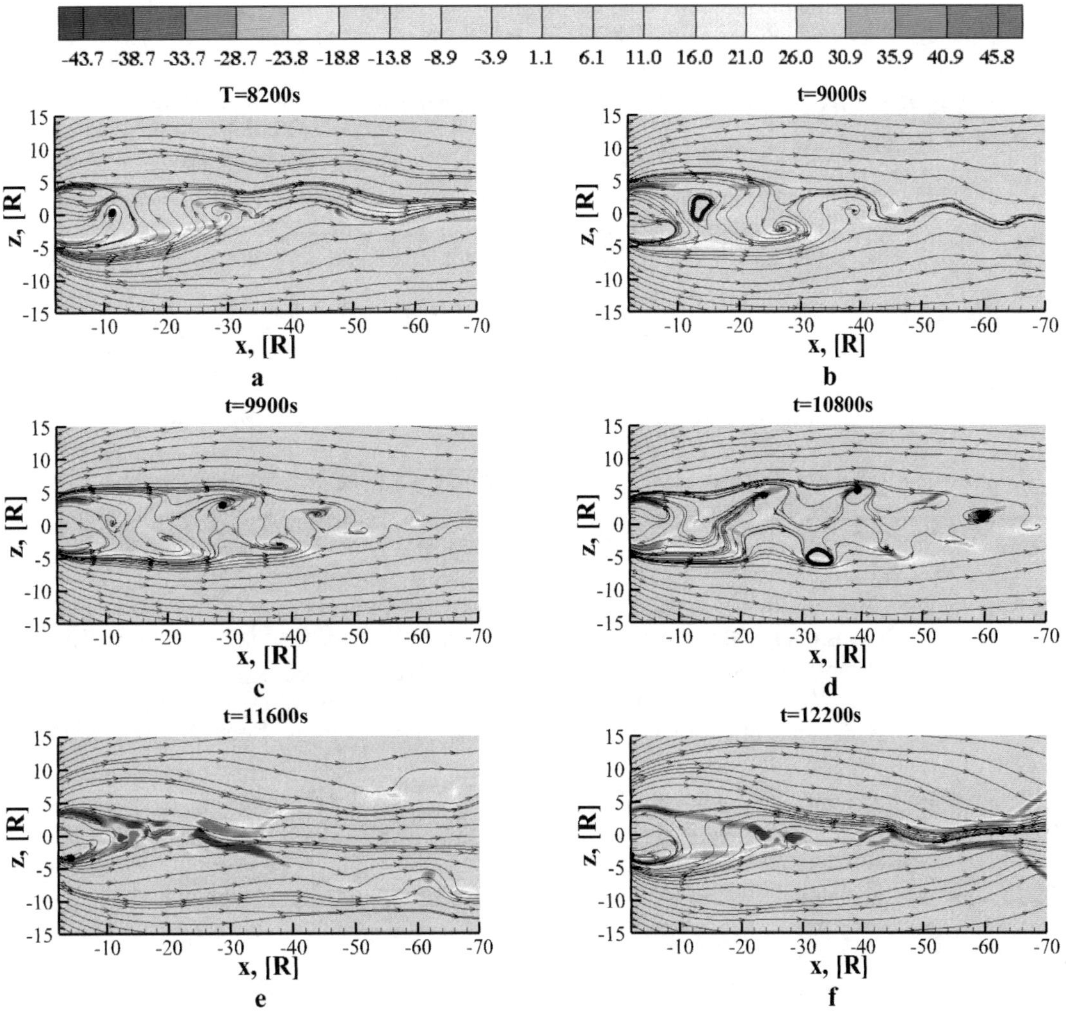

Figure 5. The plasma streamlines and the colored vorticity for the tooth cycle of 8.2-12.9 ks.

When in the recovery phase, Figure 5e,f, the magnetic island separates and the dipole sector shrinks the vortex train is quenched by the magnetic field shear. The only two vortices are attached to the body. Their extension will produce the vortex splitting and shedding in the next tooth. The frequency of vortex formation depends on the phase of the sawtooth event: it is near 185 c in the growth phase, near 165 c in the expansion phase, and near 120 s in the recovery phase.

5. Conclusions

Solar wind resistivity controls magnetospheric tail response modes, the SMC and the sawtooth events, under conditions of the rest solar wind parameters being of the same values. The MHD modeling resistivity might be evaluated from the measured width of the bow shock and from the computed anomalous collision frequency. In computations the two values of resistivity are accepted. At high resistivity of 10^6 Ω·m the simulation presents the SMC flow. At low resistivity of 10^4 Ω·m the tail sawtooth events are realized.

The SMC evolution of topology of the magnetic field and the flow velocity illustrates the dayside and the night side flow dynamics. In the day side the reconnection is quasistationary. It is formed in the

Journal of Physics: Conference Series **1394** (2019) 012026 doi:10.1088/1742-6596/1394/1/012026

shock layer through the thin magnetic island with two vertically positioned X-points. Downstream the shock and the magnetic island the double vortex is generated. In the night side the sawtooth cycle sets in when the open magnetic flux arrives at minimum and starts to grow. In the near tail the dipole constituent of magnetic field is expanding. The attached fluid vortices are splitted and begin with shedding. The splitting is provided by the combined tensile impact of the ponderomotive force and the pressure gradient. In the expansion phase the vortex street is formed. The vortex street propagates within the range of the dipole and decays at the magnetic island separation. After the magnetic island separation the open magnetic flux decreases, the magnetic dipole and the attached double vortex are spreading in the near wake. This eventually accomplishes the sawtooth cycle.

6. References

[1] Priest E and Forbes T 2000 *Magnetic reconnection: MHD theory and applications* (New York: Cambridge University Press)

[2] Gonzalez W, Parker E 2016 *Magnetic Reconnection: Concepts and Applications* (Springer: Astrophysics and Space Science Library)

[3] Furth H P, Killeen J, Rosenbluth M N 1963 Finite resistivity instabilities of a sheet pinch. *Phys. of Fluids*. 1963 **6**(4). 459-84

[4] Pulkkinen T I, Goodrich C C, Lyon J G 2007 Solar wind electric field driving of magnetospheric activity: Is it velocity or magnetic field? *Geophys. Res. Letters*. **34**. L2110

[5] Partamies N, Pulkkinen T I, McPherron R L, McWilliams K, Bryant C R, Tanskanen E, Singer H J, Reeves G D, Thomsen M F 2009 Statistical survey on sawtooth events, SMCs and isolated substorms *Advances in Space Research* **44**(3). 376-8

[6] Pulkkinen T I, Palmroth M, Koskinen H E, Laitinen T V, Goodrich C C, Merkin V G, Lyon J G 2010 Magnetospheric modes and solar wind energy coupling efficiency. *J. Geophys. Res.: Space Physics* **115** (A3)

[7] Walker R J, Ogino T, Raeder J, Ashour-Abdalla M. 1993 A global magnetohydrodynamic simulation of the magnetosphere when the interplanetary magnetic field is southward: The onset of magnetotail reconnection. *J. Geophys. Res.: Space Physics* **98** (A10) 17235-49.

[8] Gordeev, E, et al. 2015 Assessing the performance of community-available global MHD models using key system parameters and empirical relationships. *Space Weather* **13** (12) 868

[9] Ashour-Abdalla M, et al 2010 Observations and simulations of a highly structured plasma sheet during northward IMF *J. Geophys. Res.* **115** A10227

[10] Pembroke A, et al 2012 Initial results from a dynamic coupled magnetosphere-ionosphere-ring current model *J. Geophys. Res.* **117** A02211

[11] Pulkkinen A, et al. 2011. Geospace environment modeling 2008–2009 challenge: Ground magnetic field perturbations *Space Weather* **9**, S020049

[12] Honkonen I, et al 2013 On the performance of global magnetohydrodynamic models in the Earth's magnetosphere, *Space Weather* **11** 313–326

[13] Treumann R A 2001 Origin of resistivity in reconnection. *Earth Planets Space* **53**(6) 453-62

[14] Bale S D, Mozer F S, Horbury T S. 2003 Density-transition scale at quasiperpendicular collisionless shocks. *Phys. Rev. Lett.* **91**(26) 265004.

[15] Harten A 1983 High resolution schemes for hyperbolic conservation laws *J. Comp. Phys.* **49** 2 357-393.

[16] Gubanov E V, Likhachev A P, Medin S A 1999 MHD effects in the interaction of a meteoroid with planetary envelopes. Preprint 3-428 JIHT of RAS Moscow (in Russian)

[17] Boghosian M E, Cassel K W 2016 On the origins of vortex shedding in two-dimensional incompressible flows *Theor. Comput. Fluid Dyn.* **30** 511–527.

Journal of Physics: Conference Series **1394** (2019) 012027 doi:10.1088/1742-6596/1394/1/012027

Erosion of the sputtered electrodes in the surface barrier discharge

I Selivonin[1,2], A Lazukin[2], I Moralev[1,2], S Krivov[2] and I Roslyakov[3]

1. Joint Institute for High Temperatures of the Russian Academy of Sciences, Izhorskaya 13 Bldg 2, Moscow 125412, Russia
2. National Research University "Moscow Power Engineering Institute" Krasnokazarmennaya 14, Moscow 111250, Russia
3. Lomonosov Moscow State University, Leninskie Gory 1, 119991, Moscow, Russia

E-mail: inock691@ya.ru

Abstract. The paper is devoted to a phenomenological study of the modification of sputtered electrodes in a surface dielectric barrier discharge in atmospheric air. A decrease in the power supplied to the discharge in the case of using platinum and aluminum electrodes and an increase in the case of using a copper electrode were demonstrated. It is shown that during long-term operation of the electrode system, a decrease in its cold capacitance is observed. These effects are assumed to be due to the degradation of the electrode material, transfer of erosion products to the surface of the dielectric, as well as morphological changes at the electrode edge.

1. Introduction

The dielectric barrier discharge (DBD) is widely used in various applications, for example, surface treatment [1,2], plasma chemistry [3–5], plasma medicine and biology [6,7], plasma aerodynamics [8–11]. In a variety of technological problems, two basic types of electrode configurations are used: with a pair of dielectric-covered electrodes (coplanar barrier discharge) and with one electrode exposed directly to the discharge (surface barrier discharge) [12].

One of the most important requirements for devices based on DBD is the long-term stability of their operational characteristics. The papers dedicated to the development of long-operated devices based on DBD describe two main processes of degradation of the electrode system: destruction of the dielectric barrier [13,14], and the erosion of electrodes, the latter taking place in surface configuration (sDBD) [15–17].

As was shown in [16,17], at long exposures foil electrodes significantly change their morphology. It was demonstrated that the degradation of the electrodes leads to a change in the integral characteristics (energy dissipated in the discharge, input power), determining the performance of the devices based on the barrier discharge. More subtle characteristics, such as the mode of the discharge operation and microdischarge statistics, are affected as well.

The rate and scenario of electrode degradation strongly depends on electrode thickness. In [15], the processes of degradation of the thin film deposited electrodes were considered. It has been shown that even with the use of oxidation-resistant materials, intense destruction of electrode edges is observed, which leads to the to a change in the capacity of the discharge cells and dissipated power level.

Thus, the correct choice of the material of the exposed electrode and the method of its manufacture is an important task in the development of devices based on sDBD, regardless of their intended purpose. Research in this direction will allow to create stable long-operated sDBD arrangements or, at least, to predict changes in their properties during the long-term operation.

2. Experimental setup and measurement methods

In this work, the classical configuration of the edge-plane surface barrier discharge was studied. The configuration and measurements organization are shown in Figure 1. The sputtered strips of aluminum, copper, and platinum were used as corona electrodes. The thickness of the electrodes was 0.5–2 μm. Before the exposure began, the edge quality was monitored visually using an MBS-10 optical microscope with a magnification of 100x. Alundum ceramics with a thickness of 1 mm and a dielectric constant of 10.4 was used as a dielectric barrier. Earlier experiments have shown that dielectric material is immune to plasma treatment in a barrier discharge. The buried electrode was sectioned in order to more accurately measure the electrical characteristics of the discharge.

The electrode system was powered by AC sinusoidal voltage with a frequency of 100 kHz and an amplitude of 3.4 kV. Exposure was carried out for 6 hours in room air at atmospheric pressure. Venting of the volume was organized to remove the chemical products of the discharge.

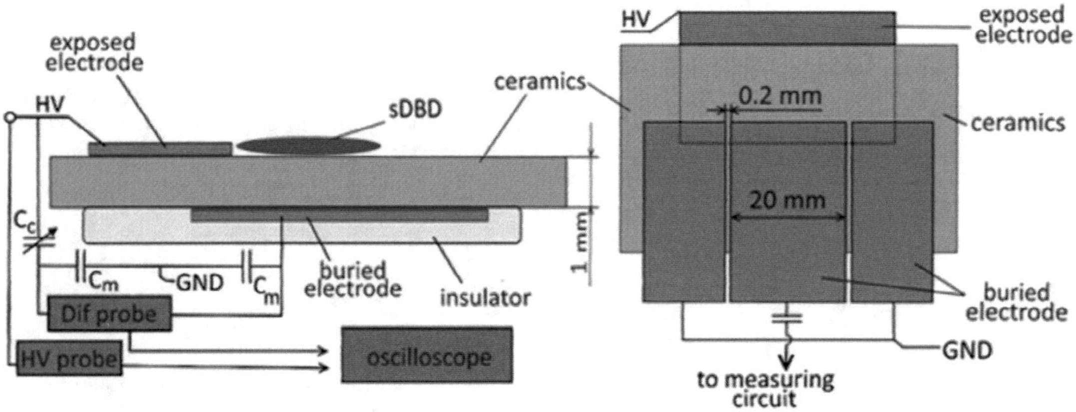

Figure 1. Discharge cell and measurements organization

To measure the electrical characteristics of the DBD, the volt-coulomb curves (VCC) were analyzed. The method was described in detail in [18]. To exclude capacitive current through the electrode system from the measurements, a Wheatstone capacitor bridge was used, in the upper arm of which a tunable vacuum capacitor C_c =3–50 pF was installed. The bridge was balanced at the low voltages prior to the experiments. Measurement of the voltage on a capacitor C_m, proportional to the charge transferred onto the dielectric barrier, was made with a differential voltage probe Pintek DP-150 (accuracy 5%, bandwidth 150 MHz). The high voltage measurement was made using a Tektronix P6015A high voltage probe (5% accuracy, 75 MHz bandwidth). The data were collected using an oscilloscope Tektronix TDS2024b with a bandwidth of 200 MHz. Recorded data were averaged across 128 periods of the supply voltage.

The energy dissipated in the discharge during one voltage period can be determined by estimating the VCC area. The cold capacitance of the discharge arrangement can be calculated as $C_{cold} = dQ/dU$ in a silent part of the VCC when the discharge does not operate. Compensation procedure before the start of exposure led to exclusion of the displacement current, associated with C_{cold}, from the measurements. Thus, the VCC had a zero slope at the silent part of the cycle (Fig. 2).

This method allows to obtain the values of the power dissipated into the discharge with an accuracy of 1 mW/mm, as well as to control changes in the discharge arrangement capacity without the discharge (cold capacitance) with an accuracy of 0.1 pF.

To obtain pictures of the morphology of the electrode edges, Olympus Lext OLS4000 laser confocal microscope with a spatial resolution better than 500 nm was used.

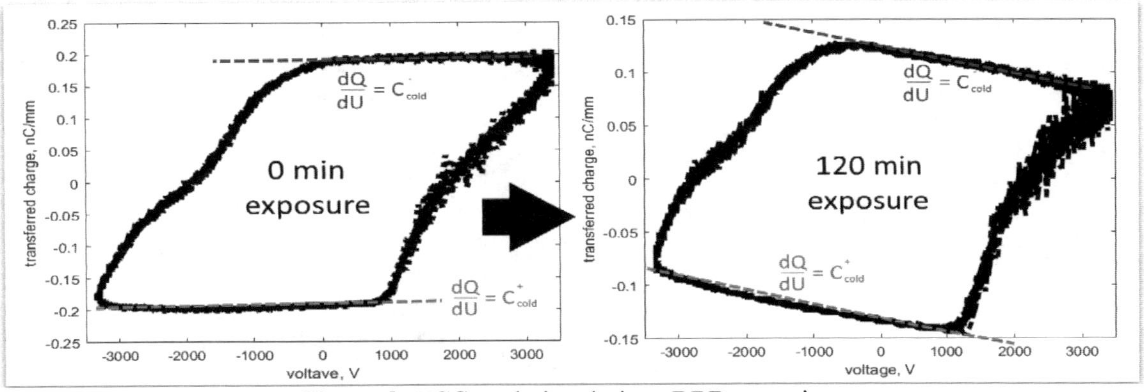

Figure 2. VCC evolution during sDBD operation

3. Results and discussion

During the first hour of operation, a decrease in the discharge power to the level of ~ 60% of the initial value is observed on the aluminum electrode (Fig. 3, left). Such a change in power is accompanied by a significant decrease in the cold capacity of the discharge cell (Fig. 3, right). The reduction of the cold capacity reaches 0.013 pF/mm after 2 hours of cell operation while the initial capacity of the electrode system is 0.5 pF / mm electrode length. In the case of using a platinum electrode, similar trends are observed, although less pronounced. In the case of a copper electrode, an increase in the power deposited in the discharge reaches 30-50% of the initial value. No change in the capacity of the discharge system is observed.

While the power dynamics qualitatively looks the same as in the case of foil electrodes [16,17], a change in the capacity of the discharge cell is a new effect, appearing only for thin exposed electrodes. The decrease in the cold capacity of the electrode system is obviously associated with a decrease in the area of the exposed electrode as a result of the effect of the sDBD plasma. Knowing the change in the capacitance of the electrode system and assuming that electrode material is removed from the surface or replaced by the oxide layer, one can estimate the width of the oxidized strip. To do this, one can use the formula:

$$C = \frac{\varepsilon\varepsilon_0 S}{d},$$

where C is the reduction in the capacitance of the electrode configuration, ε and d are the dielectric constant and thickness of ceramics, S is the width of the oxidized strip.

The result of this estimate is shown in Figure 4 on the left. In the case of using an aluminum electrode, an edge degradation is observed at distances up to 140 μm for 2 hours exposure. To verify this result, images of the edge were taken using a confocal microscope (Figure 4, right). They clearly show a blackened oxide band along the entire edge. At the 2 hours exposure time, the width of this band is about 250 μm, that somehow exceeds the estimate obtained from a decrease in the capacity of the discharge system. Obviously, in the blackened area, the electrode oxidizes unevenly; in some areas, oxidation does not occur across the entire thickness of the electrode, and there is a conductive metal layer under a thin oxide layer. This fact is also confirmed visually - during long-term discharge operation, the attachment points of the microdischarges move from the edge of the electrode, however, they are localized in the far region of the oxide band.

In the case of using a platinum electrode, edge degradation can also be counted and observed in images. At 2 hours of exposure, it is about 100 microns. However, the edge patterns obtained with a microscope show that the erosion process of the electrode differs from the processes on the aluminum edge. In addition to the oxidation of the strip, active transfer of erosion products to the surface of the barrier in the discharge region is observed. As a result of these processes, the properties of the surface of the dielectric change, which also leads, apparently, to the difficulty of developing the discharge.

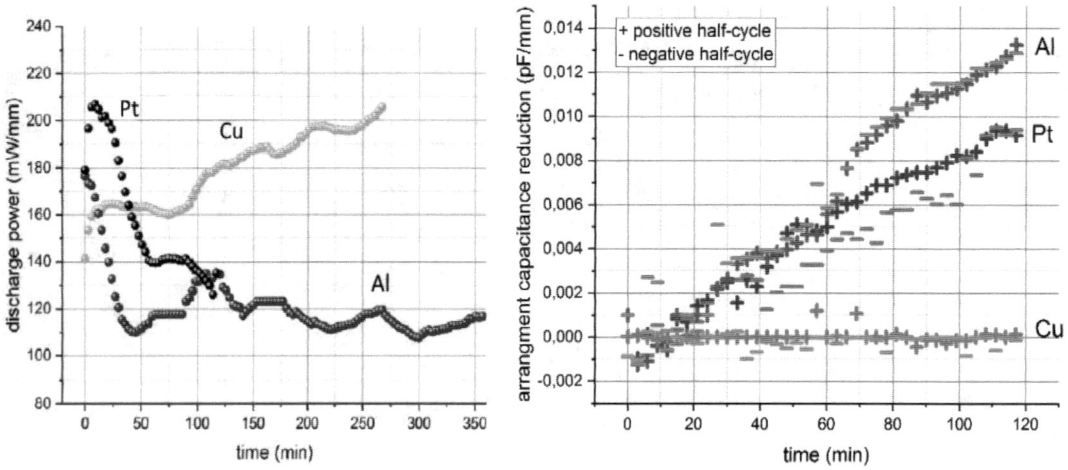

Figure 3. The dynamics of the discharge power on the electrodes of various materials (left) and the reduction of electrode system capacitance (right)

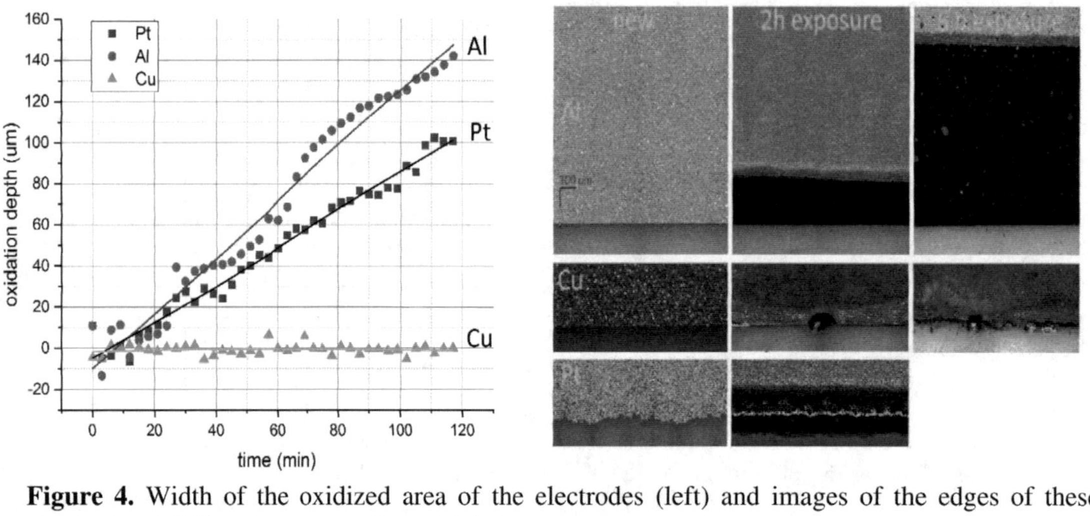

Figure 4. Width of the oxidized area of the electrodes (left) and images of the edges of these electrodes at different exposure times (right)

In the case of a copper electrode, no bulk degradation of the electrode is observed. On the electrode edge, craters with a size of the order of 100 μm, evenly distributed along the edge, are formed. These craters correspond to the conservative positions of microdischarges. In the vicinity of these craters, the intensive formation of copper oxides is observed. However, inside the craters the clean nonoxidized electrode material can be seen. Previous studies have shown that cathode erosion cleaning at the locations of stationary microdischarges can be responsible for the formation of the removal of the oxides front the crater center. The initial increase in discharge power seems to be related to the formation of the crater and stabilization of the microdischarges positions. In general, the behavior of the copper sputtered electrode is qualitatively similar to the behavior of the copper foil electrode, which was described in [16].

Summarizing the above, we can describe the processes of modification of the sputtered electrodes as follows.

In the first tens of minutes, the edge of the aluminum electrode is modified by discharge. At the same time, there significant decrease in power occurs. At the completion of this stage, the edge acquires a certain structure with a gradient of the thickness of the oxide strip. Further, this once formed structure moves into the depth of the electrode. This process leaves the power nearly constant, while reduction of the electrode area reduces the cold capacitance of the system. Obviously, the decrease in power and degradation rate of the edge are determined by electrode thickness.

The modification of the platinum electrode can also be divided into 2 stages - a fast one, in which a structure with an oxide band on the edge and contamination by products of erosion of the barrier surface are formed; and slow, in which the electrode spots of the microdischarges move deeper into the electrode from the edge. Accordingly, during the first stage, the dynamics of discharge power is observed, and during the second one it is stable.

Microdischarges on the copper electrode are initially practically stationary. In the places of attachment, the formation of craters is observed, with their inner surface intensively cleaned by discharge from the oxide layer, which leads to an increase in power. The formation of such a structure leads to the establishment of an even more stable quasi-filamentary mode of discharge operation.

4. Conclusions

An unexpected conclusion of this work is that due to the peculiarities of the processes of oxidation, noble metals are not definitely the best choice of material for the corona electrode.

The durability of the aluminum and platinum electrodes is limited by their oxidizing by the discharge products. In addition, it is necessary to pay attention to the reduction in the capacity of the discharge cell. When using resonant power sources, reducing the load on several pF may be sufficient to significantly change the power supply mode of the cell. Copper appears to be the most resistant of the deposited electrodes. However, it is worth noting that the features of the discharge mode on such an electrode (quasi-filamentation) may strongly affect the applications where the discharge homogeneity is required.

Acknowledgments

The reported study was funded by RFBR according to the research project № 18-38-00668.

References

[1] Šimor M, Ráhel' J, Vojtek P, Černák M and Brablec A 2002 *Appl. Phys. Lett.* **81** 2716–8
[2] Borcia G, Anderson C A and Brown N M D 2003 *Plasma Sources Sci. Technol.* **12** 335–44
[3] Mikeš J, Pekárek S and Soukup I 2016 *J. Appl. Phys.* **120**
[4] Bednar N, Matović J and Stojanović G 2013 *J. Electrostat.* **71** 1068–75
[5] Radacsi N, Van Der Heijden A E D M, Stankiewicz A I and Ter Horst J H 2013 *J. Nanoparticle Res.* **15** 1445
[6] Park Y, Oh K S, Oh J, Seok D C, Kim S B, Yoo S J and Lee M-J 2016 *Plasma Process. Polym.* **15** 1600056
[7] Ambrico P, Šimek M, Morano M, De Miccolis Angelini R, Minafra A, Trotti P, Ambrico M, Prukner V and Faretra F 2017 *J. Phys. D. Appl. Phys.* **50** 305401
[8] Moreau E 2007 *J. Phys. D. Appl. Phys.* **40** 605–36
[9] Wang J-J, Choi K-S, Feng L-H, Jukes T N and Whalley R D 2013 *Prog. Aerosp. Sci.* **62** 52–78
[10] Leonov S B, Adamovich I V and Soloviev V R 2016 *Plasma Sources Sci. Technol.* **25** 063001
[11] Kriegseis J, Simon B and Grundmann S 2016 *Appl. Mech. Rev.* **68** 020802
[12] Brandenburg R 2017 *Plasma Sources Sci. Technol.* **26** 053001
[13] Pons J, Oukacine L, Moreau E and Tatibouët J M 2008 *IEEE Trans. Plasma Sci.* **36**
[14] Hanson R E, Houser N M and Lavoie P 2014 *J. Appl. Phys.* **115**
[15] Pescini E, De Giorgi M G, Francioso L, Taurino A and Lavoie P 2016 *54th AIAA Aerospace Sciences Meeting AIAA* pp 1–14

[16] Selivonin I, Lazukin A, Moralev I and Krivov S 2018 *Plasma Sources Sci. Technol.* **27** 085003
[17] Lazukin A V, Selivonin I V, Moralev I A and Krivov S A 2017 *J. Phys. Conf. Ser. 927* **927** 012028
[18] Kriegseis J, Möller B, Grundmann S and Tropea C 2011 *J. Electrostat.* **69** 302–12

Journal of Physics: Conference Series **1394** (2019) 012028 doi:10.1088/1742-6596/1394/1/012028

Classification of vector lines and fields by geometric and numerical methods

Burakhanov B.M.

Joint Institute of High Temperature RAS Izhorskaya 13 bld.2, Moscow, 125412, Russia

burbm@rambler.ru

Abstract. Geometric and numerical methods of classification of vector lines and vector fields are proposed. The division of vector lines of arbitrary vector fields into potential and spiral lines is carried out. The division of vortex vector fields into layered and spiral vector fields is carried out.

It is shown that any vector field can be represented as a superposition of layered vector fields.

Conditions of existence of so-called forceless spiral vector fields are established.

The mathematical conditions to be satisfied by the magnetic fields of magnetic clouds of the solar wind are formulated.

Introduction

A significant part of all modern courses of continuum mechanics is devoted to a detailed description of the mathematical apparatus used in these courses. This shows the deep relationship of mathematics with applied disciplines. The reason for the existence of this relationship is that the general properties of mathematical objects used in the description of natural phenomena, as a rule, have their physical counterparts, and vice versa. In this paper, we discuss the classification of vector lines and vector fields, which, among other things, allows us to establish the «mathematical» conditions for the existence forceless magnetic fields of magnetic clouds of the solar wind.

The current classification of vector fields is based on the analysis of the consequences of the use of rotation and divergence operators on these fields.

This analysis allows us to divide arbitrary vector fields into potential and vortex fields, as well as to distinguish among them solenoid vector fields [1-2].

Considering that the basis of this classification is the use of differential operators of rotation and divergence, this method of classification can be characterized as an analytical method of classification.

However, in this analytical method of classification remain unclaimed some important geometric characteristics of vector lines and vector fields. The proposed additional classification of vector lines and vector fields is based on the possibility of effective use of these geometric characteristics.

This additional geometric classification allows to divide vector lines of arbitrary vector fields into potential and spiral ones, and also to divide vortex vector fields into layered and spiral ones.

If the geometric classification of vector lines is based directly on their qualitative topological difference from each other, the geometric classification of layered and spiral vector fields is based on the qualitative topological difference of their vector lines.

It is very likely that the main advantage of layered vector fields is the ability to represent any vector field in the form of their superposition. While, most likely, the main advantage of spiral vector fields is the possibility of selecting among them a set of vector fields, vector lines of which are parallel to the vector lines of their vortex fields.

Vector fields of this type are used in various branches of physics and in space magnetohydrodynamics. Because of their physical counterparts are the so-called a forceless magnetic field [3].

The impetus for the use of these fields in space magnetohydrodynamics is the practical importance of studying the magnetic clouds of the solar wind. Since the forceless magnetic fields of the magnetic clouds of the solar wind significantly affect the processes of the earth's high-latitude geomagnetic activity [4].

1 Potential vector fields

Potential or gradient fields $\mathbf{a}^{pot}$ are vector fields that meet the condition of the form

$$\mathbf{a}^{pot} = \nabla\Phi, \tag{1.1}$$

where Φ is a differentiable scalar function.

The expression (1.1) can be rewritten as

$$\mathbf{a}^{pot} = |\nabla\Phi|\mathbf{e}_{pot} = A^{pot}\mathbf{e}_{pot} \tag{1.2}$$

here $A^{pot} = |\nabla\Phi|$ - module of the vector $\mathbf{a}^{pot}$, and $\mathbf{e}_{pot}$ - directing unit vector or ort of the vector $\mathbf{a}^{pot}$.

Potential of a vector field $\mathbf{a}^{pot}$ can be determined as well as the vortex-free vector field that satisfies the condition

$$rot\,\mathbf{a}^{pot} \equiv rot(A^{pot}\mathbf{e}_{pot}) = 0. \tag{1.3}$$

This paper proposes an alternative definition of potential vector fields. This alternative definition is based on the number of scalar functions that must be specified to define a potential vector field $\mathbf{a}^{pot}$.

According to this alternative definition, potential vector fields $\mathbf{a}^{pot}$ are arbitrary vector fields **a**, to set which it is enough to set one scalar function.

2 Potential vector lines

Let us recall that vector lines are lines, tangents to which are parallel to vectors of the considered vector field [1].

It follows from this definition, that to set a family of vector lines L^{pot} of potential field $\mathbf{a}^{pot}$ enough to set the field of its unit vectors

$$\mathbf{e}_{pot} = \nabla\Phi/|\nabla\Phi|. \tag{2.1}$$

It is essential that the vector fields $\nabla\Phi$ and $\mathbf{e}_{pot}$ are two autonomous vector fields that have an identical set of vector lines L^{pot}. It is also significant that the unit field $\mathbf{e}_{pot}$, associated with the potential vector field $\nabla\Phi$ does not have to be a potential vector field $\mathbf{a}^{pot}$. Because the rotor of this field may not be zero

$$rot\,\mathbf{e}_{pot} \neq 0. \tag{2.2}$$

However, despite this, below for the name of vector lines L^{pot} describing vector fields $\mathbf{a}^{pot}$ and vector fields $\mathbf{e}_{pot}$ will be used the general term – potential vector lines L^{pot}.

In other words, in this case, a potential vector lines L^{pot} are vector lines of potential vector fields $\mathbf{a}^{pot}$.

Potential vector lines L^{pot} is a vector line to set which it is enough to set one scalar function Φ.

3 Layered vector fields

By definition, layered vector field $\mathbf{a}^{lay}$ is the vortex vector field $\mathbf{a}^{rot}$ with potential vector lines L^{pot}.

It follows from this definition that this field must meet the conditions of the form

$$\mathbf{a}^{lay} = A^{lay}\,\mathbf{e}^{lay}\,, \qquad (\mathbf{e}_{lay} = \mathbf{e}_{pot} = \nabla\Phi/|\nabla\Phi|) \tag{3.1}$$

$$rot\,\mathbf{a}^{lay} \neq 0. \tag{3.2}$$

It is possible to give alternative definitions of layered vector fields $\mathbf{a}^{lay}$.

For example, a layered vector field $\mathbf{a}^{lay}$ can be defined as a vortex vector field $\mathbf{a}^{rot}$ whose vector lines are orthogonal to the vector lines of its vortex field $rot\ \mathbf{a}^{lay}$.

This means that [2]

$$\mathbf{a}^{lay} \cdot rot\ \mathbf{a}^{lay} = 0. \quad (3.3)$$

In addition, a layered vector field $\mathbf{a}^{lay}$ can be defined as a vortex vector field $\mathbf{a}^{rot}$, which can be written as [2]

$$\mathbf{a}^{lay} = \Psi\nabla\Phi. \quad (3.4)$$

Finally, the layered vector field $\mathbf{a}^{lay}$ can be defined as a vortex vector field $\mathbf{a}^{rot}$, for which it is necessary and sufficient to set two scalar functions Φ and Ψ.

4 Spiral vector fields

There is a theorem [2] proving that any vector field **a** can be represented as

$$\mathbf{a} = \Psi\ \Phi + \nabla F, \quad (4.1)$$

where Ψ, Φ and F are differentiable scalar functions.

Where does it follow that any vector field **a** can be written as a superposition of layered $\mathbf{a}^{lay} = \Psi\ \nabla\Phi$ and potential $\mathbf{a}^{pot} = \nabla F$ fields, that is, in the form

$$\boldsymbol{a} = \mathbf{a}^{lay} + \mathbf{a}^{pot}. \quad (4.2)$$

Below is the name of an arbitrary vector field **a** for which fair conditions are

$$\mathbf{a} \cdot rot\ \mathbf{a} \neq 0, \quad (4.3)$$

$$rot\ \boldsymbol{a} \neq 0, \quad (4.4)$$

it is proposed to use the term – spiral vector fields $\mathbf{a}^{spir}$.

Spiral vector fields $\mathbf{a}^{spir}$ can be defined as vortex vector fields for which it is necessary and sufficient to set three scalar functions Ψ, Φ and F.

We briefly discuss now an important distinguishing feature of the unit fields $\mathbf{e}_{spir}$ of spiral vector fields $\mathbf{a}^{spir}$. To set this attribute, you must first rewrite expression (4.2) as

$$\mathbf{a}^{spir} = \mathbf{a}^{lay} + \mathbf{a}^{pot}\ , \quad (4.5)$$

and then divide both parts of the expression (4.5) by the module $|\ \mathbf{a}^{spir}\ |$ of the vector $\mathbf{a}^{spir}$.

By doing this, we get

$$\mathbf{e}_{spir} = \mathrm{A}^{lay}\ \mathbf{e}_{lay} + A^{pot}\ \mathbf{e}_{pot}, \qquad (\mathbf{e}_{lay} \neq \mathbf{e}_{pot}) \quad (4.6)$$

where $\mathbf{e}_{spir}$ is the unit vector of the spiral vector field $\mathbf{a}^{spir} = A^{spir}\mathbf{e}_{spir}$, and $\mathbf{e}_{lay} \neq \mathbf{e}_{pot}$ are the unit vectors of the vectors $\mathbf{a}^{lay} = (\,|\,\mathbf{a}^{lay}\,|\,/\,|\,\mathbf{a}^{spir}\,|\,)\ \mathbf{e}_{lay}$ and $\mathbf{a}^{pot} = (\,|\,\mathbf{a}^{pot}\,|\,/\,|\,\mathbf{a}^{spir}\,|\,)\ \mathbf{e}_{pot}$.

Thus, it is possible to consider that a unit vector field $\mathbf{e}_{spir}$ of spiral vector fields $\mathbf{a}^{spir}$ always is the superposition of layered vector fields.

5 Spiral vector lines

In the previous paragraph, it was found that in order to set a unit vector field $\mathbf{e}_{spir}$, you must first be set two layered vector fields $\mathrm{A}^{lay}\ \mathbf{e}_{lay}$ and $A^{pot}\ \mathbf{e}_{pot}$.

Where does it follow that to set a unit field $\mathbf{e}_{spir}$ must take into account the specifics of the method of setting unit fields $\mathbf{e}_{lay}$ and $\mathbf{e}_{pot}$.

It was established above, to set each unit field $\mathbf{e}_{lay} = \nabla\Phi/\,|\,\nabla\Phi\,|$ and $\mathbf{e}_{pot} = \nabla F/\,|\,\nabla F\,|$ is necessary and sufficient to set a one scalar function, in this case the functions Φ and F.

Knowing the unit vector field $\mathbf{e}_{spir}$ it is not difficult to construct a family of vector lines $\{L^{spir}$ $(\mathbf{e}_{spir})\}$. It is obvious that to find the family of vector lines $\{L^{spir}\ (\mathbf{e}_{spir})\}$, as well as to set the unit vector field $\mathbf{e}_{spir}$, you must to set two scalar functions Φ and F.

By definition, spiral vector lines L^{spir} are vector lines, for which you need to set two scalar functions.

It is also essential that spiral vortex vector fields $\mathbf{a}^{spir}$ can also be defined as vortex vector fields with spiral vector lines L^{spir}.

6 Conjugate layered vector fields

It is easy to show that any potential vector field ∇F can always be represented as a superposition of two layered vector fields.

To do this, it is enough to use a vector identity having the form

$$\nabla F_{1,2} = \nabla(F_1 \cdot F_2) \equiv F_1 \nabla F_2 + F_2 \nabla F_1, \quad (6.1)$$

where $F_{1,2}, F_1$ and F_2 are differentiable scalar functions.

It is obvious that for each fixed function $F_{1,2}^0$ there are infinitely many functions F_1 and F_2 satisfying the conditions $F_1 \cdot F_2 = F_{1,2}^0$.

It follows that any fixed potential field $\nabla F_{1,2}^0$ can be represented as a superposition of any number of pairs of layered vector fields $F_1 \nabla F_2$ and $F_2 \nabla F_1$.

By definition, the conjugate vector fields are any two layered vector fields $F_1 \nabla F_2$ and $F_2 \nabla F_1$, the superposition of which degenerates into the potential of vector field $\nabla F_{1,2}$.

Given the validity of the expression (6.1), it is not difficult to formally divide all layered vector fields $\mathbf{a}^{lay} = \Psi \nabla \Phi$ into two types of conjugate vector fields. Formally assuming that one of these types are vector field $\Psi \nabla \Phi$ and the other $\Phi \nabla \Psi$.

It is easy to check the validity of expressions of the form

$$rot\ (F_1 \nabla F_2) = \nabla F_1 \times \nabla F_2, \quad (6.2)$$

$$rot\ (F_2 \nabla F_1) = \nabla F_2 \times \nabla F_1. \quad (6.3)$$

Whence it follows that

$$\nabla F_1 \times \nabla F_2 = -\nabla F_2 \times \nabla F_1. \quad (6.4)$$

It is easy to see that the expression (6.4) is no more than a well-known vector identity.

Given the validity of expressions (6.2) – (6.4), it is not difficult to formulate two alternative definitions of conjugate layered vector fields:

- conjugate layered vector fields $F_1 \nabla F_2$ and $F_2 \nabla F_1$ are layered vector fields $\mathbf{a}^{lay}$, vortex fields which $rot\ (F_1 \nabla F_2)$ and $rot\ (F_2 \nabla F_1)$ equal in magnitude and opposite in sign;
- conjugate layered vector fields $F_1 \nabla F_2$ and $F_2 \nabla F_1$ are layered vector fields $\mathbf{a}^{lay}$, vortex fields which rot $(F_1 \nabla F_2)$ and rot $(F_2 \nabla F_1)$ have identical vector lines.

7 Layered vector field as the basis vector subspace of the space of vector fields

In mathematics, the basis of vector space is understood as an ordered set of vectors.

It is assumed that any vector of vector space can be represented in the form of a linear combination of vectors from this set.

In this paper, the set of layered vector fields $\mathbf{a}^{lay}$ is interpreted not as the basis of the space of arbitrary vector fields **a**, but as some basic subspace of this space.

It is assumed that any vector field **a** can be represented as a superposition of layered vector fields $\mathbf{a}^{lay}$.

The easiest way to illustrate the validity of this assumption is as follows.

First we need to use the possibility of representing an arbitrary vector field **a** as $\mathbf{a} = \Psi \nabla \Phi + \nabla F$. Then we need to use the representation of the potential field ∇F as a superposition of conjugate layered vector fields $F_1 \nabla F_2$ and $F_2 \nabla F_1$.

By doing this, we get

$$\mathbf{a} = \Psi \nabla \Phi + F_1 \nabla F_2 + F_2 \nabla F_1. \qquad (F = F_1 \cdot F_2) \quad (7.1)$$

It follows from the expression (7.1) that an arbitrary vector field **a** can always be represented as a superposition of one layered $\Psi \nabla \Phi$ and two conjugate layered $F_1 \nabla F_2$ and $F_2 \nabla F_1$ vector fields.

8 Forceless spiral vector fields

By definition forceless vector fields $\mathbf{a}_{noforce}$ are vector fields **a** parallel to fields $rot\ \mathbf{a}$.

Forceless vector fields $\mathbf{a}_{noforce}$ can also be defined as vector fields that meet the conditions of the form

$$\mathbf{a}_{noforce}^{spir} \times rot\, \boldsymbol{a}_{noforce}^{spir} = 0, \quad (8.1)$$

$$rot\, \boldsymbol{a}_{noforce}^{spir} \neq 0. \quad (8.2)$$

Significantly that forceless vector fields $\mathbf{a}_{noforce}$ can only be a spiral vector field $\mathbf{a}^{spir}$. Since neither potential $\mathbf{a}^{pot} = \nabla F$ nor layered $\mathbf{a}^{lay} = \Psi\nabla\Phi$ vector fields can not simultaneously satisfy the conditions (8.1) and (8.2).

The fact is that the potential vector fields $\mathbf{a}^{pot}$ can not satisfy the condition (8.2) because their vortex fields $rot\, \mathbf{a}^{pot}$ are zero ($rot\, \mathbf{a}^{pot} = 0$) .

While layered vector fields $\mathbf{a}^{lay}$ cannot satisfy the conditions (8.1) because $\mathbf{a}^{lay} \times rot\, \mathbf{a}^{lay} \neq 0$ (due to the orthogonality of vectors $\Psi\nabla\Phi$ and $\nabla\Psi \times \nabla\Phi$).

9. The conditions for the existence forceless spiral vector fields

It can be shown that the conditions of existence of forceless spiral vector fields $\mathbf{a}_{noforce}^{spir}$ have the form

$$|\mathbf{a}_{noforce}^{spir}| = |\nabla F_{noforce} \cdot \mathbf{i}_{noforce}^{spir}|, \quad (9.1)$$

$$|\Psi_{noforce} \cdot \nabla\Phi_{noforce}| = |\nabla F_{noforce} \cdot \mathbf{i}_{noforce}^{lay}|, \quad (9.2)$$

$$\mathbf{i}_{noforce}^{spir} = \nabla\Psi_{noforce} \times \nabla\Phi_{noforce} / |\nabla\Psi_{noforce} \times \nabla\Phi_{noforce}|, \quad (9.3)$$

where $\mathbf{i}_{noforce}^{spir}$ - ort of the vector rot $\mathbf{a}_{noforce}^{spir}$, $\mathbf{i}_{noforce}^{lay}$ - ort of the vector $\nabla\Phi_{noforce}$.

This means that the module $|\mathbf{a}_{noforce}^{spir}|$ of the forceless spiral vector field $\mathbf{a}_{noforce}^{spir}$ must be equal to module of the projection $|\nabla F_{noforce} \cdot \mathbf{i}_{noforce}^{spir}|$ of the potential field $\nabla F_{noforce}$ on the direction of the vector $\mathbf{a}_{noforce}^{spir}$.

While the module $|\Psi_{noforce}\nabla\Phi_{noforce}|$ vortex component $\Psi_{noforce}\nabla\Phi_{noforce}$ of a forceless spiral vector field $\mathbf{a}_{noforce}^{spir}$ must be equal to module the projection $|\nabla F_{noforce} \cdot \mathbf{i}_{noforce}^{lay}|$ of the potential field $\nabla F_{noforce}$ on the direction of the vector $\nabla\Phi_{noforce}$.

The result is that the existence of a forceless spiral vector field $\mathbf{a}_{noforce}^{spir}$ entirely depends on will or will not be able the corresponding projection of the potential field $\nabla F_{noforce}$ to delete the vortex component $\Psi_{noforce}\nabla\Phi_{noforce}$ of vector field $\mathbf{a}_{noforce}^{spir}$.

It is also essential that the vectors $\mathbf{a}_{noforce}^{spir}$, $rot\, \mathbf{a}_{noforce}^{spir}$, $\Psi_{noforce}\nabla\Phi_{noforce}$ and $\nabla F_{noforce}$ should be coplanar vectors.

10 The main geometric difference between divergent and divergent-free vector fields

The concept of divergence-free field, in essence, is a refinement of the concept of a solenoid vector field. Because this concept clarifies that a solenoid vector field is a vector field whose divergence is zero.

By definition, divergent-free vector fields $\mathbf{a}_{nodiv}$ are arbitrary vector fields $\mathbf{a}$, satisfying the condition of the form

$$div\, \mathbf{a} = 0. \quad (10.1)$$

Accordingly, divergent vector fields $\mathbf{a}_{div}$ are arbitrary vector fields $\mathbf{a}$, satisfying the condition of the form

$$div\, \mathbf{a} \neq 0. \quad (10.2)$$

Note that we can show [2] that the vector field $\mathbf{a}$ is a divergence-free field $\mathbf{a}_{nodiv}$ if and only if the flux of the vector through an arbitrary inner surface is zero.

The main geometric difference between divergent $\mathbf{a}_{div}$ and divergence-free $\mathbf{a}_{nodiv}$ vector fields is that the families of vector lines of divergence-free vector fields have a tubular structure.

Journal of Physics: Conference Series **1394** (2019) 012028 doi:10.1088/1742-6596/1394/1/012028

Recall that vector tubes are called closed vector surfaces formed by vector lines passing through the points of a closed contour L.

It is essential that the value of the vector flux through the cross section of vector tubes of divergence-free vector fields $\mathbf{a}_{nodiv}$ remains constant along these tubes.

Therefore, vector tubes of divergence-free vector fields cannot end inside the field [1,2].

11 The relationship of a divergence-free potential and layered vector fields and harmonic functions

In mathematics, the harmonic functions Φ^{Δ} are called scalar functions Φ, satisfying the condition of the form [1]

$$Div(\nabla\Phi) = 0 \tag{11.1}$$

If we consider the expression (11.1) as a definition of harmonic functions, we can talk about the existence of a relationship between harmonic functions and divergent-free potential vector fields $\mathbf{a}_{nodiv}^{pot} = \nabla\Phi$.

It was found above that to set an arbitrary potential field $\mathbf{a}^{pot}$ it is necessary and sufficient to set one scalar function Φ. In this paragraph it is shown that to set a divergence-free vector field $\mathbf{a}_{nodiv}^{pot}$ is necessary and sufficient to set a one scalar harmonic function Φ^{Δ}.

In the applied aspect, you can say that each divergence-free $div\,\mathbf{E} = 0$ electric field $\mathbf{E} = -\nabla\varphi$, where φ is the electric potential, can be put into correspondence with garmonic electric potential φ^{Δ}. While each vortex-free rot $\mathbf{H} = 0$ magnetic field $= rot\,\mathbf{A}$, where $\mathbf{A}$ is the vector magnetic potential, can be supplied in conformance scalar garmonic magnetic potential A^{Δ}.

Let us now discuss the geometrical method for allocating a among the multitude of harmonic functions Φ^{Δ} of a subset of the so-called ortbigarmonic functions $\Phi^{\pm\Delta}$.

By definition, the ortbigarmonic function $\Phi^{\pm\Delta}$ is a scalar function whose values are equal to the product of two harmonic functions $\Phi^{+\Delta}$ and $\Phi^{-\Delta}$, the vector lines of the gradient fields of which $\nabla\Phi^{+\Delta}$ and $\nabla\Phi^{-\Delta}$ are orthogonal to each other.

This means that in this case the expressions of the form

$$\Phi^{\pm\Delta} = \Phi^{+\Delta} \cdot \Phi^{-\Delta}, \tag{11.2}$$

$$\nabla\Phi^{+\Delta} \cdot \nabla\Phi^{-\Delta} = 0 \tag{11.3}$$

are fair.

It is easy to verify that the divergence-free gradient field $\mathbf{a}_{nodiv}^{pot} = \nabla\Phi^{\pm\Delta}$ of the orthbigarmonic function $\Phi^{\pm\Delta}$ is a superposition of the divergence-free conjugate layered vector fields $\mathbf{a}_{nodiv}^{+lay} = \Phi^{+\Delta}\nabla\Phi^{-\Delta}$ and $\mathbf{a}_{nodiv}^{-lay} = \Phi^{-\Delta}\nabla\Phi^{+\Delta}$.

So orthbigarmonic function $\Phi^{\pm\Delta}$ formally valid to consider as ort biharmonic potential conjugate layered vector fields $\mathbf{a}_{nodiv}^{+lay}$ and $\mathbf{a}_{nodiv}^{-lay}$.

12 Mathematical conditions to be satisfied by magnetic fields of magnetic clouds of the solar wind

It is well known that the "mathematical" vector field **a**, considered as a physical object, can have a different physical meaning. For example, this field can be a velocity field **v**, vortex field **ω**, electric field **E** and magnetic field **H**.

In addition, the "mathematical" vector field **a** can be formally assigned physical properties.

For example, in the case where the "mathematical" vector field **a** is associated with the "physical" magnetic field **H**, it makes sense to say that this "mathematical" vector field, considered as a physical object, may has the property of freezing.

This suggests that the "mathematical" vector fields a may have a hypothetical dualistic mathematical – physical nature.

However, it is obvious that any "physical" vector field must satisfy the restrictions imposed by mathematics on this field, considered as a mathematical object. Therefore, it makes sense to say that any physical vector field has a hypothetical dualistic physical - mathematical nature.

This dualistic physical - mathematical nature of vector fields should be taken into account in the study of any physical phenomena, in the mathematical description of which vector-valued functions are used.

For example, this dualistic nature should be taken into account when studying the magnetic fields of magnetic clouds of the solar wind. More specifically, in the study of these magnetic fields is necessary to consider that from the point of view of mathematics they are divergent-free forceless spiral vector fields.

It follows that the vector lines of forceless magnetic fields, considered as forceless spiral mathematical fields $\mathbf{a}_{noforce}^{spir}$, must coincide with the lines of intersection of the surfaces of the level $\Psi = const$ and $\Phi = const$ scalar functions Ψ and Φ, which determine the vortex component $\Psi\nabla\Phi$ field $\mathbf{a}_{noforce}^{spir}$.

In addition, note that the size of the module $|\mathbf{a}_{noforce}^{spir}|$ of a spiral field must be equal to the size of the module $|\nabla F_{noforce} \cdot \mathbf{i}_{noforce}^{spir}|$ of projection of the harmonic potential $\nabla F_{noforce}$ on the direction $\mathbf{i}_{noforce}^{spir}$ of vector field $\mathbf{a}_{noforce}^{spir}$.

Note that you must also take into account the limitation imposed on a spiral vector field $\mathbf{a}_{noforce}^{spir}$ by the divergence-free property.

This restriction has the form

$$div\,(|\nabla F_{noforce} \cdot \mathbf{i}_{noforce}^{spir}| \cdot \mathbf{i}_{noforce}^{spir}) = 0. \quad (12.1)$$

where $F_{noforce} \equiv A^{\Delta}$, here A^{Δ} is the scalar harmonic magnetic potential.

Summary

Geometric and numerical methods for classifying vector lines and vortex vector fields are discussed.

The division of vector lines into potential and spiral, and vortex vector fields into layered and spiral is carried out.

It is shown that to each layered field can be assigned an associated the conjugate layered vector field.

It is shown that any vector field can be represented as a superposition of one layered field and two conjugate layered vector fields.

It is shown that the set of layered vector fields can be considered as a basic subspace of the space of vector fields.

The conditions for the existence forceless spiral vector fields are established.

The mathematical conditions to be satisfied by the magnetic fields of magnetic clouds of the solar wind are established.

References

[1] L. I. Sedov. Mechanics of continua. vol. 1, Nauka, Moscow, 1973,pp.535.
[2] N. E. Cochin. Vector calculus and the beginning of tensor calculus. Publishing USSR Academy of Sciences, Moscow,1961,pp.426.
[3] J. Shercliff. Course of magnetic hydrodynamics. Moscow, Mir, 1967, 320 pp.
[4] N. A. Barkhatov, A. E. Levitin, E. A. Revunov , A. B. Vinogradov. Magnetic clouds solar wind as a cause of high-latitude geomagnetic activity. "Physics of Auroral Phenomena", Proc. XXXVIII Annual Seminar, Apatity, pp. 83-86.

Journal of Physics: Conference Series **1394** (2019) 012029 doi:10.1088/1742-6596/1394/1/012029

Study of the microwave interaction with supersonic plasma sheets and jet generated by Magneto-Plasma Compressor

V G Brovkin[1], P V Vedenin[1], I Ch Mashek*2*, N M Ryazanskiy[1]

1 Joint Institute for High Temperatures of the Russian Academy of Sciences, Izhorskaya 13 Bldg 2, Moscow 125412, Russia
2 Saint-Petersburg State University, Universitetskaya Naberezhnaya 7/9, Saint-Petersburg, 199034, Russia

E-mail: brovkin47@mail.ru

Abstract. In this paper, we present investigation of a method for converting a supersonic plasma axisymmetric jet, created by a Magneto-Plasma Compressor, into flat plasma layers with various thicknesses and electron concentration. This kind of plasma layer has comparatively homogeneous structure and may use for experimental modelling of interaction of X-band radiation with high-speed plasma flow. The results of experimental studies of the microwave radiation transition through these layers are presented. The analysis of the received results within the one-dimensional (1D) approximation is carried out and the plasma concentrations are calculated.

1. 1.Introduction

When a hypersonic vehicle moves through the atmosphere, a high-density plasma layer forms around it. The density of this plasma layer reaches values sufficient to lock the radio signal [1]. For experimental investigation of such effects in supersonic flows the using of plasma jets, generated by Magneto-Plasma Compressors (MPC) was proposed [2]. The modelling of interaction of microwave (MW) radiation with plasma layers usually bases on approximation of thin homogeneous plasma sheet with transvers dimensions essentially more than using wavelength and area of plasma-MW interactions [3,7].The verification of these models needs in experimental version of thin plasma sheets with comparatively high temperature and electron density. Usually, of a large-volume plasma (or a flat plasma layer) source bases on discharges with gridded anode or multianode systems [4, 5]. Another variant for such plate plasma creation – space distributed dielectrically barrier discharge (DBD) [6]. All these ways leads to arising regular and periodic nonuniformities in plasma density distribution and cannot be used with MW radiation with wavelength less, ore compare with it characteristics dimensions. Experiments presented in this paper, have shown that the magnetic plasma compressor with addition plate outlet nozzle can be used for creation of the flat plasma layers with comparatively high transverse dimensions, under the high pressure and space irregularities essentially less than X-band wavelength. The applications of created plasma sheets also provides an opportunity to study the interaction of radiation with layers of different configurations, inclined layers and assess the effect of the magnetic field.

2. The object and the methods of research

The magneto-plasma compressor of compact geometry is the result of further development of large-scale MPC [8, 9]. The MPC jet formed near the focal region of the converging microwave beam propagated along the electric field of the beam (Fig.1). A pulsed magnetron used to generate MW radiation with the following parameters: pulse power does not exceed the value of 700 kW; $\lambda \approx 2.3$ cm; $\tau \sim 4$ μs. Microwave radiation through the horn and lens was introduced into the vacuum chamber with a size 1m in length and 0.7 m in diameter. Using a dielectric lens, a slightly diverging linearly polarized electromagnetic beam was formed in the chamber and falls on a focusing mirror with a radius of about 0.3 m. Outside the zone of interaction was installed the microwave probe.

The PSO sensicam camera recorded through the mirror (Fig.2) the plasma jet of the MPC with a minimum exposure time of 1 μs and the ability to shift the registration time relative to the start of the process. The range of speeds of the plasma streams generated by such MPC shall be 2-10 Km/s, the jet lifetime should not be less than 50 μs, in the range of the surrounding gas (air) static pressures 10-500 Torr. The electron density reaches values of $(10^{16} - 10^{17})$ cm^{-3}, at a temperature of about 3 eV [2].

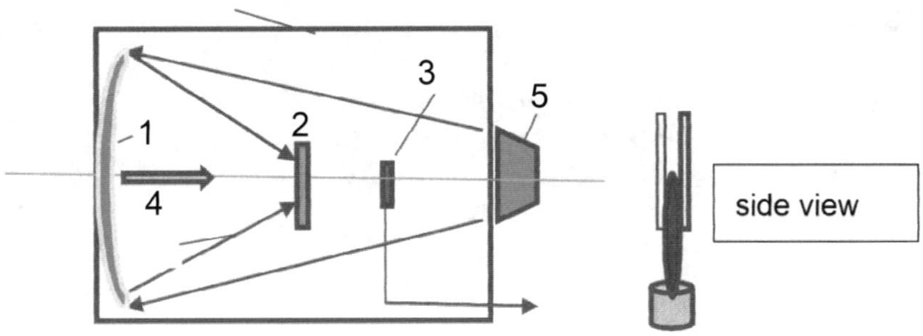

Figure 1 Scheme of the experiment (the top view). 1 - Focusing mirror; 2 – plates with a gap where the plasma jet moves; 3 – microwave probe; 4 – MW beam; 5 – horn with lens for entering microwave radiation into the chamber. The plasma flat layer was formed along the electric field vector of the wave and perpendicular to the direction of the MW beam propagation. This layer completely overlapped the focused beam.

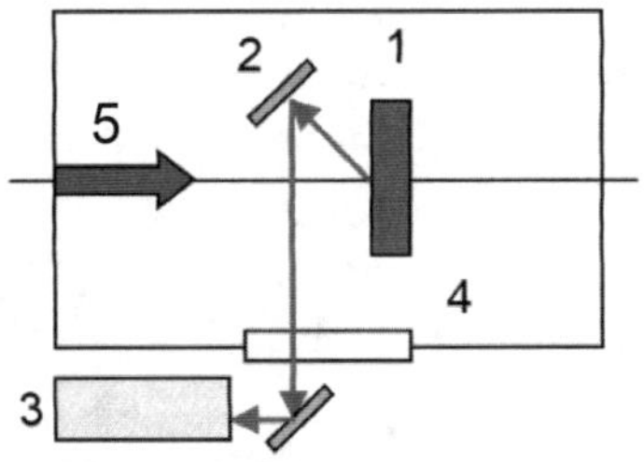

Figure 2 Scheme of the jet registration (the top view). 1 - plates with a gap where the plasma jet moves; 2 – mirrors; 3 - PSO sensicam camera; 4-quartz window; 5 - MW beam.

In this series of experiments, the dynamics of the jet was recorded at different times in order to determine its optimal size to overlap the microwave beam. A similar procedure was performed when the jet expanded between the plates.

3. The research results

The results of the experiments and corresponding calculations are presented below. On the Fig.3a shows the vertical plasma jet on 40мкс of its development. Its diameter size is close to the microwave length and reached ~ 25mm, at a height of ~ 70mm.The front of the plasma flow moves up at a speed of ~ 2 km/s to the radiation focus area, gradually blocking the microwave beam. The amplitude of the signal characterizing the transmitted radiation decreases. Later an open slot device was created in the form of two parallel radio-transparent plates (65 ×75 mm), the size of the gap has changed from 3 to 10 mm. Plates set near the electrodes of the MPC (Fig.1 - position 2). Such device transformed a cylindrical type jet into a flat layer and this makes it possible to conduct experiments on the interaction of MW radiation with plasma layers of different thicknesses. It should be noted, that there was a good repeatability of the processes. As shown by preliminary experiments, the rate of expansion of the plasma flow in the slit space was about $V\sim$ 1.5 km/s and fairly homogeneous a flat layer up to 75 mm in size was formed (Fig.3b,c).

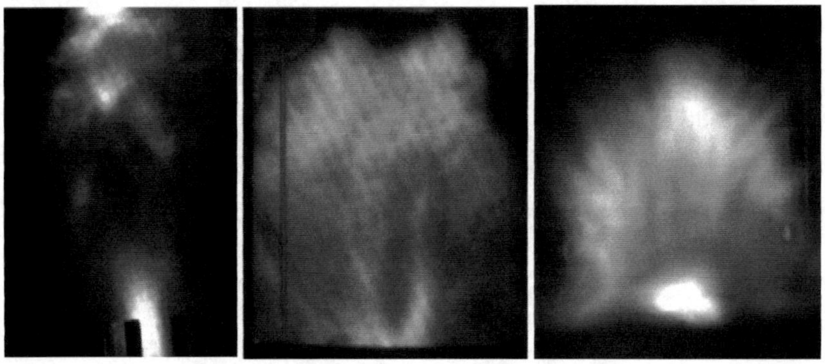

a b c

Figure 3 Plasma jet (a) at 40 µs converted into a flat layers of 10 mm (b) and 5mm (b) thickness , t exp = 3 µs, P = 25 Torr.

The transverse size of the flat plasma flow was three to four times more than the MW length, and it completely covered the incident microwave beam. In experiments plasma layers with a thickness of 3, 5 and 10 mm at air pressure of 25-30 Torr were used. It was started the study of the processes of the microwave passage (in conditions of $E/P < 40$ V/cm•Torr) through the plasma during the formation and decay of the layer. Note that the quasi-stationary phase of the flow as well as in a conventional jet is observed in the period from 30 µs to ~ 80 µs. In this interval, the signals of the microwave radiation transmitted through the layer were recorded.

Based on these data, it have been executed the estimation of plasma concentration in the layers. The analysis of the experiments results is carried out within the one-dimensional (1D) approximation. Monochromatic linearly polarized wave propagating along the z axis

$$\tilde{E}_{Wx}(t,z) = Re\{E_{W\,x}\, exp(i(\omega t - kz))\},. \quad (1)$$

where k is the wave vector, $k = \omega/c$, c is the speed of light, normally falls to infinite in the x, y directions plasma layer. Under the influence of the electric field of the wave, a high-frequency current is excited inside the plasma.

In the absence of information on the plasma distribution, we will use the approximation of the effective plasma layer with an uniform plasma density distribution

$$N_p(z) = \begin{cases} N_p, & 0 \le z \le \Delta, \\ 0, & z < 0,\ z > \Delta, \end{cases}. \quad (2)$$

In the microwave range, it can be assumed that the ions are motionless at the oscillation period. In addition, we neglect the thermal and directional movement of electrons. Under these assumptions, the current excited in the plasma with the density, $\mathbf{j}_e$, is coupled with the electric field by the expression

$$\mathbf{j}_e = \sigma \mathbf{E}. \quad (3)$$

where $\sigma = e^2 N_p / m(\nu - i\omega)$ is the high-frequency complex electrical conductivity, m is the electron mass, -e is the electron charge, ν is the effective collision frequency of electrons with molecules.

The degree of the microwave transition through the layer is characterized by the parameter

$$D = 10\,lg\left|\frac{E_W}{E_T}\right| \tag{4}$$

which is measured in decibels. Here $|E_T|$ is the amplitude of the transmitted wave. The expression for this amplitude is as follows:

$$|E_T| = 4|E_W|\left|\frac{\sqrt{\varepsilon}}{(\sqrt{\varepsilon}+1)^2-(\sqrt{\varepsilon}-1)^2\,exp(-4\pi\sqrt{\varepsilon}\Delta_*)}\right| exp(-2\pi\chi\Delta\Delta_*), \tag{5}$$

where $\chi = \sqrt{0.5\left(\sqrt{(1-N_*)^2+(N_*\nu_*)^2}-1+N_*\right)}$, $\varepsilon = 1 + i\frac{\sigma}{\omega\varepsilon_0}$ is the complex permittivity of the plasma,ε_0 is the permittivity of free space. $N_* = \frac{e^2N_p}{m\varepsilon_0(\omega^2+\nu^2)}$, $\nu_* = \frac{\nu}{\omega}$, $\Delta_* = \frac{\Delta}{\lambda}$, λ is a wavelength. Thus,

$$D = 20\pi\chi\Delta_*\,lg\,e + 10\,lg\frac{\left|(\sqrt{\varepsilon}+1)^2-(\sqrt{\varepsilon}-1)^2\,exp(-4\pi\Delta_*\sqrt{\varepsilon})\right|}{4|\sqrt{\varepsilon}|}. \tag{6}$$

From the equation (6) one can find the value of the plasma density corresponding to a given degree of the microwave transition through the plasma layer. For the layer thickness Δ= 3, 5, 10 mm, the following values of the degree of radio transparency $D \approx$1.5, 2, 2.6 dB were experimentally obtained. These values correspond to the levels of the plasma density $N_p \approx$4.6, 3.6, $2\cdot10^{12}$cm^{-3}.

4. Conclusions

The paper shows the possibilities of using plasma flow. produced by a magneto plasma compressor, in a plasma experiment. Plasma stream of MPC can be successfully applied both for the study of flow around models and for creating plasma layers. In the latter case, it is shown that the use of slit spaces allows the formation of flat plasma layers with specified properties. The first experiments were performed with (3, 5, 10)mm layers through which the microwave propagated. The analysis of the received results within the one-dimensional (1D) approximation is carried out. The corresponding plasma densities are obtained: $N_p \approx$4.6, 3.6, $2\cdot10^{12}$cm^{-3}.

This work was partially supports by grant RFBR 18-08-00707

References

[1] A E Bezmenov, V A Aleksashenko. Radiophysics and gas-dynamic problems passing atmosphere. Moscow, Mechanical engineering, 1982.

[2] B A Balakirev3, V A Bityurin1 at all, Supersonic plasma jets in experiments for radiophysical testing of bodies flow, Journal of Physics: Conf. Series 946 (2018).

[3] B A Balakirev3, V A Bityurin1 at all, Propagation of microwave radiation through an inhomogeneous plasma layer in a magnetic field, Conf. Series: Journal of Physics: Conf. Series 946 (2018)

[4] Yuan C., Tian R., Eliseev S.I., Bekasov V.S., Bogdanov, E.A., Kudryavtsev, A.A., Zhou Z. On self-sustainment of DC discharges with gridded anode, Journal of Applied Physics, Volume 122, Issue 14, October 2017.

[5] Yuan C., Yao J., Eliseev S.I., Bogdanov, E.A., Kudryavtsev, A.A., Zhou Z. Numerical simulation and analysis of electromagnetic-wave absorption of a plasma slab created by a direct-current discharge with gridded anode Journal of Applied Physics, Volume 123, Issue 11, March 2018.

[6] P. B. Repin, A. G. Rep'ev Self-Organization of the Channel Structure of a Nanosecond Diffuse Discharge in a Wire–Plane Electrode System. Technical Physics, Vol. 46, No. 5, 2001, pp. 632–634.

[7] Bityurin V A, Brovkin V G and Vedenin P V 2017 J. Phys. D: Appl. Phys. 50 275201 629–633

[8] Morozov A I 1990 Physics of Plasma 16 131–146

[9] Astashinsky B, Bakanovich G and Minko L 1980 Journal of Applied Spectroscopy 33 629–633

Influence on separation point position by rapid local gas heating with spark discharge on the surface of airfoil streamlined by supersonic airflow

A S Saveliev[1,2]

[1]Joint Institute for High Temperatures of the Russian Academy of Sciences, Izhorskaya 13 Bldg 2, Moscow, 125412 Russia,
[2]Moscow Institute of Physics and Technology, Institutskii lane 9, Dolgoprudny, Moscow Region, 141700 Russia

E-mail: fisteh@mail.ru

Abstract. Experimental investigation of interaction between gas-dynamic perturbations caused by long spark discharge and separation region on the surface of an aerodynamic model, streamlined by supersonic airflow with Mach number M=2, is presented in this paper. A NACA-airfoil with a relative thickness of 12% of the chord length was used as aerodynamic model. With the help of the Particle Image Velocimetry method airflow was visualized at various angles of attack, delays 0–100 μs relative to the moment of discharge initiation and at various values of stored electric energy 0.1–10 J, that was spent for discharge creation. It was found that the flow disturbance caused by the discharge leads to a short-term displacement of separation point downstream.

1. Introduction

The question of improving the aerodynamic characteristics and the pattern of flow around the surface of the vehicles is relevant today. When aircrafts move in the atmosphere, such regimes can be realized in which the possibility of flight or its control is threatened, for example, when an airplane is flying at supercritical angles of attack. One of the negative phenomena in the flow pattern is the emergence of separation region on the surface of the aircraft, which is characterized by nonstationarity and altered parameters, for example, an increased pressure value, temperature inside the flow separation region, as well as an increased drag. Currently there are passive or mechanical methods of controlling separation on the surface of aircraft and they are being used, and methods of active influence on the flow by initiating electrical discharges of various types near the streamlined surface are being investigated and developed. The main advantages of using electrical discharge are low response time, high peak power, the ability to use it only when it is necessary, and not during the entire flight at all altitudes. The rapid development of the high-voltage semiconductor industry past years makes it possible to speak of a lower specific weight of high-voltage equipment (per unit of power) compared with mechanical and pneumatic devices acting on the flow. However, there are also disadvantages of using an electrical discharge, primarily due to the need to use dielectric materials, because a discharge is initiated at a high electric field strength, as well as with the complexity of computer simulation of physical processes in plasma and gasdynamic processes occurring simultaneously near the surface of the aircraft, due to significantly different values of time parameters of these processes. Therefore, discharge plasma occurrence process is often considered simplistic, for example, as heating of the flow at the volume of discharge initiation. In this paper the investigation of initiation of long spark discharge of submicrosecond duration on the surface of a two-dimensional model of a wing across a supersonic airflow around this model at a positive angle of attack is presented.

With such duration of discharge it turns out that the transit time is much longer, and the discharge effect can be simulated as almost instantaneous local heating of the incoming airflow. After the rapid release of energy for local gas heating, a quasi-cylindrical shock wave is formed in the oncoming airflow, which propagates relative to this flow. Behind the shock wave front there is airflow, which has a certain speed relative to the main flow. This speed can be both subsonic at a relatively low intensity of the shock wave, and supersonic at high intensity. In this work using a spark discharge, a shock wave is generated, the recorded propagation velocity of which corresponds to a Mach number less than 2, and the air velocity behind its front has a velocity of the order of several tens of meters per second. The purpose of this study is to visualize with Particle Image Velocimetry method (PIV) the propagation of a gasdynamic perturbation caused by long spark discharge across a supersonic flow, and its interaction with the separation region at various angles of attack of airfoil and discharge energy.

2. Experimental setup

The experiments were carried out on the ST-4 supersonic wind tunnel of atmospheric-vacuum type, a detailed description of which can be found in [1]. At this facility aerodynamic models are tested in supersonic airflow. In this work the free-stream Mach number was M=2, the static pressure of the flow was p_{st}=0.15 bar, the mass flow rate through the nozzle was about 1 kg/s, and the time for maintaining a supersonic flow was one minute. The unit Reynolds number in the experiment is Re_1~10^7 m^{-1}. An aerodynamic model was placed in a supersonic airflow, which was the NACA64a212 profile with a chord length of 98 mm and a span of 150 mm, made of fused silica material and with a discharge device mounted at 38 mm from the trailing edge. A more detailed description of used aerodynamic model can be found in [2]. It is important to note that, due to its design, the discharger made it possible to obtain long spark discharge of 7 cm in length in a supersonic air flow with a Mach number M=2, while the heat release using a discharge occurred across the flow. The design of the working chamber ST-4 provides a mechanism by which it was possible to set the desired angle of attack of the wing model.

During test run the flow around a wing model was diagnosed using a Particle Image Velocimetry laser system. The working chamber of the supersonic wind tunnel is equipped with several transparent windows, which made it possible to introduce a laser sheet for visualizing the flow above the airfoil, as well as to photograph from above long spark discharge in the airflow. For measurements, the LaVision FlowMaster 2D PIV system was used, as well as an oil aerosol generator based on Laskin nozzle [3], which provides a productive airflow seeding by droplets of oil with a diameter close to 1 μm. The area visible by the PIV-videocamera was 48x35 mm^2 and it was located near the trailing edge of the profile, so that particles moving in the flow above the upper surface of the profile and the place of initiation of the discharge (discharger) were in PIV-photos. The spatial resolution of the PIV system was 0.5 mm.

High free-stream velocity, the presence of shock waves and the separation region require the usage of tracer particles with a relatively small diameter and, accordingly, high-energy lasers in a single pulse. Since the PIV system used an Nd:YAG laser pumped by a gas-discharge lamp with a pumping time of about 100 μs, which is close to the time of flight in experiments, and operational frequency less than 14 Hz, a high-voltage pulse generator was used to control the initiation of the spark discharge. A generator is an electrical circuit based on a Marx two-stage circuit. Switching in the generator took place with the help of two air spark dischargers, one of which is controlled. In [2, 4] another discharge initiation scheme was used, where the initiation of a spark discharge occurred in an uncontrolled manner, and the non-stationary flow caused by the discharge was recorded independently of this process.

Simply the high-voltage generator circuit can be characterized by a net capacitance C_0, which at the time of discharge is charged up to doubled value of the charge voltage U_0. The capacitance C_0 was set by the generator circuit (C_0 was varied from 4 nF to 50 nF) with the help of several high-voltage capacitors, and U_0 was varied in accordance with the breakdown voltage for a controlled discharger and was set with the help of a high-voltage power supply. The amplitude of the high-voltage pulse when applied to the discharger mounted in the wing model corresponded to a voltage that exceeds a breakdown value of 1 kV/cm, which was previously determined under the same experimental conditions [2].

In experiments LeCroy WaveRunner 66Zi digital storage oscilloscope was used, the first input channel of which was supplied with voltage from a high-voltage 1:1000 LeCroy PPE20KV probe connected to a high-voltage generator, and to the second input channel a sync pulse from a PIV system was applied, which was a positive TTL-pulse, the rise edge of which corresponds to the moment of PIV-measurement. Oscillograms were recorded during the entire experimental run (10 s) with a discretization time 100 ns. The oscillogram of the voltage on the capacitors of a high-voltage generator can be characterized by a periodic process of slow charging to a voltage U_0 (~10 ms) and fast discharging to a voltage U_1 due to the occurrence of a spark discharge and the release of energy into the airflow (~0.1 μs). Due to this form of oscillograms of the voltage across the capacitor of the generator and the sync pulse of the PIV system, they can be processed with computer algorithm. Thus, by recording the oscillograms and their subsequent processing using a computer program, one can determine the voltage U_0, the residual voltage U_1 after the discharge, the moments of the occurrence of the discharge and the PIV measurements. With simultaneous processing of the oscillograms and PIV photographs of single test run, the average spatial distribution flow velocity vector at the time Δt after the discharge, the average value of the electric energy E that was consumed per discharge, the average duration of the discharge τ were obtained. The amount of electrical energy was determined by the formula:

$$E = 2C_0(U_0^2 - U_1^2) \tag{1}$$

In addition, the static pressure of the flow p_{st} and pressure p_a of air which surrounds the supersonic jet in the working chamber were recorded synchronously with the PIV measurements, so that each frame could be assigned a flow regime beyond the nozzle exit section. As stated above, one experiment lasted 10 seconds, 5 seconds before the onset of the ideally expanded flow regime on the ST-4 facility and 5 seconds after. At the same time, the pressure ratio p_a/p_{st} varied in the range from 0.85 to 1.15. The number of PIV frames registered with a frequency of 14 Hz was 140.

Thus, in each experimental test run the following parameters were set: electrical energy E, which was spent on long spark discharge creation, the delay Δt between the start of the discharge and the time of the PIV measurement, the angle of attack of the airfoil. The place of initiation of the discharge (i.e. the place of energy release) was fixed on the surface of the airfoil and was specified by the position of the discharger. By varying the value of Δt in the range from 0 μs to 100 μs (the characteristic value of the transit time) with a step of 10 μs or 20 μs, it was possible to reconstruct the dynamics of the propagation of gasdynamic perturbations that occur after the initiation of the discharge. To do this, the PIV frames obtained in single test run were tested (by oscillogram processing) for compliance with the delay relative to the set value and for the occurrence of a discharge (by the frames taken from the top view of the airfoil). The frames that passed tests were averaged, and the flow pattern was obtained in the form of fields of velocity vectors for a given delay Δt.

3. Obtained results

In order to average the passed PIV frames, obtained with fixed $E, \Delta t$, it is necessary to check how wide the dispersions of these parameters from discharge to discharge in single test run are. It is also necessary to check that the discharge duration τ has a constant value, and the place of discharge initiation coincides with the location of the discharger on the surface of the wing model. To do this, the oscillograms and recorded images of a spark discharge, obtained in each test run, were processed. After such processing statistics of the energy E, the duration τ, the delay from the PIV-frame moment relative to the moment of initiation of the discharge Δt were obtained of an each individual discharge. The processing was carried out with the help of the developed computer program, which takes on its input the oscillograms of the voltage on the capacitor of the high-voltage generator and the PIV system sync pulse. To reconstruct the nonstationary flow pattern, i.e. to obtain PIV measurements at different Δt values, it is necessary to show that from one test run to another the value of E does not change significantly. Figure 1 shows the typical histograms of the distribution of delay Δt and energy E, obtained after processing the oscillograms of one experimental test run. On the presented histograms one can select the maximum and its width. Figure 2 shows the variation of the parameters $\Delta t, E$ and τ in different experimental series depending on the number of the experimental launch. Table 1 presents the parameters of each run. In table 1 and in figure 2, the values with the 'av' index is the mean values for the corresponding experimental series. The nonmonotonic dependence between discharge time τ and the net capacitance C_0 of the generator is associated with the usage of various ceramic capacitors in various experimental series. These sets of capacitors differ in their nominal capacitance (1 nF, 4.7 nF or 10 nF), in size, manufacturers. Thus, in the experimental series, not only the magnitude of the net capacitance C_0 changed, but also the inductance of the discharge circuit.

The following conclusions can be drawn from the presented statistics of the properties of discharges. The duration of any discharge τ did not exceed 1 μs in any experiment, while the average discharge duration τ_{av} did not exceed 0.5 μs. During this period, the main flow over the airfoil travels no more than 0.6 mm, which is two orders of magnitude less than the length of the wing profile chord. At the same time, the obtained photographs of the discharge from top view with an exposure of about 10 μs do not show a 'blurred' image of the spark channel. These two circumstances make it possible to consider the process of energy release into the oncoming air stream almost instantaneously.

The scatter of the magnitude of the energy E of all discharges mainly (except series 4) did not exceed ±5% of the average value of the electric energy E_{av}. In the theory of a very strong explosion [5] the velocity of a generated shock wave with a two-dimensional fast release of energy has a power dependence on E with an exponent of 0.25. The length of the energy release is fixed and coincides with the distance between the outer segments of the spark gap (this distance is 7 cm). These circumstances make it possible to assume that the intensity of the shock waves generated by the discharge in each experimental test run has a relatively stable value. At the same time, this intensity is relatively stable throughout the entire experimental series.

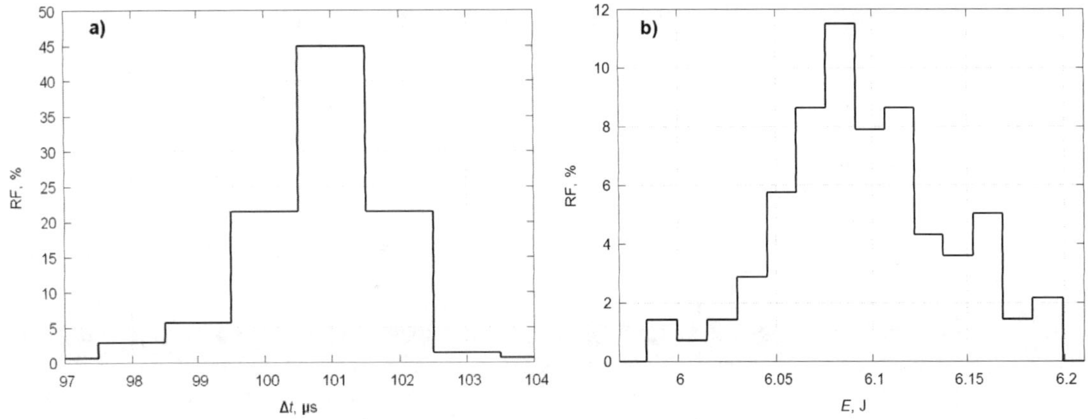

Figure 1. Typical histograms of the distribution of the delay Δt (a) and energy E (b). 'RF' is the relative frequency of the discharge occurrence with a given parameter value.

The change of the Δt value is from –10 μs to +20 μs relative to that set in the experiment test run, and this dispersion increases with increasing energy of the discharge. But it should be noted that the maxima in the Δt histograms lie in the range of ±5 μs relative to the average value, i.e. most discharges occur with 10 μs jitter.

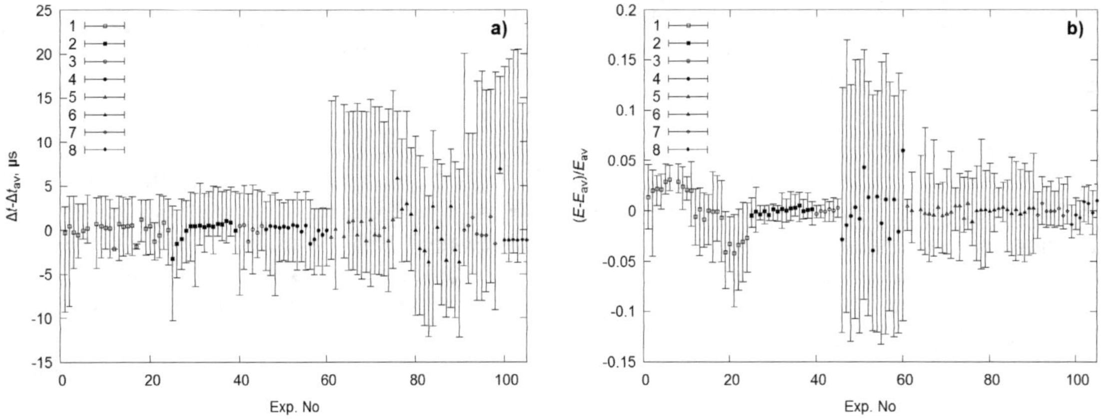

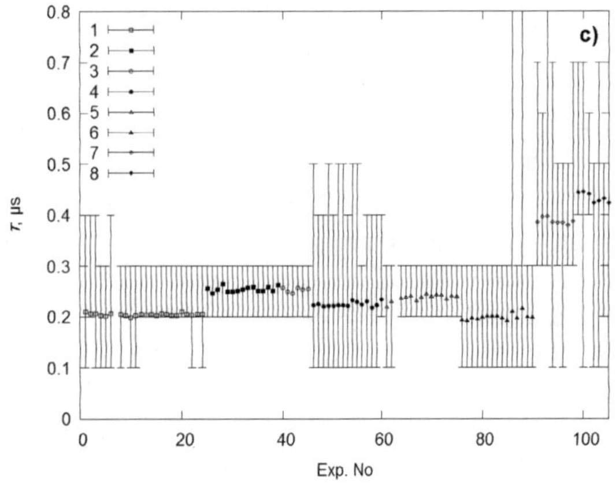

Figure 2. The dispersions of delay Δt (a), energy E (b), and discharge duration τ (c) in various experiments with different parameters. The number denotes the experimental series, the parameters of which are presented in the table 1.

Table 1. The parameters of experimental series. 'AoA'—angle of attack.

Ser. No	Exp. No	AoA, °	U_0, kV	C_0, nF	E_{av}, J	τ_{av}, μs
1	1–24	12	7.32	1.8	0.17	0.20
2	25–39	12	8.80	3.9	0.56	0.25
3	40–45	14	8.80	3.9	0.55	0.25
4	46–60	12	9.50	8.4	1.26	0.22
5	61–75	14	9.00	19.5	2.95	0.24
6	76–90	14	9.00	8.4	1.23	0.20
7	91–98	14	9.50	38.9	6.08	0.39
8	99–105	16	9.50	44.0	6.86	0.43

Figure shows the typical averaged fields of the flow velocity vector (the V_y component of the velocity vector is shown in color map) when initiating long spark discharge on the surface of the wing profile model for two experimental series (see experimental series 4 and 7 in table 1). In this case, averaging is made over approximately 100 PIV measurements. The red line shows the place of initiation of the spark discharge, the gray area is the schematically indicated airfoil, and the black area below is the shadow from this airfoil. The white area in the flow patterns actually corresponds to the separated airflow. Oncoming supersonic airflow is from left to right. The x axis is directed horizontally from left to right, and the y axis is directed vertically from bottom to top.

On the presented unsteady flow patterns for the fourth series at the time Δt=12 μs, it can be seen that a shock wave is formed after the discharge, which adds velocity to the incident flow in the direction of the y axis. At the time Δt =91 μs the separation area has the smallest size, and the distance from the separation point to the trailing edge (coincides with the right edge of the shadow from the airfoil) has the smallest value at this time point. At the time Δt =131 μs, the front of the shock wave is not registered in the flow pattern, because on the one hand the intensity of the shock wave has significantly decreased by this time, and on the other hand, the shock wave partially left the area visible by the video camera due to blow-off by the incident airflow. At this time moment the position of the separation point has almost returned to the initial one.

The flow pattern for the seventh series (figure b) shows that at the initial moment of time after discharge (Δt =2 μs), the separation region is closer to the place of initiation of the discharge (because angle of attack is greater than in fourth series of experiments), but still energy release occurs upstream relative to the separation point. After initiating discharge, a quasi-cylindrical shock wave briefly arises in the flow, which propagates downstream and interacts with the separation region. It is seen that at a certain point in time (Δt =83 μs) this interaction leads to a shift of the separation point downstream. In the middle frame (Δt =83 μs), the shock wave is still visible in the form of a relatively high value of V_y in the upper part of the flow pattern. The position of the separation point returns to its original value as the shock wave is blown off by an external supersonic and flow and its intensity decays (Δt =122 μs).

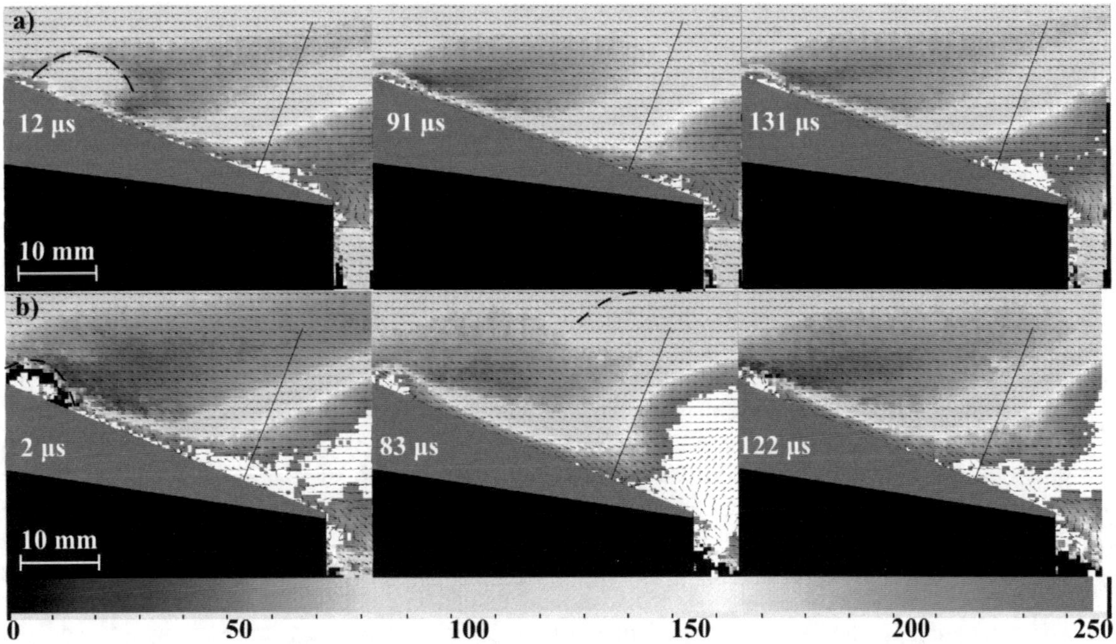

Figure 3. The averaged fields of the V_y component (in m/s) of the velocity vector at different times (indicated in the frames) after the start of the discharge for the fourth (a) and seventh (b) experimental series. Arrows indicate airflow direction. Dashed curves indicates schematically shock waves front detected by increased value of V_y.

Interaction of flow separation region and gasdynamic perturbation caused by discharge can also be shown by a set of velocity profiles along the interval perpendicular to the airfoil surface. On figure **4**a and **4**b V_x component (normalized on oncoming airflow speed V_∞ =500 m/s) profiles along the corresponding intervals are presented. For experimental series 4 (see table 1) this interval starts on the distance 10 mm from trailing edge, for experimental series 7 this distance is 11 mm. These values were chosen to demonstrate clearly the effect of interaction between gasdynamic perturbation and flow separation region. If interval is not in flow separation region then V_x is always greater than 0 m/s (see red curves), but if V_x<0 m/s then there are separation region (see black and blue curves). Velocity vector is directed almost parallel to the airfoil surface (in vicinity to it) and V_y is also can indicate separation region and separation point position. The practice shows that spatial distribution of V_y component (that is presented at figure) more clear shows separation point position at the surface of airfoil. It should be noted that on presented profiles V_x is not equal 0 m/s at l =0 mm, because start points of intervals lay slightly higher than airfoil upper surface. In experiment PIV-measurements cannot be performed in closer vicinity of this surface due to light scattering on it and spatial resolution of optical system.

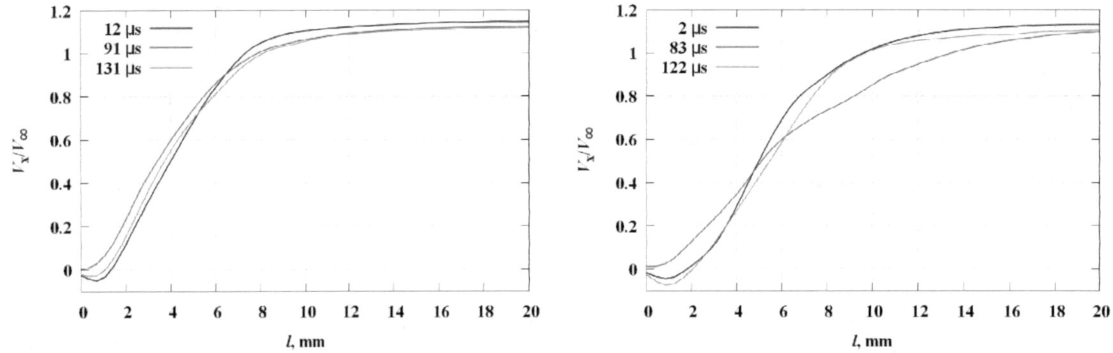

Figure 4. V_x component (normalized on incoming airflow velocity V_∞=500 m/s) profiles along intervals shown on figure . '*l*' is a distance from airfoil surface.

The experimentally observed effect can be represented as graphs of the dependence of the separation point position x_{SP} relative to the trailing edge versus time Δt after the discharge initiation. Figure 5 presents the dependences of $x_{SP}(\Delta t)$ for some experimental series 4, 5, 7, 8 (see table 1). From the presented dependencies one can conclude that in the range of electrical energy from 0.18 J/cm to 0.98 J/cm, which is spent for spark discharge initiation on the surface of the airfoil, a short-term displacement of the separation point is less than 1 cm. When the discharge energy changes by 5 times, this shift turns out to be almost the same. If one compares all the experimental series presented in the figure, it turns out that the displacement of the separation point downstream begins earlier at a larger value of angle of attack. This circumstance is due to the fact that at a larger angle of attack the position of the separation point is closer to the place of initiation of the discharge. If we compare the experimental series 5 and 7, it turns out that with an increase in the discharge energy by a factor of two, a short-term effect of early separation may occur due to an excessively strong shock wave caused by the discharge. On the graph this is expressed by a short-term increase in the x_{SP} position of the separation point.

The separation of the flow from the upper surface of the airfoil has unsteady nature, and the position of the separation point is not constant in time. Figure 5 shows the average values of separation point position, which was determined by the average velocity field. In addition to the averaged field, the fields of root mean square (RMS) of the components and the length of the velocity vector from the average were calculated in order to determine the boundaries of the separation point position. Figure 6 shows the fields of the RMS value for the length of the velocity vector for experiments corresponding to figure . To determine the magnitude of the fluctuations in the position of the separation point, a threshold of 250 m/s was selected, which corresponds to half the free-stream velocity V_∞=500 m/s. In the presented RMS fields regions with maximum values that are close to this threshold are visible. The projection of these regions onto the upper surface of airfoil determined the magnitude of the fluctuations in separation point position. It should be noted that as a result of this approach, it was found that at angles of attack 12°–16° the region of fluctuations of separation point position on the upper surface has a length of about 1 cm, and at the time when the gasdynamic disturbance interacts with the separation zone, this length is reduced to 3–5 mm.

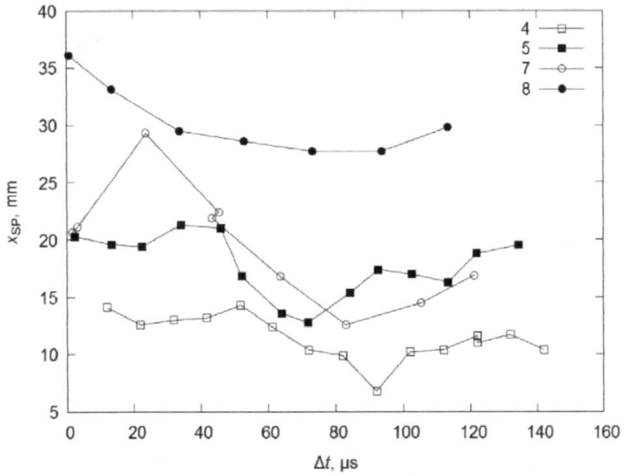

Figure 5. Distance x_{SP} between separation point and trailing edge at various time moments after discharge Δt.

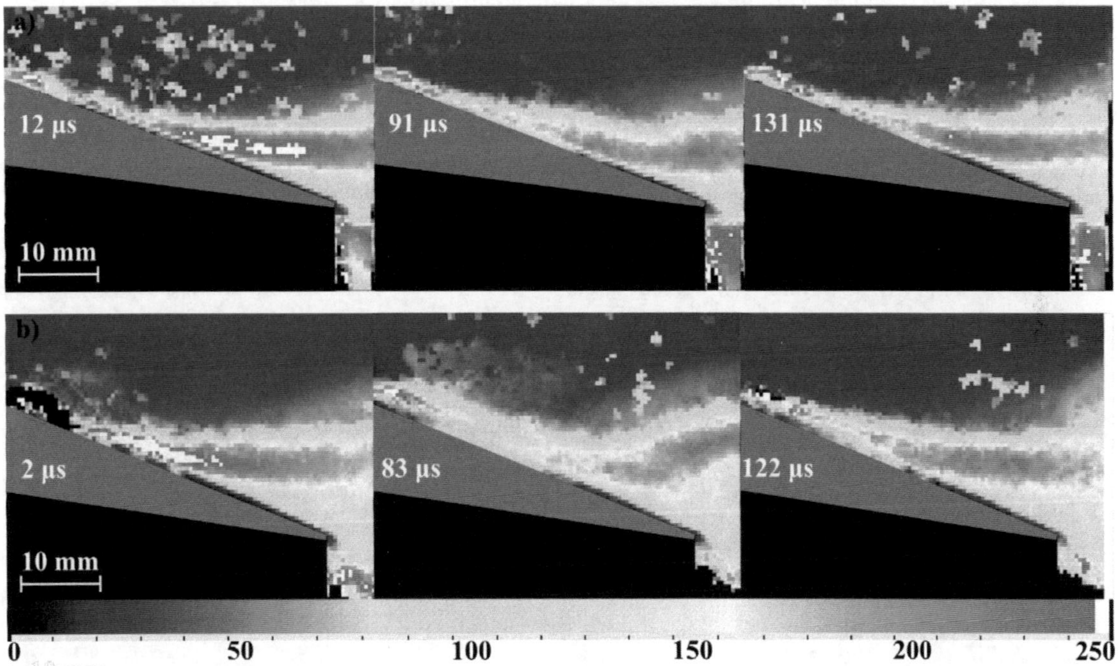

Figure 6. The RMS fields of the velocity vector length (in m/s) at different times (indicated in the frames) after the start of the discharge for the fourth (a) and seventh (b) experimental series.

4. Discussion

The main parameters of energy and time of the discharge are experimentally investigated. With the help of high speed video photography of the discharge and oscillograms of voltage of high-voltage generator, it was shown that the energy release occurs almost instantly in less than 1 µs, which is much less than the transit time. The mentioned properties of the object of investigation give reason to believe that the process of heat release is fast (instantaneous), and the unsteady flow is quasi-two-dimensional. These properties are important because they allow computer simulation of this process in a two-dimensional formulation without the use of significant computational resources. And in this case the initial local heating of the flow can be simulated as an area of increased pressure and temperature. The calculation of these values of pressure and temperature should take into account that as a result of a spark discharge only a part of the eclectic energy E is transferred to gas heat. The fraction of this energy is usually determined by comparing the experimentally obtained flow patterns with flow patterns obtained in numerical simulation. For long spark discharges, this value may be different depending on the type of discharge initiation and external conditions (flow parameters). From the literature values of this fraction differ several times from 10% [6] to 40% [7].

As a result of processing the PIV data, averaged patterns of unsteady flow were obtained at various time moments after discharge. The pictures show that as a result of rapid local heating of the incident flow with discharge, a nonstationary gas-dynamic perturbation occurs, which is an expanding hot gas region heated by a spark discharge, and a quasi-cylindrical shock wave, behind the front of which the gas has an additional velocity.

The averaged flow patterns obtained by the PIV method showed that the propagation of a gas-dynamic perturbation leads to a short-term displacement of the separation point downstream. The experimental data show that an increase in the input electrical energy by a factor of two from 0.42 J/cm to 0.87 J/cm does not lead to a significant change in the magnitude of this displacement, but at the initial moment of time the position of the separation point may shift upstream when energy value is large enough. The maximum displacement of the position of the separation point is about 1 cm in about 40 μs at angles of attack 12°–16°, which corresponds to the result obtained earlier at an angle of attack of 14° [2]. According to the data of the present and previous experiments [2, 4], carried out under the same experimental conditions, the velocity of the separation point downstream has a value (0.2–0.3 km/s) that is less than the velocity of external air flow (depends on the angle of attack, about 0.6 km/s). This circumstance suggests that the left-hand side of the gasdynamic perturbation, which moves against the oncoming airflow, has an influence on the position of the separation point. The influence mechanism can be the turbulization of the boundary layer, the generation of additional vorticity due to the presence of pressure and temperature gradients in the energy release region, the transfer of an additional impulse to the boundary layer due to the expansion of the heated region.

Relative energy deposition η can be estimated as the ratio between the consumed electrical energy to create a discharge and the enthalpy of the incident flow:

$$\eta = \frac{\gamma-1}{a_0^2}\frac{E}{QT}, \tag{2}$$

where a_0 is the speed of sound of atmospheric air (~320 m/s), γ is heat capacity ratio for air (1.4), Q is airflow mass rate through the relative (planform) area A_{rel} of airfoil ($Q=A_{\text{rel}}\rho_\infty V_\infty=1$ kg/s), T is discharge initiation period, E is electrical energy consumed by initiation of one discharge. The value of T can be estimated from the data obtained in the experiment, if we assume that the restoration of the original flow pattern occurs after about 50 μs (figure 5). Then in the range of E from 1.26 J to 6.86 J, the value of η will be from 0.1 to 0.5. This value can be reduced by initiating the discharge with the smallest necessary energy, as well as by increasing the efficiency of converting electrical energy to local rapid heating of the gas.

5. Conclusions

An experimental study was carried out about the interaction between a separated flow and a gasdynamic perturbation generated by rapid local heating of the oncoming supersonic airflow near the upper surface of the airfoil at positive angles of attack of 12°–16°. For this a method for initiating long spark discharge of 7 cm length and an electrical circuit for it initiating in a pulse-periodic mode are proposed. A study of the parameters (energy, duration, delay) of the electric discharge was made, which showed their sufficient stability from experiment to experiment with average energy input, that was varied in range 2.4–98 J/m. This circumstance made it possible to measure the velocity vector field of airflow over the upper surface of the profile using the PIV method. The obtained instantaneous (averaged) fields of the flow velocity vector at various time moments after the initiation of the discharge show that the gasdynamic perturbation, which is a quasi-cylindrical shock wave and a region of increased pressure that moves downstream, affects the separation region. This effect leads to a shift in the position of the separation point (up to about 1 cm) downstream to the trailing edge. Fields of root mean square for the velocity vector are also obtained, from which the boundaries of the position of the separation point on the upper surface of the profile are determined. At the specified angles of attack, the oscillations of separation point position were ±5 mm from average value. The experiment showed that the fluctuations in the position of the separation point decrease by 2–3 times during interaction of the gasdynamic perturbation with separation region. Thus, the region of increased pressure that caused by the discharge and propagating downstream limits the oscillation range of the separation point position, and this on average leads to its displacement to the trailing edge.

Acknowledgments
The reported study was funded by RFBR according to the research project № 18-38-00803.

References

[1] Son E E 2016 *Physical Mechanics: Laboratory Workshops* (Begell House Publishers)
[2] Saveliev A S 2018 *J. Phys.: Conf. Ser.* **1112** 012014
[3] Raffel M, Willert C E, Scarano F, Kähler C, Wereley S T and Kompenhans J 2018 *Particle Image Velocimetry: A Practical Guide* (Springer International Publishing)
[4] Golub V V, Saveliev A S, Sechenov V A, Son E E and Tereshonok D V 2010 *High Temp.* **48**(6) 903–9
[5] Sedov L I 1993 *Similarity and Dimensional Methods in Mechanics* (CRC Press)
[6] Elias P Q and Castera P 2013 *J. Phys. D: Appl. Phys.* **46** 365204
[7] Znamenskaya I A, Latfullin D F, Lutsky A E, Mursenkova I V and Sysoev N N 2007 *Tech. Phys.* **52**(5) 546–54

AUTHOR INDEX